# LEFT SIDE, RIGHT SIDE

# LEFT SIDE, RIGHT SIDE

## A Review of Laterality Research

Alan Beaton

Yale University Press
New Haven & London

To my parents, M.H. and Sue

First published in the United Kingdom in 1985 by
B. T. Batsford Limited
First published in the United States of America in 1986 by
Yale University Press

Library of Congress Catalogue Card Number: 85 – 52037

International standard book number: 0 – 300 – 03549 – 7


Printed in Great Britain

10 9 8 7 6 5 4 3 2 1

# Acknowledgements

It is a pleasure to acknowledge the help given to me by different people during the preparation of this book. I particularly wish to thank Dr Susan Clews for her careful reading of several draft chapters. She made many valuable criticisms and suggestions which are incorporated in the final text. I am grateful also to Dr Marian Annett who read the first draft of Chapter 2 and suggested a number of changes. Dr Natalie Danford was equally generous with her time in discussing with me genetic matters of interest. Dr Dennis Molfese and Dr Ruben Gur read the completed work and made a number of helpful comments and suggestions. Needless to say, responsibility for the final manuscript rests with me alone and no blame can attach to any of the aforementioned for any errors that may remain.

Tony Seward has been a most patient and understanding editor. Keith Smith also deserves mention for keeping a watchful eye on my endeavours and for providing advice and timely encouragement.

Finally, I thank Mrs Maureen Rogers who typed most of the first draft of the book and Mrs ·Christine Angus who typed the final version. The latter's devotion to the task far exceeded her professional obligations and I am greatly in her debt.

# Contents

*Acknowledgements*   v

*Preface*   xiii

1   **Introduction**   1

2   **Handedness**   6
  Hand preference and skill
  The measurement of handedness
  Correlations of handedness with other lateral asymmetries
  Hand posture during writing
  The genesis of handedness
    The birth-stress hypothesis
    Pathological left-handedness
    Genetic theories of handedness
      The Levy-Nagylaki hypothesis
      Annett's right-shift theory
      Collins's theory
    Non-genetic theories of handedness
    Socio-cultural factors influencing handedness
  Summary

3   **The split-brain studies**   34
  The role of the corpus callosum in inter-hemispheric transfer
    Visual functions
    Auditory functions
    Somesthesis
    Olfaction
    Motor functions
    Comparison with patients of the Akelaitis series
    Comparison with callosal agenesis
  Lateral specialisation in the bisected brain
    Verbal functions
      Linguistic functions of the right hemisphere
    Non-verbal functions in split-brain patients
  General observations on the effects of commissurotomy
  Meta-control in the split-brain
  Summary

4   **Hemispheric asymmetry in normal subjects: methods, findings and issues**   ·62

Tachistoscopic visual half-field studies
   Ocular dominance
   Tachistoscopic studies using verbal stimuli
   Tachistoscopic studies using non-verbal stimuli
      Form recognition
      Dot enumeration and localisation
      Perception of line orientation
      Stereopsis
      Facial recognition
   Theoretical interpretations of laterality effects
   Information processing and visual field asymmetry
      Iconic registration
      Hemisphere differences in coding
      Modes of processing
Dichotic listening
   Models of dichotic listening
   Dichotic studies with non-verbal stimuli
   Reliability and validity of dichotic listening asymmetry
Tactile perception
Electro-physiological studies
Lateral eye movements
Dual task experiments
The measurement of laterality
Summary

5   **Language and laterality**   109

Unilateral cerebral lesions
The Wada test
Electrical stimulation of the exposed cortex
Unilateral electro-convulsive therapy
Language laterality in neurologically intact subjects
   Electrophysiological studies
   Dichotic listening and visual half-field studies
The role of familial handedness
   Clinical studies
   Dichotic listening studies
   Tachistoscopic studies
   Familial sinistrality in dextrals
   Familial handedness and non-verbal performance
Inheritance of language lateralisation
Dominance proportions
Hand posture
Stuttering
Speech and motor functions of the left hemisphere
Language and the right hemisphere
   Language following hemispherectomy
   The effect of right cerebral lesions on language functions
   Recovery from aphasia
   Studies with normal subjects
      The Moscovitch model
      Lexical decisions in left and right visual fields
   Deep dyslexia
Summary

**6  Biological and comparative aspects of asymmetry  143**
Hemispheric blood flow
Regional Cerebral Blood Flow
  Techniques for measuring regional cerebral blood flow
Hemispheric activation and cerebral blood flow
Anatomical asymmetry of the human brain
Anatomical asymmetry in relation to handedness
Anatomic asymmetry in animals
Hemispheric functional asymmetry in animals
Summary

**7  The ontogeny of cerebral specialisation  155**
The effects of early brain damage
Evidence of functional asymmetry in the intact infant brain
Lateral asymmetries in perceptuo-motor response
Lateral preference in infants and young children
Tactile asymmetry in children
Dichotic listening
  Verbal stimuli
  Non-verbal stimuli
Tachistoscopic hemifield presentation
  Verbal stimuli
  Non-verbal stimuli
Laterality and second language acquisition
Environmental influences on brain lateralisation
Hemispheric specialisation in the deaf
Summary

**8  Sex differences in asymmetry  176**
Clinical investigations
Neuroanatomical asymmetry
Laboratory studies with normals
  Dichotic listening: verbal stimuli
  Interaction of sex with personal and familial handedness
  Dichotic listening: non-verbal stimuli
  Tachistoscopic half-field studies: verbal stimuli
  Tachistoscopic half-field studies: non-verbal stimuli
  Summary of dichotic listening and tachistoscopic half-field studies
  Electrophysiological studies
  Dual-task performance
  Lateral eye movements
Sex differences in handedness distribution
Ontogeny and sex differences
  Sex differences in infant asymmetry
  Studies with older children
    Dichotic listening
    Tachistoscopic half-field studies
    Tactile asymmetry
Theories of sex differences in lateralisation
Biochemical aspects of sexual differentiation
Summary

**9  Hemispheric asymmetry and reading disability  196**
Anatomical investigations of dyslexia
Cerebral dominance and dyslexia

Lateral preference and reading
Hand posture during writing
Divided visual field studies
  Verbal stimuli
  Non-verbal stimuli
Critique of divided visual field studies
Dichotic listening studies
Critique of dichotic listening studies
Electrophysiological investigations and dyslexia
Models of reading disability in relation to hemispheric function
  Satz and Sparrow's developmental lag hypothesis
  Bakker's balance model
  Annett's right-shift theory
  Witelson's model
  The Corballis and Beale model
  The Beaumont and Rugg model
  The Dunlop model
Summary

**10 Laterality of hand and brain: sinistral versus dextral** 221
Levy's model
Hardyck's model
Beaumont's model
Handedness and occupational choice
The dimension of field dependence-independence
Handedness and personality
Handedness and anxiety
Summary

**11 Emotionality** 234
Clinical observations
Experimental investigations with brain damaged patients
Studies with normal subjects
Theoretical interpretations of lateral asymmetry in emotion
Summary

**12 Asymmetry and psychopathology** 249
Neuropsychological studies
Psychophysiological studies of hemispheric asymmetry
Eye movement studies
Schizophrenia: the disconnection hypothesis
Auditory tasks and schizophrenia
  The effect of changes in clinical state
  The influence of particular symptoms
Tachistoscopic studies and schizophrenia
Electrophysiological studies
Résumé
Laterality of motor performance
  Twins, handedness and schizophrenia
Childhood autism
Hysterical phenomena
Summary

**13 Channel capacity, attention and arousal** 272
The normal brain: one or two channels?
Are two hemispheres better than one?

Information processing in split-brain animals
Information processing in human split-brain patients
Hemisphere asymmetry in attention and arousal
Summary

14  **What, why and how? – some loose ends**   285
Dichotomies of hemisphere function
The evolution of asymmetry
How should laterality be measured?

*References*   294

*Index*   361

# Preface

During the past ten years or so there has been an explosion of interest in the phenomenon of cerebral asymmetry of function. Yet when I began work on the manuscript of this book no fully comprehensive and up-to-date source of reference was available. Students and other non-specialists were at a loss to know where to begin their reading. Even for established researchers it had become difficult to keep abreast of the burgeoning literature. There was, in short, a need to draw together the different strands of laterality research within the covers of a single book that would be of value to students, practising professionals and researchers in this field. I therefore took upon myself the burden of providing such a text. Had I realised how onerous a task it would be I would have thought twice before embarking on the enterprise. Doubtless I should have done so anyway, for is there not an element of perversity in us all?

My intention in writing this book was not to develop any novel theory of laterality nor to view the phenomenon from a new perspective. My more modest aim was to record the progress that has been made in understanding the nature and importance of cerebral asymmetry of function. The work is offered as a review of findings and issues in this area spanning a period of approximately two decades from 1962 – the year in which the first of the modern split-brain studies was reported. Where the logic of exposition did not otherwise dictate, I have preserved the chronological sequence of published material in an attempt to capture both the historical and contemporary flavour of developments in this field. This should provide students with some insight into the way in which scientific progress actually evolves and give to more advanced readers an indication of which ideas are ripe for resurrection!

It is an occupational hazard of an undertaking such as this that from time to time review articles are published in the journals just as one's own review of a particular topic is completed. The conscientious reviewer should not rely on the efforts of others but sometimes their insights are more penetrating than his own. In fairness to my readers, therefore, I have been at pains to refer in the text to virtually all the serious reviews known to me even when, as was often the case, they appeared in print after I had completed my own survey of the literature.

At other points in the text citation of the relevant published papers has had to be limited due to economic considerations. In these circumstances I have cited, all else being equal, those articles which will enable the reader to consult the most comprehensive or up-to-date references. In this way the book will serve, I hope, as a useful guide to those who wish to undertake original research themselves.

I began work on this book at the suggestion of the late Professor Stuart Dimond. He had been invited to review the field himself but having other projects in hand he declined the invitation. Instead he encouraged me to take up my pen. Sadly, Stuart Dimond did not live to comment on the manuscript which would surely have benefited from the fertility of his ideas. If, however, my efforts have gone some way towards producing the book that Stuart himself would have written I shall be well pleased.

A.A.B.
*Craig-Cefn-Parc, 1984*

# 1

# Introduction

If a jellyfish could speak, the terms 'up' and 'down' would doubtless be part of its everyday vocabulary. But 'back' and 'front', 'left' and 'right' would have no meaning. Since a jellyfish is radially symmetric, the meaning of 'back' and 'front', 'left' and 'right', could not be determined by the creature in relation to its own physical structure, unlike 'up' and 'down' which are closely related to the jellyfish's asymmetry about the horizontal plane.

Unlike the jellyfish, certain animals show striking structural asymmetries. The unequal claw sizes of the fiddler-crab or the presence of a single tusk on one side of the narwhal are perhaps two of the most well-known examples. The human body also is rarely perfectly bilaterally symmetric. The limbs on one side are often shorter than those on the other, the left and right sides of the face are not exact mirror-images, in females one breast may be larger than the other, and in males one testicle usually hangs lower than its fellow (Chan, Hsu, Chan and Chan, 1960).

In addition to asymmetries of bodily structure, both animals and humans exhibit asymmetries of function. The most conspicuous example of functional asymmetry in humans is a preference for using one hand rather than the other. Most people prefer to use their right hand rather than their left for skilled activities, although this preference is by no means absolute. Some individuals prefer to use their right hand for some actions, their left hand for others. The issue of manual preference is not a simple one and is dealt with more fully in the chapter which follows. For the present it is sufficient to note that right-handedness (dextrality) is far more common than left-handedness (sinistrality).

The reason for the predominant preference for the right hand is to be found not in some peculiarity of the right hand but in some advantage which the left half of the brain has over its partner on the right. As a general rule, one half of the brain is connected primarily, though not exclusively, with the opposite (contralateral) side of the body. Thus the right hand is controlled largely by the left cerebral hemisphere. In a sense we have two brains, just as we have two eyes, two ears, two kidneys, and so on. However, the two cerebral hemispheres are linked by a large number of nerve fibres which enable the left and right sides of the brain to integrate their activities. The largest and most important of the bundles of nerve fibres running between the two sides of the brain is known as the great cerebral commissure or corpus callosum.

To the naked eye the brain appears a bilaterally symmetric organ; each half is the mirror-image of the other. But while this is true, roughly speaking, of the brain's structure, it is not true of its functions. Left and right cerebral hemispheres each have their own specialised abilities.

At one time it was common to refer to the preferred hand as the 'dominant' hand, and hence the hemisphere controlling that hand was referred to as the 'dominant' hemisphere. Given the preponderance of right-handedness, the left hemisphere has usually been considered to be the 'dominant' hemisphere. It was considered that the left hemisphere was somehow superior to its partner on the right, a view supported by clinical findings of the late nineteenth century (Broca, 1861; 1865; Jackson, 1874) indicating that speech – one of man's most distinctive faculties – is controlled by the left hemisphere in most people. The discovery that, in addition to defects of speech, certain other psychological defects more commonly follow damage to the left than the right side of the brain (Gerstmann, 1927) reinforced the belief that all important psychological functions are subserved by the left hemisphere. Thus there arose the concept of left- and right-'brainedness', as distinct from 'handedness'.

The pre-eminence of the left half of the brain remained more or less unchallenged throughout the first half of the present century. In 1962, however, certain events occurred which marked the start of an era of energetic research which has necessitated radical revision of the rather loose idea of 'cerebral dominance'. (For a history of this idea see Benton, 1965). The first of these events was the publication of the proceedings of a conference concerning 'Cerebral Dominance and Inter-hemispheric Relations' (Mountcastle, 1962). What emerged from the research reported at this conference was that the right half of the brain, for so long accorded only minor status, was not only the equal of the left hemisphere, but for certain non-verbal functions was actually superior.

The second event of 1962 to stimulate interest in the relationship between the two sides of the brain was publication of a report on a patient with extensive damage to the corpus callosum. This patient could write with his right hand but could make no attempt at all with his left. He was also unable to name correctly objects placed in his left hand, although capable of doing so if they were placed in his right hand. The interpretation which the authors (Geschwind and Kaplan, 1962) placed on these findings was that the callosal lesion prevented information passing from the right half of the brain to verbal centres located in the left half.

Although the possible role of the corpus callosum in inter-hemispheric transfer of information had been suggested (e.g. Dejerine, 1892) long before publication of the paper by Geschwind and Kaplan (1962), the relevant case histories had concerned individuals with damage to areas of the brain in addition to the callosum. Geschwind and Kaplan's patient was the first recorded case where the lesion was confined almost exclusively to the callosum itself. Experiments by Myers and Sperry in the 1950s had conclusively demonstrated the crucial role of the callosum in inter-hemispheric exchange of visual information in cats. In contrast, research with humans by Akelaitis and his colleagues during the preceding decade had failed to show any convincing psychological consequence of surgically dividing the callosum when this procedure was carried out on a few patients as a part of their medical treatment. Geschwind and Kaplan's paper reopened the question of the role of the corpus callosum in humans.

As it turned out, the year 1962 also saw the publication of the first of an extensive series of now-celebrated papers by Sperry, Gazzaniga, and their colleagues on the effects of surgically dividing the callosum in humans. This operation, known as cerebral commissurotomy, was revived as a treatment for otherwise intractable epilepsy by the neurosurgeons Bogen and Vogel. The research of Sperry and Gazzaniga with commissurotimised patients entirely vindicated the findings of the animal experiments in demonstrating the crucial role of the corpus callosum in transmission of information between the two sides of the brain. Furthermore, these studies of so-called split-brain patients (reviewed in detail in Chapter 3) revealed clear differences in function between the left and right cerebral hemispheres. This research was undoubtedly the most important single factor underlying the subsequent upsurge of interest in cerebral asymmetry, or left-right differences, in the human brain.

The most clearly established difference between the left and right sides of the brain concerns speech and language generally. In almost all right-handers, and in a good proportion of left-handers, it is the left hemisphere which is specialised for language (see Chapter 5). Among right-handers, at least, the neural centres for control of preferred hand function are close to the centres involved in articulate speech. Such proximity of the language centres to centres for preferred hand function may not be accidental.

It has been suggested on a number of grounds (Hewes, 1973) that from a phylogenetic point of view language has its origins in gestural activity of the hands (see Chapter 5). It is arguable that human language represents the highest level of evolutionary achievement. While certain other animals communicate vocally, they do not do so verbally. Even if it is granted that some apes can be taught rudimentary language skills, there is no doubt that these are trivial in comparison with the sophistication of human language. Thus both a sophisticated system of language and a predominance of the right over the left side of the body are unique to man. Do manual and cerebral asymmetry therefore represent some crucial phase in evolutionary development?

If left-hemisphere control of language, together with right-handedness, represents a significant evolutionary adaptation, there is a problem in accounting for the not inconsiderable proportion of the population who are not clearly right-handed. Historically, such individuals have often been viewed with suspicion and even hostility. This has no doubt been compounded by findings that among certain clinical populations there are more left-handers than would be expected purely on the basis of their frequency in the general population (see Chapter 2). From this observation it is but a short, though illogical, step to considering all cases of left-handedness as pathologically abnormal. Regardless of the flaws in this line of reasoning, it is of interest to investigate the consequences of being left-handed and, in particular, to ask how these relate to functional brain asymmetry.

Although research over the past 20 years has established that the two sides of the brain are specialised for different functions, it is not yet clear how best to classify those functions for which each half-brain is specialised (see Chapter 14). With regard to right-handers (dextrals), it is common to ascribe verbal functions to the left hemisphere and spatial, especially visuo-spatial, functions to the right hemisphere. While this may not be the most appropriate characterisation of left and right hemisphere functions, it is certainly the case that, whatever description is most appropriate, left-handers as a group do not simply show the reverse pattern of cerebral asymmetry to that exhibited by right-handers. In other words, it is not true to say that among left-handers (sinistrals) the right hemisphere is specialised for language and the left hemisphere for visuo-spatial skills. Some left-handers show the same pattern as right-handers, some left-handers show the reverse pattern, and some appear to show little or no cerebral asymmetry. Indeed, it is probably wrong to think of cerebral asymmetry as an all-or-none phenomenon, even in right-handers. It is more a question of *degree* of asymmetry (Zangwill, 1960), one side of the brain having a relatively greater or lesser propensity to subserve a particular function or set of functions than the other side.

If it is true that evolution has favoured asymmetry, the prediction can be made that those individuals endowed with symmetrically-functional brains should in certain respects be at some disadvantage in comparison with their brethren who have a more asymmetric cerebral organisation. The issue is far more complex than this suggests, but there is evidence that degree of cerebral asymmetry is indeed related to individual and group differences in cognitive functioning. The nature of this relationship is explored in the following chapters of this book.

Research into hemispheric specialisation represents part of the wider attempt to unravel brain-behaviour relationships. But as well as satisfying purely theoretical ends, knowledge of the specialised functions of the left and right sides of the brain may have practical application. For example, a patient's performance on psychological tests bearing a known relationship to lateralised brain function may prove to have localising significance in the diagnosis of cerebral disease or in indicating the prognosis for recovery. Further, knowing the functional specialisation of the left and right hemispheres of the brain may be valuable in devising suitable rehabilitation programmes. After damage has occurred to one half of the brain the most efficacious therapy might be that which attempts to capitalise on the known specialisation of the intact hemisphere, as in the form of speech therapy known as melodic intonation therapy which attempts to engage right hemisphere mechanisms in improving the speech of patients who have been rendered aphasic by left-hemisphere strokes.

Asymmetry of brain function has been implicated in such diverse matters as reading ability, choice of occupation, emotional reactions, psychiatric illness, and even aesthetic preference and classroom seating arrangements. The very ubiquity of brain asymmetry as a correlate (determinant?) of behavioural response invites speculation as to the nature and significance of the phenomenon. Unfortunately, this has led to widespread exaggeration and misunderstanding, especially in popular accounts of laterality (left-right) differences in the human brain. The implications of certain of these accounts are totally misleading, and it is therefore important to be clear just what the facts are and to distinguish these from both theoretical argument and downright error. The aim of this book, then, is to bring together the scientific findings concerning asymmetry of brain function and to present the theories that have been devised to explain them.

During a period of approximately 20 years, from 1962, scientists from a wide range of disciplines, notably psychology, psychiatry, neurology, anatomy, and biology, have been investigating laterality differences in the brains of a variety of different species of animal, including man. The results of this research are published in a large number of different journals, and it is difficult to keep up to date with the enormous literature. By bringing together the most important facts and theories this book should make it easier for individuals, researchers, and others to follow the developments that have taken place in different areas of the field as a whole. The ultimate aim of research into the brain and its functioning must, of course, be a greater understanding of ourselves. This book documents the progress that has been made in this direction as far as laterality of the brain is concerned.

# 2

# Handedness

The most striking example of behavioural asymmetry in humans is a preference for using one hand rather than the other, the majority of people preferring the right hand. This bias in favour of the right hand appears to have existed throughout recorded history (Coren and Porac, 1977) and even during prehistoric times, as far as can be judged from drawings on Egyptian tombs (Dennis, 1958) or inferred from examination of fossil skulls (Dart, 1949).

## HAND PREFERENCE AND SKILL

Although it is clear that most people are right handed, the question of how handedness is to be defined and measured is not easily answered. Two aspects of handedness can be distinguished – preference and skill. Is it the case that the preferred use of one hand is invariably correlated with a higher degree of skill? Annett (1967; 1970a) used a peg-moving task, in which the subject had to place pegs into close-fitting holes, to investigate this problem. She found that differences in the time taken by the two hands correlated with the degree of preference assessed by questionnaire. By contrast, Todor and Doane (1977) found a decreasing degree of correlation between degree of manual preference and difference in left-right hand proficiency scores as the difficulty of the experimental task increased. Specifically, Todor and Doane required their subjects to tap two adjacent targets with a stylus as quickly as possible. The level of difficulty was manipulated by varying the size of the targets. However, changing the size of the targets presumably alters the nature of the task. With small targets, accuracy is clearly more important than with large targets. The motor programming for the required movements may therefore differ and there is no a priori reason to suppose that different kinds or components of movement control show the same degree of asymmetry between the left and right hands.

Flowers (1975) distinguished between 'ballistic' movements of the hands – which once made continue without feedback control – and 'corrective' movements which require a feedback loop in order to monitor and modify movements. He found that strongly lateralised individuals, either dextral or sinistral, were better at corrective movements using the preferred than the non-preferred hand, but in individuals whose manual preference was not so distinct the non-preferred hand was superior. The performance of the non-preferred hand was at an equal level for both groups of subjects. Thus 'ambilaterals' showed comparatively poor control over the preferred hand without any compensatory increase in control over the non-preferred hand. On a 'ballistic' task, ambilaterals performed significantly better with the preferred than non-preferred hand whereas strongly lateralised subjects showed no such difference. Thus the relative superiority of one hand over the other depended upon degree of manual preference and upon whether the required movement was 'ballistic' or 'corrective'.

Differences between the hands, then, are not fixed but depend upon a number of factors including type of movement, task complexity (Steingruber, 1975) and degree of practice. For some tasks the effect of practice is to diminish the difference (Provins, 1967) while for other tasks the asymmetry remains constant (Hicks, 1975; Peters, 1981). But what is the nature of the superiority of one hand over the other in different tasks? This was investigated by Annett, Annett, Hudson and Turner (1979) who analysed the effects of changing the width of the hole in Annett's peg-moving task. They found that the width of the hole was related to the difference between the speed of the hands, a result in agreement with that found for changing the size of the targets in a tapping task used by Peters (1980). Annett et al. filmed the performance of their (well-practised) subjects and subsequently analysed the movements of the hands in detail. The relative slowness of the non-preferred hand was seen to be due not to the speed of the initial phase of moving the hand to the target but to the positioning phase (that is actually placing the peg in the hole) in which the non-preferred hand made more small corrective movements. Accuracy of aiming by the non-preferred hand was more variable than by the preferred hand, which added to the inferior overall performance of the non-preferred hand.

While practice no doubt confers some additional advantage on the preferred hand for certain skills, the initial difference between the hands is presumably of neural origin. Nakamura and Saito (1974) took electromyographic (EMG) recordings, that is electrical measurements of muscle activity, from the biceps of left and right handed subjects during flexion or supination (straightening) of the arms in response to a tone signal. Reaction times were calculated as the interval from the onset of the tone to the initiation of electrical activity. Reaction times were faster for the preferred than for the non-preferred hand in supination but the reverse was the case for flexion. Since the same group of muscles performs these two actions the opposite asymmetries of reaction time in the two actions convinced Nakamura and Saito that the two movements are represented separately in the brain.

It is easy to forget that a single hand is composed of five different digits and that the relationship between the fingers may be both subtle and complex. Although an individual may prefer to use the right hand, it does not necessarily follow that all tasks are performed best by this hand. Kimura and Vanderwolf (1970) and Parlow (1978) found among right handers that flexion of individual digits was carried out more effectively by the fingers of the left hand.

In view of these findings it is clear that the blanket attribution of superiority to one or other hand is an over-simplification. The relationship between the hands will depend upon the way in which the different digits are co-ordinated in different tasks to produce a finely programmed sequence of movements (Peters, 1977; Kelso, Southard and Goodman, 1979). As yet, little experimental work has been done to examine the co-ordinated performance of the hands in delicate or intricate tasks.

## THE MEASUREMENT OF HANDEDNESS

For research purposes, handedness has been measured in one of several different ways. These are (i) self-report – subjects simply state whether they consider themselves left- or right-handed or ambidextrous, (ii) writing hand – the hand used for writing is taken as the preferred hand, (iii) observation of the hand used in a number of unimanual activities – the hand used most often is noted, and (iv) questionnaire – the subject fills in a questionnaire concerning hand usage.

There are obvious problems with all of these methods. Regarding the first method, individuals differ in their criteria for assigning the labels 'left' and 'right'. Regarding the second method, the writing hand may not be representative of the hand used for other activities, especially in cases where natural left-handers have been taught to write with their right hand. The third method, observation of the hand used in a number of activities, is clearly the best indicator but the question arises as to how many activities should be sampled and how often. This applies also to the fourth method, questionnaires, and there is in addition the question of whether subjects' written responses correlate with their actual behaviour.

Typically, the items of a handedness questionnaire ask the subject to state whether he usually uses his left hand, his right hand, or either hand for each of a number of specified activities. The more questions that are asked, the more likely it is that some deviation from a consistent preference for one hand will emerge. This has theoretical implications which have rarely been addressed, even though statistical techniques are available for assessing the suitability or otherwise of particular items of a questionnaire (but see Oldfield, 1971). Many investigators have simply chosen an arbitrary number of items for their 'questionnaire' without carrying out the sorts of procedures which would be considered mandatory in the construction of such questionnaires as intelligence tests or personality inventories. The two most important features of any test instrument are its reliability and its validity. Reliability refers to the consistency of responses given to the questionnaire – does it always give the same result when administered on different occasions? Validity refers to whether the questionnaire measures what it purports to measure. In the case of handedness, the issue is whether scores on a pencil-and-paper test correlate closely with actual hand usage. Do people who say they use a particular hand for a particular activity in fact do so (item validity)? And does the questionnaire overall distinguish between individuals who on other grounds might be said to differ in handedness? At the time of writing only one questionnaire (Annett's) has been adequately validated against behavioural observations.

A final problem that arises in connection with questionnaire data is how to define left-handedness, right-handedness, ambidexterity, or whatever. How consistent must subjects be in their hand preference to be classified in a particular way? In the absence of any compelling theoretical criteria, the decision is purely arbitrary. This has probably accounted for much of the contradictory evidence in the literature concerned with handedness.

Until fairly recently it was usual to consider that people were one of two or three basic biological types – dextral, sinistral or mixed handed – but it is now clear that as descriptions of hand preference the terms left, right and mixed-handed do not do justice to the full range of manual asymmetry which may be observed.

Annett (1970a) used a questionnaire which samples 12 behaviours. Some of the items concern unimanual activities (such as writing or throwing) whereas others could be considered bimanual although one hand may play only what is literally a supporting role (as, for example, in dealing cards or unscrewing the lid of a jar). Responses to the questionnaire take the form of indicating whether the left, right or either hand is usually used for a particular item. On the basis of the responses of a large number (2321) of subjects to her questionnaire, Annett distinguished 23 types of hand preference which emerged from a treatment of the data known as association analysis. She argued that handedness should be considered not as a discrete variable (left handed, right handed and mixed-hand preference) but as a continuous distribution ranging from strongly right handed to strongly left handed. Peters and Durding (1978) have argued similarly on the basis of their finding of a linear relationship among 513 children between a laterality ratio, based on the comparative use of the right and left hands for 7 everyday activities, and manual asymmetry on a tapping task.

It is probably true to say that few, if any people, now dispute that handedness is continuously rather than discretely distributed but the implications of this are only just becoming appreciated among researchers interested in psychological differences between handedness groups. The point is that if a variable is continuously distributed the choice of a cut-off criterion for classifying different groups of subject (say into left or mixed handers) is entirely arbitrary. This may well have accounted for various discrepancies in the literature concerned with handedness in relation to other attributes.

As well as Annett's questionnaire, another in common use is that devised by Oldfield (1971) and referred to as the Edinburgh Handedness Inventory. It consists of 10 items, selected from an original sample of 20, giving a laterality quotient from $+100$ (extremely right handed) to $-100$ (extremely left handed). Oldfield provides a frequency distribution of the laterality quotients and presents data showing a sex difference in responses to individual items making up the inventory. It is worth noting, though, that the inventory was not validated against a behavioural measure, responses from subjects being taken at face value. Also, the entire standardisation sample of 1128 individuals were all first-year psychology undergraduates who are an unrepresentative sample of the total population. However, Annett (1970a) in her study found no conspicuous difference in responses made by undergraduates and servicemen from one of the armed forces, so for practical purposes this is perhaps not a very serious criticism.

The construct validity of the Edinburgh Inventory has been examined using the technique of Factor Analysis by White and Ashton (1976) and by Bryden (1977). In both studies a single major factor was extracted which was identified with handedness, and a minor factor (White and Ashton) or factors (Bryden) not associated with handedness. The stability of the handedness factor extracted in a similar study by McFarland and Anderson (1980) was found to be high across both age and sex and over a four week test-retest period. Thus the inventory as a whole may be considered to have reasonable construct validity even though one or two particular items are unstable.

Annett (1970a) validated her own questionnaire against an independent criterion, that of relative differences in speed between the hands shown by 300 subjects on her peg-moving task. The relative magnitude of the hand difference for each of 8 preference groups, derived from 23 groups originally yielded by an association analysis, was predicted by the degree of hand preference shown by each group.

Annett's questionnaire allows for different degrees of manual preference to be recorded for a particular item only in so far as the respondent has to indicate whether he habitually uses the left hand, the right hand or either for a particular activity. Briggs and Nebes (1975) modified the questionnaire to take greater account of expressed strength of hand preference for each item. This does not necessarily represent an improvement since no behavioural validation of the modified scoring system has been published (Beaton and Moseley, 1984). Loo and Schneider (1979) used this version of the questionnaire to assess handedness in 51 male and 59 female students. On the basis of a factor analysis Loo and Schneider concluded that the inventory taps three factors which they identify as power, skill and rhythm. By contrast, Richardson (1978), who factor analysed the responses of 160 students to a 6-item questionnaire, maintained that handedness represents a single dimension. (Writing and throwing were the activities which most clearly defined the factor extracted in his analysis.)

With regard to reliability, McMeekan and Lishman (1975) obtained a Kappa coefficient of 0.80 for Annett's questionaire with a test-retest interval ranging from 8 to 26 months (mean 14.5). Almost half of the 73 subjects (46.6%) changed response to one or more items but only 12 (16.4%) changed group classification as a result. On the Edinburgh Inventory individuals with an initial laterality quotient of between − 1 and − 99 yielded a reliability coefficient (product-moment) of 0.75 and those with initial quotients between 0 and + 99 gave a coefficient of 0.86. (Combining the two subsets had yielded a spuriously high coefficient of 0.97 due to the fact that considerable shifts in the scores of subjects with negative laterality quotients were counteracted by equally large shifts in subjects with positive quotients). Other investigators have also shown that though particular items may not be highly reliable (see e.g. Raczowski, Kalat and Nebes, 1974) by and large questionnaires provide a reasonably reliable method of assessing handedness (Bryden, 1977; Coren and Porac, 1978). Certainly the Annett and Edinburgh inventories are improvements on earlier (Humphrey, 1951; Crovitz and Zener, 1962) questionnaires in terms of their test construction.

It has sometimes been claimed that left handers are less consistently sinistral in their manual preference than are right handers (Humphrey, 1951; Crovitz and Zener, 1962; Benton, Meyers and Polder, 1962). This, however, may only be an artefact of the criterion used for classifying left and right handed groups. Individuals calling themselves right handed are very likely to show strong dextral preference when tested objectively, whereas among persons calling themselves left or mixed handed, some will be found who in point of fact preferentially use the right hand for a number of activities (Humphrey, 1951; Friedlander, 1971; Bryden, 1977). If self-report is taken as the criterion, then, so-called left-handers as a group will show less consistency in their use of the right hand. A more strict criterion of left handedness would result in a reduction of the apparent inconsistency in hand usage among sinistrals. However, this does not explain the lower *within* subject consistency in hand usage for *particular items* among left handers (Raczkowski, Kalat and Nebes, 1974).

## CORRELATION OF HANDEDNESS WITH OTHER LATERAL ASYMMETRIES

The question arises as to the relationship between handedness and other lateral asymmetries. Does a dominant right hand imply a preference for the right foot, right ear and right eye? There is some confusion in the literature as to the degree of association between handedness and eyedness, in particular. This is because the term eyedness has been used in a number of different ways to refer to (i) the eye having greater acuity, (ii) the eye used for sighting (as in looking through a key-hole), (iii) the eye whose stimulus dominates perception in experiments on binocular rivalry and (iv) the eye by which visual direction is specified during normal binocular viewing.

(For reviews of the topic of eye dominance see Friedlander, 1971; Money, 1972; Porac and Coren, 1976).

In general, correlations between preferred hand and foot appear to be higher than between hand-eye; hand-ear; eye-foot; ear-foot (Coren and Kaplan, 1973; Porac and Coren, 1975; 1976; 1978; 1979a). Porac, Coren, Steiger and Duncan (1980) factor analysed the responses of 962 subjects and concluded that there were three factors or dimensions (i) hand and foot, (ii) eye, (iii) ear. They therefore suggested that laterality of sensory or motor function is not produced by a single cause. That being said, it is entirely possible that there is some super-ordinate factor − perhaps involving attentional strategies − that may bias an individual towards the consistent use of a limb or organ on one side of the body over and above the particular mechanisms underlying a specific preference. This would help to explain why it is that, though not perfect, there is a statistically significant association between hand and eye dominance in the population. Merrell (1957) reported that 70.5 per cent of 464 right handers (defined by the writing hand) had a dominant right eye and 29.5% were left-eyed. Of 33 sinistrals, 60.6% were left-eyed and 39.4% were right-eyed.

Levy (1981) refers to findings, unpublished at that time, which indicate an increased incidence of left eyedness among (strong) right handers who had suffered birth stress as reported by the subjects' mothers. If it is assumed that all right handers are naturally right-eyed, and if birth stress can cause a change not only in the preferred hand (see below) but also in the dominant eye, then relatively minor damage to one side of the brain during labour might cause a shift in both eyedness and/or handedness.

It might be asked why eye dominance may be subject to change as a result of minor *unilateral* injury, since each eye is connected to both halves of the brain. The answer may be that eye dominance is in some way a function of the phylogenetically older crossed pathways (Polyak, 1957) as opposed to the more recently evolved uncrossed pathways. This would link a particular eye with the contralateral hemisphere and damage to this hemisphere might affect the dominance or otherwise of this eye. In support of this proposition it may be noted that following hemispherectomy monkeys (Kruper, Patton and Koskoff, 1971) and human patients (Smith, 1972) show a shift in preference to the eye opposite the intact hemisphere.

Despite the lack of close correlation between hand preference and eye dominance in the normal population, interest in so-called mixed or crossed-dominance (in which the preferred hand and eye are on opposite sides) has been slow to wane, particularly among researchers investigating reading disability in children (such studies are dealt with in Chapter 9). This is particularly surprising in view of the fact that adequate theoretical grounds for expecting a difference between crossed and uncrossed dominant subjects are hardly, if ever, specified. The foregoing discussion provides such a rationale but appears to have been little considered in the literature.

Studies comparing crossed-dominant with uncrossed-dominant subjects have used a variety of tasks. Dawson (1973), for example, found among a

West African tribe living in a jungle or forest environment that mixed dominant subjects were less susceptible to certain visual illusions, but his measures of handedness would now be considered totally inadequate. Other researchers have noted effects of eye dominance, sometimes interacting with handedness, on tasks involving the use of a tachistoscope, a device for presenting visual information for very brief durations (Hayashi and Bryden, 1967; Bryden, 1973; Kershner, 1974; Porac and Coren, 1979b; McKinney, 1967). With such brief durations, differences between nasal and temporal hemiretinae of each eye (Overton and Weiner, 1966; Markowitz and Weitzman, 1969; Neill, Sampson and Gribben, 1971) together with the effect of a dominant eye may be sufficient to account for the results. In any event, the findings obtained are sufficiently diverse as to preclude any firm statement at the present time as to the effect or otherwise of eye dominance in any particular experimental set-up.

## HAND POSTURE DURING WRITING

As well as differing in the hand used for writing, individuals differ with regard to the position of their hand during writing. Some people write with the hand more or less in line with the fore-arm, the wrist being straight (the so-called normal position) while others bend their wrist to a greater of lesser degree giving an inverted or hooked posture. It is sometimes suggested that inversion of the left hand is an adaptation to the left-to-right direction of written English; inversion is said to prevent smudging what has just been written. If so, then one would expect a relatively high proportion of Israeli right handers to use the inverted position since Israeli script is written and read in a right-to-left direction. However, there is not a greater frequency of inversion among Iraeli than American right handers, but neither is the incidence of inversion among left handed Israelis as high as that found for American left handers (Shanon, 1978). The inverted writing posture is more common among males than among females (Peters and Pedersen, 1978; McKeever, 1979; Coren and Porac, 1979). McKeever (1979) also claims that inversion is more frequent among those with sinistral relatives although this was not found to be the case by Searleman (1980).

Although control of the upper limbs is generally held to be undertaken by the contralateral cerebral hemisphere, there is evidence that this holds true more for fine movements carried out by the distal musculature (the fingers), as in writing, than for gross movements of the whole limb. Motor pathways from the cortex of one hemisphere descend to the spinal cord via the brain stem where the majority of fibres cross the midline and continue downwards as the lateral corticospinal tract. A proportion of fibres, however, do not cross and constitute the ventral corticospinal tract. It has been shown in monkeys that these uncrossed tracts terminate in the region of the spinal cord that controls gross movements of the limbs (Brinkman and Kuypers, 1973). This is consistent with clinical evidence in man which suggests some measure of ipsilateral cerebral control of gross motor movements.

It has been suggested that the upright or normal hand posture might indicate that writing is controlled by the contralateral cerebral hemisphere whereas the inverted posture indicates ipsilateral cerebral control (Levy and Reid, 1976). Levy and Reid (1976) use the term 'control' in this context to refer to the hemisphere specialised for language in its written forms. Certainly, there is evidence that writing and speech may depend upon opposite hemispheres in the same individual (Heilman, Coyle, Gonyea and Geschwind, 1973; Heilman, Gonyea and Geschwind, 1974). It is conceivable, then, that purely motor control of the hand during writing is undertaken by one hemisphere while the linguistic aspects of what is written are determined by the opposite side of the brain through trans-commissural mediation.

Levy and Reid (1978) hypothesise that in cases of inversion of the left hand during writing, it is the left hemisphere that is in control. The mechanism of control, as Levy and Reid point out, may be either by way of ipsilateral fibres or by means of transcommissural mediation from the left hemisphere influencing the right hemisphere, which then controls the left hand via the contralateral motor pathway. Herron, Galin, Johnstone and Ornstein (1979) have argued on the basis of EEG (electroencephalograph) data that the right hemisphere is *always* engaged during writing by the left hand, regardless of the posture adopted, but the possibility that their data reflect afferent rather than efferent processes cannot be discounted.

The idea that a not inconsiderable number of the population have their writing hand controlled (in the sense of sensori-motor control) by the ipsilateral cerebral hemisphere might be thought to conflict with the observation that the majority of pyramidal fibres (forming part of the main motor pathways) cross over from one side of the brain stem to the other (at the level of the medulla) in almost all recorded cases (Yakovlev and Rakic, 1966; Mizuno, Nakamura and Okamoto, 1968). In one study the fibres from the left side were found to cross before those of the right in 73 per cent of 158 cases including four left handers (Kertesz and Geschwind, 1971) but the functional significance of this, if any, is not known.

If it is true that inverted writers have ipsilateral hemispheric motor control for tasks other than writing, then it might be hypothesised in such cases that reaction times to stimuli presented to the side of the body opposite the responding hand would be faster than to stimuli on the same side. This is based on the assumption that faster responses will be obtained when the hemisphere in direct receipt of the stimulus is also the one that controls the motor response (as compared with the situation in which the hemisphere responding is not that which receives the stimulus). It is normally found (i.e. with presumed contralateral motor control) that reaction times are faster to stimuli on the *same* side of the body, for example a right hand response to a stimulus presented on the right of the field of vision.

Moscovitch and Smith (1979) applied the logic of this neuro-anatomical model to the case of individuals with presumptive ipsilateral neuro-motor control (i.e. inverters) and predicted faster responses to stimuli presented on the side of the body opposite the response hand than to stimuli presented on the same side. That is, Moscovitch and Smith predicted the opposite results to those normally found. Their findings were that inverted (ipsilateral control?) and non-inverted (contralateral control?) subjects showed the usual pattern of results when signals were presented in the auditory and tactile modalities but with visual stimuli (see also Smith and Moscovitch, 1979) the results were in line with the predictions. Using a simple unimanual reaction time task, McKeever and Van Hoff (1978) also found non-inverted left handers to show the usual ipsilateral advantage but inverted left handers showed no difference between conditions. On the Levy and Reid hypothesis there should have been a contralateral advantage.

In the experimental situation employed by Moscovitch and Smith (1979) and McKeever and Van Hoff (1978) any effect due to length of the anatomical routes between stimulus reception and response output is confounded with effects due to what is known as stimulus-response (S-R) compatibility. Responding with the right hand to a stimulus on the right is said to be a compatible relationship while responding with the left hand to a stimulus on the right represents an incompatible relationship (Fitts and Seeger, 1953). Attempts have been made to deconfound effects due to stimulus response compatibility from those due to length of anatomical pathway by requiring subjects to respond with their forearms crossed so that the right hand is on the left side of the body and the left hand on the right side of the body. Under these circumstances, an advantage for the 'same' hemisphere condition over the 'different' condition can be demonstrated, at least for simple reaction time tasks (Anzola, Bertolini, Buchtel and Rizzolatti, 1977; Berlucchi, Crea, Di Stefano and Tassinari, 1977). However, the difference between conditions is much smaller than that typically found in situations where the forearms are not crossed. In the latter situation, then, where S-R compatibility and anatomical effects are concordant the greater effect is presumably due to a compatible S-R relationship. (The fact that the S-R effect is 'over-ridden' when the forearms are crossed might be attributed to the subjects' lack of experience with the unusual stimulus-hand relationship). Thus in the experiment by Moscovitch and Smith it is perhaps not surprising that inverted and non-inverted writers performed similarly in two out of the three modalities because whatever the sensori-motor control, ipsilateral or contralateral, in the two groups of subjects the S-R relationship was always compatible.

Moscovitch and Smith saw their results as indicating not a difference in sensori-motor organisation between inverted and non-inverted writers but a difference in visuo-motor organisation, since a difference between the two groups of subjects was found only for the visual modality. Yet a difference between left handed inverters and non-inverters has been observed for purely motor tasks (though performed under visual control) by Parlow and Kinsbourne (1981). Non-inverters tended to show the mirror-image asymmetry of right handers. Unlike right handers, non-inverted sinistrals showed an advantage for the right hand in finger flexion, and left hand superiority in the same direction as non-inverters but reduced in magnitude. This suggests that differences in hand posture may indeed reflect some difference in sensori-motor organisation as hypothesised by Levy and Reid (1976). Further discussion of their hypothesis will be found in Chapter 5. For the present it may be noted that differences between left-inverters and dextrals in comparison with sinistral non-inverters have been found in a cerebral blood flow study (Halsey, Blauenstein, Wilson and Wills, 1980) and on one item of a neuropsychological battery of tests (Gregory and Paul, 1980).

## THE GENESIS OF HANDEDNESS

Given the preponderance of right handedness in all cultures that have been studied (Hécaen and Ajuriaguerra, 1964) it was common at one time to view right handedness as the universal norm and departures from full dextrality as due either to some abnormal influence on brain development or to an unusually perverse resistance to socio-cultural pressure since at one time left handedness incurred considerable social disapproval.

It is not easy to explain on purely cultural grounds why it is the right hand that is normally preferred. Attempts have been made but they are unconvincing. One well-known idea is that, among warfaring peoples, holding a shield in the left hand protected the heart, thereby increasing an individual's chances of survival in battle. If the shield was held by the left hand, then the right hand was used for wielding a sword. Natural selection thus favoured holding a sword in the right hand, which led to a preference for this hand in other skilled activities. For a review of early theories of handedness the interested reader is referred to Harriss (1980).

### The Birth Stress Hypothesis

The most recent proponent of the view that left handedness is abnormal in a sense other than the purely statistical is Bakan. He noted that left handedness is more frequent in males than in females and in twins compared with singletons and, further, that male gender and twinning are both associated with increased infantile mortality. He therefore wondered whether left handedness was a consequence of stressful pre-natal and/or birth conditions. Since birth stress is more frequent in primiparous and older mothers, Bakan (1971) compared the numbers of left and right handed writers among first born and fourth and later born students (high risk birth ranks) with the frequencies in birth ranks 2 and 3 (low risk). Among males, but not females, there was a significant association between handedness and birth order; there were comparatively more left handers among the high stress birth ranks. Hubbard (1971), however, reported the opposite effect, namely more sinistrals among low risk births, without distinguishing between the sexes.

In an attempt to improve upon the classification of handedness used in previous studies of the birth stress hypothesis, Bakan, Dibb and Reed (1973) distributed a handedness questionnaire among students. They found that left and mixed handers reported birth stress significantly more often than did right handers and that significantly more non-dextrals were first born to mothers aged 30 years or over. These findings were supported by Bakan (1977) who found a significant excess of sinistrals in high-risk birth ranks among post-secondary students. In contrast, Teng, Lee, Yang and Chang (1976), Schwartz (1977), Hicks, Pellegrini and Evans (1978) and Dusek and Hicks (1980) failed to find any association between handedness and presumed birth stress.

Hicks, Evans and Pellegrini (1978) compiled data from five published studies and found a near zero correlation in the pooled data between handedness and birth order. The weight of the evidence, then, suggested that birth stress was not responsible for an increased incidence of left handedness. Even maximising the chances of finding a positive effect by differentiating between births to mothers who were less than 20 and over 30 years of age as a high risk group, in comparison with mothers between the ages of 20 and 30, failed to reveal any significant association (Hicks, Elliot, Garbesi and Martin, 1979).

There are a number of methodological factors that might contribute towards the inconsistency in the findings of different investigators. In the studies by Bakan, Dibb and Reed (1973) and Schwartz (1977) birth stress was reported by the subjects themselves who, since they cannot recall their own births, may not be a reliable source of information. Annett and Ockwell (1980) obtained relevant reports directly from the mothers of their subjects and found no evidence of birth stress being more common for left handers, although there was a slight trend towards a higher proportion of first and fourth rank births among left handed daughters but not among sons, mothers or fathers. Tan and Nettleton (1980) also obtained reports direct from mothers but found no increase in left handedness among fourth or later borns. Indeed, among first borns, there was a decrease in left handedness even though first births were reported as more stressful than others.

One of the most conspicuous discrepancies in the studies reviewed here lies in the different criteria used to classify handedness. Most investigators have used questionnaires of one kind or another but have been inconsistent in their choice of inventory and criteria for classification. One way of avoiding the problem of defining cut-off points for left handedness and mixed handedness is to examine not proportions of subjects in particular subgroups but the entire distribution of handedness. Hicks, Dusek, Larsen, Williams and Pellegrini (1980) used the Briggs-Nebes modification of

Annett's (1970a) questionnaire to assess handedness in 1,501 students. Of this sample, 181 reported birth complications and this group showed a small shift in their scores away from the dextral end of the distribution. A criticism that might be levelled at the majority of the studies quoted is that the respondents have all been university students. If birth stress is sufficiently severe to 'cause' all left handedness then it might be expected that this would militate against the chances of university entry. Thus it would not necessarily be expected that any relationship between birth stress and handedness would emerge among student populations. Indeed, it might be expected on this view that left handers would be under-represented in such samples. There is, however, no evidence that this is the case (Oldfield, 1969; Byrne, 1974).

University students undoubtedly represent a social group unrepresentative of the population as a whole. Birth stress complications might be expected to be more frequent in socially disadvantaged mothers and thus any association between birth stress and handedness would be more likely to be revealed in families whose socio-economic status is relatively low. (There is, of course, the complicating possibility that families differing in socio-economic status also differ in their tolerance of 'deviant' hand usage). However, Leiber and Axelrod (1981a) found no difference in non-dextrality among student respondents as a function of the occupational (or educational) level of their parents but the incidence of low occupational status groups in this study was very low, which may have contributed to the negative finding. Hardyck, Petrinovich and Goldman (1976) and Silverberg, Obler and Gordon (1979) also failed to find any effect of socio-economic class. Interestingly, in the Leiber and Axelrod study, among the parents of the respondents, the proportion of left handedness was significantly higher among fathers (and in the same direction but not significant for mothers) whose occupational or education level was high compared with fathers of low occupational or educational level. This might be explained by assuming that high-status fathers come from families who, two generations ago, were more tolerant than low-status families of transgressions of the cultural norm. In the more relaxed attitude of today, there may be no difference between families of high and low status, thus accounting for the negative finding for Leiber and Axelrod's student respondents.

Leiber and Axelrod (1981b) using a combination of Annett's and Oldfield's questionnaires found a higher incidence of non-dextrality among high compared with low risk birth ranks among males but the reverse was the case among females. The finding for males was marginally ($p < 0.10$) significant but for females the effect reached a conventional level of significance ($p < 0.05$). Reported birth complications were associated with increased non-dextrality, but only among male university staff and not female staff nor student respondents of either sex. An interaction between gender and generational effects could explain these results.

In a very large scale analysis McManus (1981) used data from the

National Child Development Study – a prospective study of all children born in England, Scotland and Wales during a single week in 1958. Treating handedness as dichotomous (on the basis of the hand used for writing at age 11 years) he found no relationship between sinistrality and either social class, birth order or a large number of variables indicative of, or associated with, birth complications.

In summary, there is little firm evidence that birth stress is associated with non-dextrality but that, if so, this is more likely to be found among males than among females (Bakan, 1971; Leviton and Kilty, 1976). Such a sex effect may relate to a greater risk of peri-natal distress among males than females, a possibility suggested by the higher rate of foetal and peri-natal death among males (Taylor and Ounsted, 1972). It may also relate to differential hemispheric vulnerability to damage (Taylor, 1969).

Whatever the evidence in favour of the birth-stress hypothesis, it is clearly dangerous to argue from an increased incidence of left handedness in high-risk births to the conclusion that all sinistrality is due to peri-natal complications such as anoxia (lack of oxygen). Nonetheless, even in the absence of frank obstetric complications there are indications that the birth process is related in some way to later hand preference. Churchill, Igna and Senf (1962) and Grapin and Perpère (1968) noted an association between handedness and the part of the baby's head first presenting at birth. In the Churchill et al. study, 93 children were said to be left handed at 2 years of age, of whom 62.4 per cent presented at birth in the right occipito-anterior position (ROA); of 836 right handers 57.3 per cent presented in the left-occipito-anterior position (LOA) and of 173 ambidextrous children 50.9 per cent presented in the LOA and 49.1 per cent in the ROA positions. The association between handedness and birth presentation position is statistically significant. However, such an association clearly cannot account for the distribution of hand preference which overwhelmingly favours right handers. Even if birth presentation were totally correlated with handedness the question would arise as to why the baby's head presents in a particular position, usually the LOA.

## Pathological left handedness

A higher than expected incidence of left handedness has been reported for a wide variety of different clinical populations including stutterers (Hécaen and Ajuriaguerra, 1964), autistics (Colby and Parkinson, 1977), epileptics (e.g. Silva and Satz, 1979) and the mentally retarded (Hicks and Barton, 1975). This may have helped to promote the idea that left handedness, per se, is abnormal and may explain the suspicion that has attended the left (sinister?) hand since Biblical times (see *Judges* II:15 and XX:16).

A raised incidence of sinistrality among diverse clinical groups can be explained by assuming that in the case of very early brain damage lesions affecting only one side of the brain induce a shift from pre-determined handedness. If unilateral lesions affect left and right hemispheres with equal probability, then, given that most people are initially destined to become dextral, a larger number of pre-determined right handers will become 'shifted' sinistrals than pre-determined left handers will become 'shifted' dextrals. Thus the population of normal left handers becomes inflated to a proportionally greater extent (Satz, 1972).

Given an estimate of the base rate of left handedness in the general population, it is possible to calculate the expected proportions of left and right handers among brain damaged populations given the assumption that a left or right sided lesion is equiprobable. Satz (1972, 1973) found a good fit between the observed incidence of left handedness among brain injured patients and expectations based on his model of pathological left handedness. Further support came from studies of EEG abnormalities in relation to handedness in retarded and epileptic populations (Silva and Satz, 1979; Satz, Baymur and Van Der Vlugt, 1979). It is noteworthy, however, that a raised incidence of left handedness was observed not only in the presence of (mainly left sided) unilateral EEG abnormalities but also in association with bilateral EEG abnormalities. Although the incidence of sinistrality in the latter case was not as high as in the former case, there is no reason according to the model of Satz (1972) to suppose that there should be any increased sinistrality with bilateral brain damage. Silva and Satz (1975) therefore suggested that 'the probability of being right-handed decreases with poorer levels of cortical functioning. This postulate is based on two assumptions: 1. that persons having bilateral EEG abnormalities function at lower cognitive levels than those whose EEG recordings are within normal limits and 2. that the depressed level of functioning in retarded persons with normal EEGs is not severe enough to influence handedness' (p. 15). They saw the findings of Hicks and Barton (1975) that preference for the right hand was less common among severe than mild cases of mental retardation as consistent with their suggestion. A relationship between left handedness and severe, but not minimal, brain damage may also explain the failure of McManus (1980a) to find an increased incidence of sinistrality among children with relatively mild epilepsy.

The pathological left-handedness model has recently been extended to a normal population of school children. Bishop (1980a) argued that if a shift from pre-determined handedness has occurred as a result of early brain damage, then performance of the non-preferred hand should be noticeably impaired. Consequently, she predicted that there would be a raised proportion of left handers among children with particularly poor performance of the non-preferred hand (the target group). This prediction was confirmed. In addition there was a significant tendency for children in this group to be fourth or later born in comparison with control children. The target group children also had a lower mean IQ than controls, implying some degree of cognitive impairment. The implication of Bishop's findings is that it is fruitful to consider the *relative* performance of the non-preferred and preferred hands in attempting to discover relationships between putative brain damage and hand preference or cognitive impairment.

Applied to the data for birth order and handedness this approach might go some way to resolving some of the inconsistency in the literature.

## Genetic Theories of Handedness

The implausibility of explaining all sinistrality in terms of pathological influences on the brain leads naturally to a consideration of whether variation in human handedness might be accounted for in genetic terms. Arguments which support a genetic influence on handedness have been summarised by Levy (1976a) and include the following:

1.   A number of functional and behavioural asymmetries present at or around birth (see Chapter 7), and hence not explicable in terms of learning, have been found to correlate with handedness.
2.   Handedness has been related to anatomical asymmetries of the brain (see Chapter 3), and nose (Sutton, 1963) and to dermatoglyphic (finger-print) patterns (Rife, 1940). It would be surprising if these asymmetries were under genetic control and yet handedness was not so influenced.
3.   Family studies of handedness suggest a genetic factor (Rife, 1940; Annett, 1973a): in particular, the fact that 80 per cent of left handers have two dextral parents is not easy to explain on environmental grounds. The influence of familial sinistrality on recovery from aphasia (Luria, 1970) and on various perceptual asymmetries (see Chapter 4) also suggests a genetic mechanism affecting laterality.
4.   Children adopted by left handed parents are not more often sinistral than those adopted by right handers (e.g. Carter-Saltzman, 1980). Intrafamilial learning therefore seems to play little or no part in manual preference.

Early genetic theories dealt with handedness in isolation from the problem of hemispheric representation of speech (e.g. Chamberlain, 1928; Rife, 1940; Trankell, 1955). Yet the fact that cerebral dominance is related to handedness, albeit imperfectly, implies that both manual and cerebral asymmetry should be considered together. Two recent genetic theories which attempt to do so are those of Levy and Nagylaki (1972) and Annett (1972, 1975, 1978). These will be briefly dealt with in turn. It should be borne in mind, however, that theorists championing the influence of genes in the determination of handedness do not thereby exclude the possible influence of environmental factors on the final expression of hand preference.

### The Levy-Nagylaki Hypothesis

Levy and Nagylaki (1972) proposed a two gene model in which one gene coded for language representation (left or right hemisphere) and the other determined whether the preferred hand was contralateral or ipsilateral to this hemisphere. The alleles for language representation were denoted L (left hemisphere) and 1 (right hemisphere) and the alleles for handedness denoted C (contralateral) and c (ipsilateral). One allele within each pair was considered dominant (L and C) and the other recessive (l and c). Full expression of the gene for hemispheric representation of speech was assumed to depend upon the presence of the dominant alleles, either singly or in pairs, at each of the two gene loci. In order to make predictions concerning recovery from aphasia, it was assumed that the homozygote LL (i.e. those individuals who receive the dominant allele from each of their parents) does not have the capacity to develop speech in the right hemisphere following injury to the left, while the heterozygote Ll (i.e. individuals who receive one dominant and one recessive allele) does have this capacity.

Levy and Nagylaki tested their model against Rife's (1940) family handedness data and data on recovery from aphasia in different handedness groups collected by Zangwill (1960) and Goodglass and Quadfasel (1954). An excellent fit between the observed data and the values predicted by their model was reported by Levy and Nagylaki despite the fact that Rife classified as left handed any subject who carried out any one of ten tasks with the left hand while to be classified as right handed a subject had to perform all ten tasks with the right hand.

The model proposed by Levy and Nagylaki (1972) and Nagylaki and Levy (1973) was criticised by Hudson (1975) on a number of grounds. Firstly, it failed to predict new data on handedness collected by Annett (1973) and to fit a re-worked sample of Chamberlain's (1928) data, rejected by Levy and Nagylaki as a possible test of their model because of various inconsistencies in the presentation of Chamberlain's data. Secondly, there is no allowance in Levy and Nagylaki's model for two empirical observations. The first of these is the sex difference in the incidence of handedness; females have generally but not universally been reported to be less strongly or less often sinistral than males (see Chapter 8). The second observation is of a difference in the incidence of left handedness between generations.

The ability of Levy and Nagylaki's model to predict Rife's (1940) data, but not Hudson's re-working of Chamberlain's data nor the data collected by Annett, cannot be attributed to differences in the criteria of left handedness adopted by these different investigators. Levy and Nagylaki specifically state that their model is relatively insensitive to differences in criteria of left handedness. Although Levy (1977a) has replied to Hudson's criticisms it is difficult to accept a model whose expected values do not fit the empirical data. Conversely, it is difficult to see the force of Hudson's arguments concerning the gender and generational differences in the incidence of left handedness since Levy and Nagylaki would not wish to argue (see Levy, 1976a) that genetic mechanisms explain the entire variation in handedness; social factors may also have a part to play. A more serious criticism of the Levy and Nagylaki model is that, like previous models, it treats handedness as a dichotomous variable with no provision for mixed handedness. Supporters of Levy and Nagylaki might argue that mixed handers consist of left handers who have yielded to social pressure to use the right hand. However, a problem with this argument is that Gillies, MacSweeney and Zangwill (1960) reported that mixed handers are more likely to use the right hand for activities less subject to social control than those for which they use the left hand (but see Lee Teng, Lee, Yang and Chang, 1976).

A further problem of Levy and Nagylaki's model is that it predicts that monozygotic (MZ) twins should not be discordant for handedness. The data do not bear out this prediction. Nagylaki and Levy (1973) argue that twins do not provide a fair test of genetic theories because of the high incidence of pathology among twins. McManus (1980b) attempts to refute this and other arguments put forward by Nagylaki and Levy and suggests, on the contrary, that any genetic model must attempt to explain 'a significant but non-binomial degree of concordance in MZ twins'. The model proposed by Levy and Nagylaki cannot do so (Corballis and Beale, 1976).

*Annett's Right Shift Theory*

Annett (1964) proposed a model in which hand preference is determined by two alleles, D (right handed) and R (left handed), D being dominant and R being recessive. Individuals of the DD genotype are right handed (and left hemisphere dominant for speech), the RR genotype is left handed (and right hemisphere speech dominant) and the heterozygote DR may use either hand preferentially and have speech in either hemisphere depending upon the degree of penetrance of the incompletely recessive gene. Assuming random assortative mating in the population the model predicts that left, right and mixed handers will occur in approximately binomial proportions (r²:2rl:l²). This prediction was confirmed by Annett (1967). However, despite support for certain features of the model from the findings of Satz, Fennell and Jones (1969) the model is inadequate since it makes predictions concerning the inheritance of handedness within families and those predictions were not fulfilled in the data collected by Annett. She has therefore proposed an alternative model which makes the radical proposal that whereas right handedness is, in some sense, genetically programmed, left handedness is not. To understand this model it is necessary to consider certain features of the data collected by Annett (1970a).

Annett (1970a) analysed responses to a handedness questionnaire and discerned 23 different categories into which respondents could be placed according to which hand they used for a number of activities. There were no sharp boundaries between these categories implying that the number of categories possible was limited only by the number of questions asked. Furthermore, on a timed peg-moving task a frequency distribution of the difference in time (skill) between the hands was approximately normal in shape with, as one would expect, a mean score favouring the right hand.

The fact that this distribution was not bimodal, with peaks at the left and right hand ends, indicates that hand differences in skill lie on a single continuum. This approximately normal distribution of differences between the hands in *skill* is not inconsistent with a J-shaped distribution of hand *preference* given that the mean of the skill distribution is to the right of zero. Those people whose skill strongly favours the left hand, making them consistently left-preferent, constitute a small proportion of the skill distribution; those whose skill favours the right hand constitute a much larger proportion of this distribution giving rise to high frequencies of right-preferent individuals. Those for whom the difference in skill is very small (their scores fall either side of zero, the point of equality between the hands) will tend to show degrees of preference intermediate between those whose skill strongly favours one hand or the other and they will be spaced out on the preference distribution between consistent left and consistent right handers. The J-shaped preference distribution thus derives from the normal distribution in skill. The latter distribution is therefore taken as the basis for what follows.

As Annett has pointed out, what needs to be explained is (a) the variation in the distribution of differences between the hand in skill and (b) the dextral bias of this distribution, that is, the fact that the mean is to the right of zero. Variability is characteristic of other species which show a preference for one paw or the other, such as mice (Collins, 1977), rats (Peterson, 1934), monkeys (Lehman, 1978) and chimpanzees (Finch, 1941). Such limb preference has not been shown to be under genetic control. By the same token, it is unreasonable, according to Annett, to consider *variations* in human handedness as due to genetic influences. It is the *bias* towards dextrality that is peculiarly human, other species preferring the use of the left or right forelimb in approximately equal proportions.

Annett (1972, 1975, 1978) proposes that the *variation* is due to chance (i.e. accidental influences operating on early physical growth) and the *bias* is produced by a genetic factor accentuated by cultural influences. Since Man is the only animal to show a population bias towards preferential use of one limb, Annett considers this genetic factor to be unique to Man. Given the uniqueness of Man's faculty for speech, and the association between handedness and cerebral dominance for speech, she postulates a genetic factor which favours left hemisphere representation of speech and incidentally loads the dice in favour of the right hand. That is, the genetic factor biases the factors influencing handedness in favour of a dextral outcome. Given the evidence that the left hemisphere is specialised for sequential motor behaviour (see Chapter 5) one may speculate that this is the common denominator underlying both fluent speech, which requires very finely modulated sequences of movements in the vocal tract, and the greater dexterity of the right hand. It should be noted, however, that this suggestion is not made by Annett herself who prefers to leave the nature of the genetic factor unspecified, except to suggest that it might have something to do with the early development of language. This factor, which she terms the 'right-shift' factor, since it biases the handedness distribution toward dextrality, is not thought to be *necessary* for speech but only to confer some advantage on the left compared with the right hemisphere.

Annett, then, postulates a single allele which she identifies as whatever it is that, when present, influences the left hemisphere towards subserving speech and which incidentally shifts the mean of the distribution of hand differences in skill away from zero towards right hand superiority. The alternative allele at the same locus she considers to be uncommitted with regard to hemispheric speech representation and hence for handedness. In the absence of the right shift factor, which might equally be termed a left shift factor if one prefers to think in terms of cerebral as opposed to manual asymmetry, the directions of both hemispheric speech dominance *and* manual preference are thought to develop according to chance influences and independently of each other. In other words, individuals without the right shift factor (rs − −) may have speech in either hemisphere and be either left or right handed according to chance variation. Thus this single gene theory, unlike other genetic models, does not posit one gene for right handedness and another for left handedness; a single gene influences the individual's chances of becoming right handed.

Considering the distribution of differences between the hands in skill, Annett supposes this to be a composite of two underlying distributions, that for individuals without the right-shift factor on either chromosome (rs − −) and that for individuals in whom the right-shift factor is present on one (rs + −) or both (rs + +) chromosomes (see Figure 1). (Since it is the mere presence of the right-shift factor in either single or double dose that is important for present purposes these two distributions are simply referred to as RS− and RS+.) By definition the mean of the RS− distribution is zero. (In the absence of the RS factor there is an equal probability of having a superior left or right hand, chance variation determining the degree of superiority in relation to zero, the point of equality between the hands). The RS+ distribution has a mean somewhere to the right of zero − since the presence of the right shift factor predisposes an individual towards right handedness.

Assuming equal variance of the RS− and RS+ distributions, the relative height of each can be calculated from estimates of the relative frequencies of the genotypes in the population. The best estimate we can make of the size of the RS− distribution is the proportion of the population who are right brained for speech. For this, Annett took the combined series of patients reported in the literature as showing dysphasia following a unilateral cerebral lesion where entry to the series was independent of brainedness or handedness. Of these patients, 9.2 per cent were dysphasic and had a right sided lesion. Since individuals lacking the RS+ gene will be right brained in half the number of cases and left brained in the other half of cases it follows that $9.2 \times 2 = 18.4$ per cent of

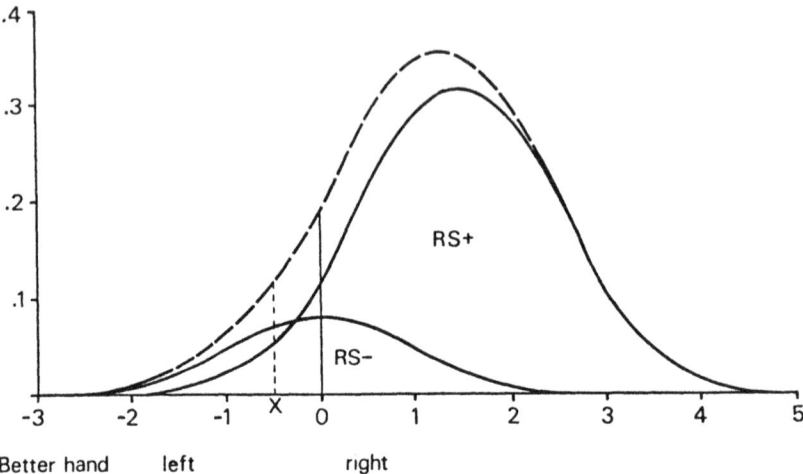

**Fig. 1** A schematic representation of the hypothesised distributions of L-R skill in the population. In the RS+ (*rs*+ + and *rs*+ − genotypes) the chance distribution of L-R skill is shifted toward the right. In the RS− (*rs*− − genotype) there is no systematic bias to either side. The threshold for expression of left handedness, *X*, can vary over a wide range to either side of 0.

dysphasics belong to the RS − distribution. RS + individuals thus comprise 81.6 per cent of the dysphasic population. Using these values as the best estimate of the proportions of RS − and RS + individuals in the general population, the relative heights of the RS − and RS + distributions must be made such that the areas under the two curves represent 18.4 and 81.6 per cent respectively of the total area under the curve representing the sum of the two distributions combined.

Fundamental to Annett's model is the notion that the position of the cut-off point defining left handedness is essentially arbitrary. It follows from the overlap of the RS − and RS + distributions that different cut-off points, or thresholds of sinistrality, give rise to different proportions of RS − individuals among both left and right handers relative to the total population. As the threshold is moved rightwards from position x in Figure 1 to include a greater proportion of the population to the left of threshold (designated left handed), the *proportion* of RS − individuals within both left and right handers falls. (This is because left handed RS − individuals are being included in a larger segment of the total population and RS − right handers, those to the right of the threshold, constitute a decreasing proportion of the RS − distribution). Since RS − individuals turn out to be right brained in half the number of cases, the percentage of right brainedness falls as the threshold is moved rightwards. This concept of the sinistrality threshold thus enables the model to account for the different incidences of right brainedness among left and right handers that have been reported by different investigators.

Given the notion of the sinistrality threshold, the question arises as to how far to the right of zero is the mean of the RS+ distribution.

From the published data Annett (1975) calculated that of 533 dysphasic patients, 3.4 per cent were said to be left handed with a right cerebral lesion. Consequently, $2 \times 3.4 = 6.8$ per cent may be considered left handed and RS− (since only half of the sinistral RS− population will be right brained for speech). Left handers constituted 7.7 per cent of the total dysphasic population and hence $7.7 − 6.8 = 0.9$ per cent are left handed and RS+. Similarly, Annett calculated that 5.8 per cent of dysphasics had a right cerebral lesion and were right handed. Thus $2 \times 5.8 = 11.6$ per cent of the population may be thought of as RS− and right handed. The proportion of the population who are RS+ and right handed is therefore $100 − 6.8$ (RS−, sinistral) $− 0.9$ (RS+, sinistral) $− 11.6$ (RS−, dextral) $= 80.7$ per cent.

The proportion of all RS+ individuals who are left handed is $0.9 \times 100/ (80.7 + 0.9) = 1.10$ per cent. Of all RS− individuals $6.8 \times 100/ (6.8 + 11.6) = 36.95$ per cent are left handed. Converting these values to Z scores (standard deviation units), 36.95 per cent would arise from a cut-off point placed at 0.34z to the left of the mean of the RS− distribution (i.e. the area under the curve to the left of this point makes up 36.95 per cent of the total area under the curve). Similarly, on the RS+ distribution 1.10 per cent arises from a cut-off placed 2.28z to the left of the mean of this distribution. For the cut-off points on the RS− and RS+ curves to coincide, therefore, since both cut-offs define left handers, the mean of the RS+ distribution must, assuming equal variance of the two distributions, be $2.28 − 0.34 = 1.94Z$ to the right of zero, the mean of the RS− distribution.

Given the fixed parameters of the model (the relative heights of the RS− and RS+ distributions, their distance apart and the assumption of equal variance) the model can be used to generate precise quantitative predictions. From estimates of the incidence of left handedness in a particular sample of individuals can be calculated the criterion or threshold of sinistrality for that sample in relation to the hypothesised RS− and RS+ distributions. From this it is possibe to estimate the relative frequencies of the genotypes rs++, rs+− and rs−− for sinistrals and dextrals. From this in turn the expected frequency of left handed offspring born as a result of R × R, R × L and L × L matings can be calculated. Annett (1979) reports that her model successfully predicts family handedness data for three generations from incidences of parental left handedness as different as 3.6 and 23.7 per cent. This ability to predict the proportion of left handed children as found by investigators differing widely in their definition of left handedness constitutes powerful evidence in favour of Annett's theory.

Two other features of the reported data on handedness can also be readily accommodated by the right-shift theory. The first is the raised incidence of sinistrality in twins compared with singletons; the second is the common but not universal finding that females are less likely to be strongly left handed than males.

It is possible to account for the raised incidence of sinistrality among twins by assuming that pre-natal pathological influences have caused a shift in pre-determined handedness (see above). However, Annett (1975) prefers to argue that the RS+ factor is less effective in twins than in singletons and thus for twins the mean of the RS+ distribution is not as far to the right of the RS− distribution as it is for singletons. In other words, the extent of the bias towards dextrality is not as great among twins. The effect of bringing the RS− and RS+ distributions closer together is that for any criterion of sinistrality there will be a higher proportion of left handers than for the same criterion applied to the more widely separated distributions for singletons. Annett (1978) shows how her theory can predict successfully the numbers of pairs concordant and discordant for handedness among both dizygotic and monozygotic twins. In this connection it should be noted that the existence of monozygotic twins discordant for handedness proves an embarrassment for any theory postulating a gene controlling left handedness. This is not a difficulty for Annett's theory since right handedness is not presumed determined by the RS factor but only made more likely. The theory, in other words, is probabilistic not deterministic. Variation in handedness is considered due to chance (accidental causes); only the bias towards dextrality is induced by the RS factor. Thus, even in the presence of the RS factor it is possible for one twin to become left handed and the other to turn out right handed. (For a review and critique of studies claiming a higher incidence of sinistrality in monozygotic than dizygotic twins, see McManus, 1980b).

The postulate that the RS+ distribution in twins is closer to the RS− distribution than it is among singletons has been applied in reverse to explain the often reported sex difference in manual preference. Relatively fewer females report themselves to be strongly left handed (Oldfield, 1971; Newcombe, Ratcliffe, Carrivick, Hiorns, Harrisson and Gibson, 1975; Thompson and Marsh, 1976; Bryden, 1977; Hardyck, Petrinovich and Goldman, 1976; Heim and Watts, 1976 but see Birkett, 1981 for contrary view). Annett (1979) argues that the right shift is expressed more strongly in females than in males. She sees this as consistent with the fact that girls acquire speech earlier than boys and are less at risk for delay in learning to speak and to read. It might be argued that the reported sex difference in handedness is artefactual, resulting from a reluctance in females to give 'extreme' responses on questionnaires (Bryden, 1977). However, Oldfield (1971) observed that the excess of male left handers was not confined to extreme left handers but was distributed throughout the sinistral part of the range of laterality quotients derived from his questionnaire. Two further considerations argue against an interpretation of the sex difference as artefactual and imply that the greater extent of right shift in females is real. First, Annett (1979) has reported that the mean of the distribution of differences between the hands in *skill* is further to the right of zero for females than for males, even in a sample where there were not proportionately fewer left *preferent* females than males. Secondly, the assumption that the right shift is greater for females gives rise to the prediction that a larger proportion of left-handed children will be born if the mother is left handed than if the father is left handed; this prediction is confirmed by empirical observation (Falek, 1959; Annett, 1973, 1979).

Although some people might feel uncomfortable at the probabilistic nature of the basic postulates of Annett's right-shift theory, its predictive power highlights the heuristic value of considering handedness not as a dichotomy but as a continuum.

*Collins's Theory*
Whereas Annett sees variation in human handedness as due to accidental, or chance, influences affecting early physical growth, and the bias towards dextrality as determined genetically, Collins (1977) argued the opposite case. He sees the variation as under genetic control but not the bias. He has carried out some elegant experiments (Collins, 1970; 1975; 1977) with mice in the course of which he found that the preferential use of one paw or the other can be changed by bringing up mice in a world biased in the opposite direction to that of their natural lateral preference. Specifically, mice are brought up in cages in which the feeding tube is flush against the left or right wall of the cage. To get food, the animals have to put one paw through the tube. The majority, but not all, left-pawed mice will change to using the right paw if brought up in a cage with the feeding tube on the right and similarly for mice with an initial preference for the right paw. The fact that the mice differ in readiness to change to the reversed orientation supports the idea that there are natural lateral biases but this does not necessarily mean that they are genetic in origin.

Collins has also carried out selective breeding experiments in which mice of the same paw preference are mated together for several generations. This procedure does not result in any change in the incidence of left and right paw preference among the offspring, which it would be expected to do if there are separate genes coding for left and right paw preference.

The results of the selective breeding experiments and the finding that the lateral preference of most animals can be modified if they are brought up in a world biased in the opposite direction has led some commentators to assert that Collins is a proponent of an environmentalist view of human handedness, opposed to any genetic argument. This is not true. What Collins does suggest is that direction of hand preference, left or right, is not determined genetically but arises out of what he terms an 'asymmetry lottery' which is a seemingly random process. However, genetic variation may determine the *degree* to which lateral preference is expressed.

## Non-Genetic Theories of Handedness

The idea that right and left handedness is determined genetically, one or more allele coding for left handedness and another or others coding for right handedness, as suggested, for example, by Levy and Nagylaki (1972), has been disputed by Morgan (1977) and Corballis and Morgan (1978) – see also Corballis and Beale (1976). These authors argue that genes can not, in effect, tell left from right – they are 'left-right agnosic' – and that the biological factors governing handedness are specified in the spatial structure of the oocyte (unfertilised ovum) rather than in the genes. Left cerebral dominance they see as one example of a fundamental biological asymmetry which often, but not always, favours the left side among chordates.

In Man, and other mammals, the heart is normally displaced to the left of the body midline: the left claw of the fiddler crab is longer than the right, the narwhal has a tusk only on the left, and so on. However, the very fact that many asymmetries favour the right side (the liver is on the right, the larger claw of the stone crab is usually the right one) implies that the left sided phenomena are so far from universal that it is not very helpful to look for a single, fundamental left-right gradient which might underlie left hemisphere dominance in Man as well as anatomic and behavioural asymmetries in other species (cf. Boklage, 1978 and other commentators on Morgan and Corballis, 1978). This, however, is the speculative side of the thesis offered by Morgan and his colleagues. More pertinent is their claim that no clear example of genetic control over the direction of any asymmetry in chordate species has been demonstrated. Consequently, they believe there is no need to invoke genetic mechanisms to explain human handedness, which they prefer to think of as determined by asymmetry in the cell cytoplasm during early embryonic development. A critique of their view is offered by Levy (1977b).

## Socio-Cultural Factors Influencing Handedness

Even those who argue in favour of genetic determination of handedness are willing to acknowledge that the degree and/or direction of hand preference can be modified through learning. Such learning may be mediated through the culture in which an individual lives and/or through his immediate family environment. The fact that cross-cultural differences in the distribution of handedness have been observed (Silverberg, Obler and Gordon, 1979) suggests a sociocultural influence, although cross-cultural differences in the gene pool logically can not be ruled out.

The role of familial learning was examined by Leiber and Axelrod (1981a,b) among 2257 subjects. It was hypothesised that having sinistral first degree relatives would influence the direction and/or degree of hand preference expressed by the respondents. The data did indeed show an increased incidence of sinistral preference among individuals with left handed relatives, the incidence being higher with a sinistral parent compared with sinistral siblings and slightly greater when the mother as compared with the father was left handed. With regard to degree of consistency of hand preference, the prediction was upheld for right handers in so far as there was a small shift away from full dextrality in individuals with sinistral relatives. For left handers, however, having sinistral relatives was associated with a decrease in degree of left handedness. This apparently paradoxical finding was attributed by Leiber and Axelrod to the fact that people calling themselves left handed are less consistent in their use of the left hand than are self classified right handers with respect to the right hand. Consequently, in families with left handers there is a less consistent model to follow and this might increase the variability of hand preference among respondents. McKeever and van Deventer (1977a), using a much smaller sample, found no significant influence of familial sinistrality on degree of left or right handedness as measured by a shortened version of the Edinburgh Handedness Inventory.

Since in their study the degree of handedness was shifted only slightly in the presence of familial sinistrality, the inference which Leiber and Axelrod drew from their findings was that intra-familial learning plays only a very minor part in influencing manual asymmetry. They concluded that there are 'definite limits set ... by biological determinants to the influence which training can exert on human handedness'. They do not, however, explain the increased incidence of left handedness in respondents with sinistral relatives. They presumably regard this as reflecting an inherited, rather than learned, tendency for left handedness to run in families. Clearly, an equally strong case could be advanced for the contrary argument. A compromise is suggested by the finding that the *incidence* of sinistrality among respondents is greater with a left handed mother compared with a left handed father, which might reflect a sex-linked biological determinant, whereas the presence of a sinistral father was associated with an increased *degree* of left handedness in sinistral respondents and decreased dextrality in right handed respondents. This effect of paternal sinistrality might reflect the influence of early learning.

The issue of whether handedness is determined primarily by intra-familial learning, or is largely inherited, can only be decided by adoption studies. Hicks and Kinsbourne (1976) found in a partial cross-fostering study that the hand preference of their student respondents was associated significantly with the writing hand of their biological parents but not with that of their step parents. However, as the mean length of time the students had spent with a step parent was $7.24 \pm 3.12$ years and the mean age of the students was 20.18 years it would be the case that most of the respondents would have been old enough for hand preference to have become firmly established (Palmer, 1964) before they acquired a step-parent. Thus the results reported by Kinsbourne and Hicks are not surprising. This problem was not present in a more recent larger-scale study by Carter-Saltzman (1980) who compared handedness distributions in the children of biological

and adoptive parents (of babies before their first birthday) as a function of
parental handedness. Her results indicated that handedness is determined
more by biological factors than by upbringing.

While patterns of hand preference at an individual level may be more
subject to genetic determination than to intra-familial learning, population
studies of handedness indicate that the influence of culture is not negligible.
Silberberg, Obler and Gordon (1979) studied 1171 Israeli high school
students and found that the distribution of handedness patterns differed
significantly from those reported in the literature for Taiwanese (Teng,
Lee, Yang and Chang, 1977) and British populations (Oldfield, 1971). The
Israeli sample contained larger proportions of subjects at both ends of the
distribution than did either of the other two samples.

Shanon (1979a) observed Americans and Israelis while they were
drawing various characters. He noted that sinistrals tend to execute lines in
a right-to-left direction whereas dextrals do so in a left-to-right direction.
(Both handedness groups, it may be noted, move their preferred hand from
the midline outwards.) Since English script is written left-to-right, American right handers have their natural tendency in accordance with the
direction of reading and writing in their culture. For Israeli right handers
the opposite is the case since Hebrew is written from right-to-left. Shanon
found that right handers of both cultures execute lines predominantly in
the direction of their natural tendency whereas left handers were less
consistent. Where the natural tendency and the direction of reading and
writing were in conflict (i.e. for American left handers) some individuals
tended to draw in their natural direction whereas others conformed to the
direction in which English script is written. Shanon therefore argued that
left handers are less resistant than right handers to cultural pressure.

It is possible that there has been a relaxation of cultural pressure to use
the right hand over the last few decades. An increase in dextral preference
with increasing age has been reported among normal (e.g. Annett, 1973a;
1979; Carter-Saltzman, 1980) and psychiatric subjects (Fleminger, Dalton
and Standage, 1977). Levy (1976) calculates that the frequency of left
handed writing in the United States has risen from 2.2 per cent in 1932 to
over 11 per cent in 1972, which she regards as the asymptotic frequency of
sinistrality in the absence of cultural pressure to use the right hand (Levy,
1976a). This may not be the only factor influencing manifest handedness.
Brackenridge (1981) found that left handed writers increased from an
estimated 2 per cent or so of the populations of New Zealand and Australia
in 1880 to approximately 13 per cent in 1969 by which time the sinistral
proportion had levelled off. He supposes that differential perinatal survival
rates as well as relaxation of cultural pressure have acted together so as to
promote the increase in left handedness in recent years.

## SUMMARY

This chapter has reviewed the literature concerning genetic, pathological, cultural and familial influences on the expression of hand preference. The validity and reliability of methods of measuring handedness and the association between handedness and other aspects of motor laterality were also considered. The most important point to emerge was that handedness should not be considered as a dichotomy but as a continuum: pure left and right handers lie at opposite ends of a continuous distribution of manual preference. The implications of this for the study of handedness in relation to other variables have not yet been fully appreciated.

Writing posture may also be an important aspect of handedness in distinguishing between those with ipsilateral and contralateral motor control.

Attempts to relate handedness to language lateralisation are discussed in Chapter 5; differences between left and right handers are dealt with in Chapter 10.

# 3

# The split-brain studies

## SOME BASIC NEUROANATOMY

Put very simply, the brain consists of the brain stem, a central core (the mid-brain) and the cerebral cortex which is above and around the mid-brain. In addition, the cerebellum, or hindbrain, is located behind the brain stem beneath the level of the cortex (Figure 2).

In phylogenetic terms, the cerebral cortex represents the most recent development of the brain, the ratio of cortex to sub-cortical tissue increasing as the phylogenetic scale is ascended. So great is this development in man that, in order to be accommodated by the cranium, the cortex has had to fold over on itself. This is what gives rise to the corrugated pattern of humps and troughs which form such a distinct feature of the brain's appearance. Technically, a hump is referred to as a 'gyrus' or 'convolution' and a trough as a 'fissure'.

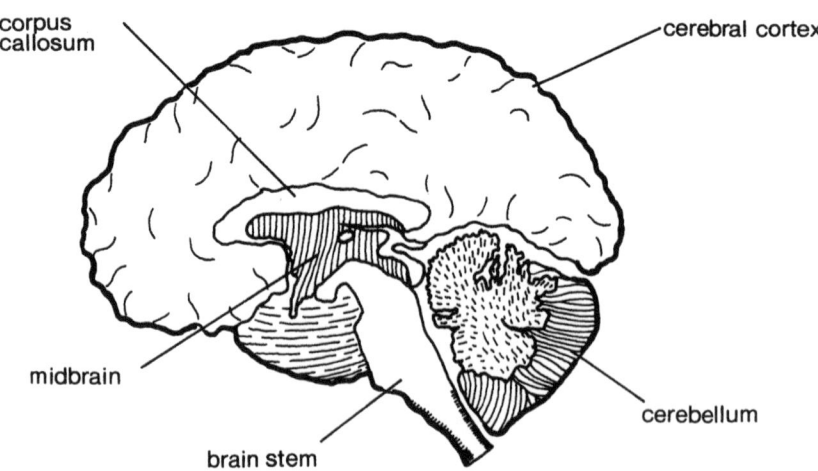

**Fig. 2** Diagram of brain divided down the middle (mid-saggital view)

The exposed surface of the brain represents only a proportion of the total mass of cortical tissue, the walls and floor of each fissure making up the remainder. The precise pattern of convolutions is different for each individual but there is sufficient general uniformity for certain fissures and gyri to serve as landmarks on the cortical surface and to be assigned names. The most important of these landmarks are shown in Figure 3.

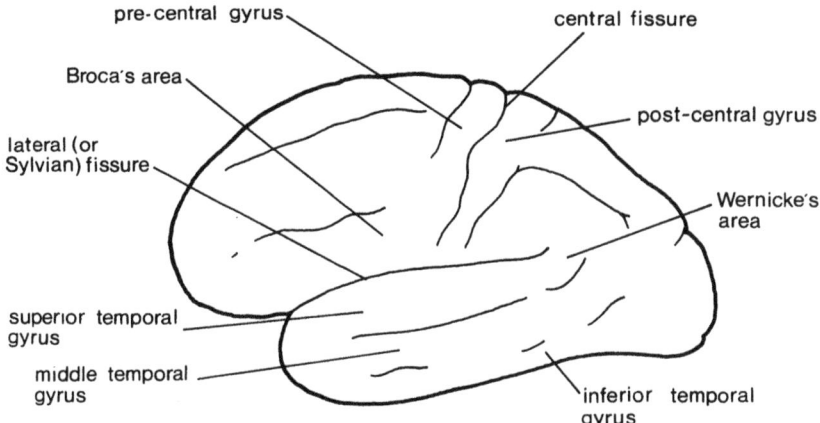

**Fig. 3** Diagram showing main fissures and gyri of the left hemisphere

The cortical mantle around the mid-brain is not a single blanket of tissue but consists of two distinct and apparently separate halves, the left and right cerebral hemispheres. The left and right halves of the brain are joined by a large band of fibres known as the corpus callosum. Fibres connecting the two sides of the brain are known as commissures. In addition to the corpus callosum there are other smaller bands of fibres, notably the anterior, posterior and hippocampal commissures, connecting the cortex on the two sides. Below the level of the cortex, at the mid-brain, most structures are duplicate in form in that they have a definite left and right side. Again there are commissures running between the two sides. Thus although the brain is bilateral in its structure there are extensive connections between its two halves.

It is customary to divide the cortex of each hemisphere into a number of regions or lobes. The simplest division is into frontal, parietal, temporal and occipital lobes as shown in Figure 4. Roughly speaking, the temporal lobe is concerned with auditory functions and the occipital lobe with seeing, although some visual functions are represented in the temporal lobe.

As a general rule, one side of the brain is concerned primarily, though not exclusively, with the opposite (contralateral) side of the body. Damage

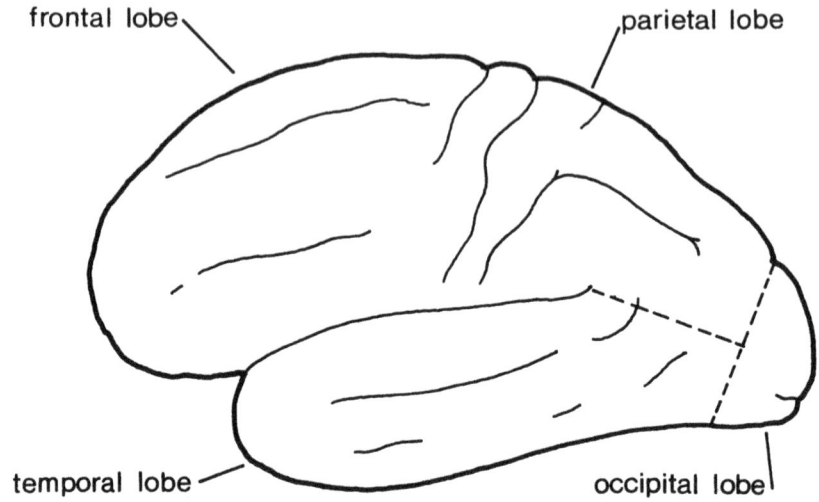

frontal lobe

parietal lobe

temporal lobe

occipital lobe

**Fig. 4** Diagram showing division of cerebrum into four lobes

to the sensori-motor cortex on the left side of the brain, for example, causes sensory impairment or paralysis of the limbs on the right side of the body.

Consistent with the general rule of crossed or contralateral nervous control, the left and right halves of the total field of vision are each represented in the occipital cortex and the opposite cerebral hemisphere. Light reflected from an object in the right half of the field of vision, the right visual hemifield, is transmitted to the left half of the brain. Similarly, light from the left half of the visual field is transmitted to the right half of the brain (see Figure 5, p. 40).

Although each visual hemifield is represented only in the contralateral cerebral hemisphere we are not consciously aware of this any more than we are aware of the blind spot in our retina. In our conscious experience the two hemifields are perceived as one continuous field of vision. There is, of course, no physical 'division' in the real world corresponding to the boundary – referred to as the vertical meridian – between the left and right visual hemifields. Thus we perceive no dividing 'line' between the two half fields of vision.

The fact that crossed neural representation applies to vision as well as to sensori-motor functions means that the right half of the brain both receives visual information from the left half of the visual field and controls the motor output of the limbs on the left side. The same hemisphere, in other words, both perceives a visual stimulus and controls a motor response on that side. This may be an evolutionarily adaptive way of dealing with spatially distributed events in the environment (Dimond, 1972). However, since the neural centres corresponding to each body-half are located at opposite sides of the brain some cross-talk between the hemispheres must normally take place in order for an integrated course of action to be adopted. If it were otherwise, each body-half would always act independently.

The so-called split-brain operation was first employed by Van Wagenen in the early 1940s as a treatment for otherwise intractable epilepsy (Van Wagenen and Herren, 1940). The idea was to prevent the spread of seizures across the brain. Epileptic convulsions arise as a result of abnormal activity at one point in the brain spreading to involve the entire cortex. Cutting the connections between the two sides of the brain confines the abnormal electrical discharge to one hemisphere thereby reducing the frequency and severity of fits. Total forebrain commissurotomy, to give the operation its technical name, involves division of the anterior and hippocampal commissures as well as section of the corpus callosum and, when present, the mass intermedia. Partial commissurotomy refers to incomplete section of the callosum. Despite the term 'split-brain' operation, commissural section does not completely divide the brain, which remains unified below the level of the cortex. Descriptions of the surgical procedures adopted are given by Bogen and Vogel (1962); Bogen, Fisher and Vogel (1965) and Wilson, Reeves, Gazzaniga and Culver (1977).

Van Wagenen's patients were studied by Akelaitis who found little or no impairment in psychological functions which could be attributed solely to the effects of the operation (Akelaitis, 1941; 1942; 1943; 1944; 1945; Akelaitis, Risteen, Herren and Van Wagenen, 1942). This stood in contrast to the effects observed with naturally occurring lesions in or near the callosum (Dejerine, 1892; Liepman and Maas, 1907; Trescher and Ford, 1937; Geschwind, 1965; 1969; 1970) and led to considerable speculation as to the functions of this structure. It was jocularly suggested that it was present for purely mechanical reasons, namely to prevent the hemispheres from sagging (Lashley, 1929). However, in the 1950s and 1960s the studies carried out on cats and monkeys by Myers and Sperry demonstrated unequivocally in these animals the crucial role of the corpus callosum in the transfer of information from one side of the brain to the other (Myers and Sperry, 1953; Sperry, Stamm and Miner, 1956; Stamm and Sperry, 1957; Sperry, 1958; Myers, 1960; 1962; 1965). Briefly, the technique was to cut the callosum longitudinally and then, by sectioning the optic chiasm and presenting stimuli to one eye only, to restrict visual input to one side of the brain. Animals were trained using one hemisphere to perform various visual discriminations and, once these were learned, the animal was tested with the stimuli presented to the untrained side of the brain. With the callosum sectioned prior to training, the animal required as many trials to learn the discrimination as it had originally taken. This demonstrated that callosal section prevented transfer of learning to the untrained hemisphere. In another condition of the experiment animals were trained through one eye with the optic chiasm sectioned but the callosum left intact. Following section of the callosum after training, the discrimination was usually performed correctly when stimuli were presented through the untrained eye. Thus the presence of an intact callosum during training led to both hemispheres learning the task (e.g. Butler, 1968).

Since the early 1960s the revival of commissurotomy as a treatment for debilitating convulsions (Bogen and Vogel, 1962; Bogen, Fisher and Vogel, 1965; Wilson, Reeves, Gazzaniga and Culver, 1977) has made available a number of patients for investigation. The now classic results confirmed the findings of the animal research with regard to the functions of the corpus callosum and provided a sharp impetus to the study of cerebral lateralisation of function. The most well known series of patients to have undergone split-brain surgery (of varying degrees) are those operated on by Bogen and Vogel in the 1960s and studied by Sperry and his collegaues at the California Institute of Technology. More recent studies have been carried out by Gazzaniga with patients operated on by Wilson working on the East side of the United States.

Although brain-bisection affords a singular opportunity to assess the abilities of each hemisphere disconnected from the other, it needs to be borne in mind that the patients studied by Sperry, Gazzaniga, and their collaborators have all, without exception, suffered severe and long-standing epilepsy. They have also suffered the trauma of the surgery itself and have usually been under anti-convulsant medication for extended periods of time. The level of cognitive functioning of split-brain patients is generally lower than that of neurologically intact subjects. Response times, for example, are much slower in split-brain patients, and as a rule IQ scores are lower than average. This reduced level of functioning may influence the results of experimental tasks in such a way as to give spuriously low levels of performance under certain conditions.

The particular pattern of symptoms observed after commissurotomy depends upon a number of factors including the age, intellectual status, pre-operative history, post-operative experience and extra-callosal brain damage of the patient concerned. In particular, the age at which the original brain damage was sustained and the age at which the surgery was carried out are each significant. The younger the age of the original damage and the surgical insult, the greater the opportunity for some degree of cerebral reorganisation of function to occur. The extent to which the results from one patient can be generalised to other patients, far less to neurologically intact subjects, is also limited by the probability that the surgical lesions are not identical in all cases. There is usually no independent verification of the extent of the surgery – and hence of the extent of any remaining commissural fibres. Not surprisingly, then, there are considerable individual differences in symptomatology, both in the immediate post-operative period and during long-term recovery. This chapter summarises those findings which can be considered to apply to all patients tested, though it should be realised that much of the published literature is based on a relatively small number of unusually well-practised individuals. Reviews of this research are also provided by Sperry (1974); Sperry, Gazzaniga, and Bogen (1969); Bogen (1969a,b; 1979); Gazzaniga (1970); Levy (1974a,b); Nebes (1974).

It is convenient to consider split-brain research under two main headings, dealing firstly with the issue of inter-hemisphere transfer, and secondly with lateral specialisation of function.


## THE ROLE OF THE CORPUS CALLOSUM IN INTERHEMISPHERIC TRANSFER

**Visual Functions**

The anatomical pathways from the eyes to the brain are shown schematically in Figure 5. The optic fibres from the outer (temporal) half of each retina pass to the lateral geniculate nucleus on the same side of the brain. Here they synapse with other cell bodies whose fibres travel to the visual cortex. (A small number of fibres also pass from the geniculate nucleus to the superior colliculus). The inner (nasal) fibres from each eye, unlike the temporal fibres, cross over at the optic chiasm and terminate in the lateral geniculate on the opposite side of the brain. The effect of this hemi-decussation of optic nerve fibres is that stimuli presented to the right of where a subject is looking impinge on the left temporal and right nasal hemiretinae and are thus conducted in each case to the left visual cortex. Similarly the left side of visual space is represented in the right visual cortex. Thus, provided that a subject does not move his eyes during stimulus exposure, it is possible to direct visual information to one side of the brain or the other by requiring the subject to steadily fixate a particular spot. In order to ensure that information is transmitted to only one side of the brain, it is usual to present stimuli at very brief exposure durations that are less than the time required for the eye to turn towards a stimulus that is presented off-centre. This limitation on the time for which material can be presented is a problem, given that split-brain patients almost invariably require longer exposure durations than normal subjects in order to perform recognition or reaction time tasks. Such very short exposure durations may interact with memory limitations so as to seriously impair the ability of patients in some circumstances to perform the required tasks. Recently, however, a technique has been devised which allows patients to view stimuli for an extended period of time while maintaining the visual input to a single hemisphere. Results using this technique are indicated at the appropriate point in the review which follows.

Tests show that each half of the brain is capable of independently perceiving and responding to stimuli presented tachistoscopically in its own half field of vision (Gazzaniga, Bogen and Sperry, 1965). The usual procedure is to ask the subject to indicate, by pointing or by some other response, that he has understood what was shown to him. Using such methods it has become clear that, while each hemisphere has the capacity to perceive and remember non-verbal stimuli and to perform cross-modal matches, inter-hemispheric integration of pattern or colour is not usually possible. The patients cannot indicate, for example, whether two stimuli flashed one on either side of fixation are the same or different although such

LEFT                                                    RIGHT

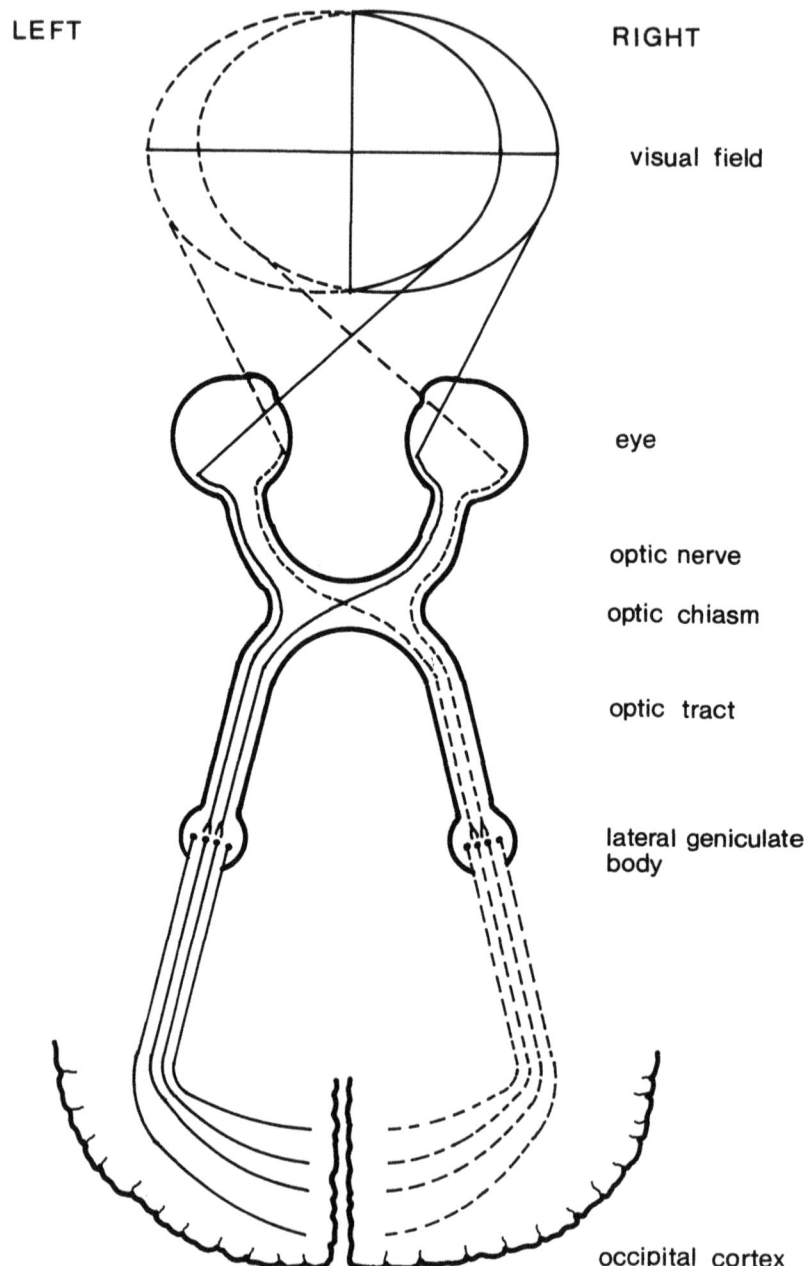

visual field

eye

optic nerve

optic chiasm

optic tract

lateral geniculate
body

occipital cortex

**Fig. 5** Diagram of the human visual system

a comparison can readily be made if both stimuli are exposed within the same half-field. These and similar tests confirm that the forebrain commissures effect an interchange of visual information between the hemispheres, each of which analyses input from the contralateral half of visual space.

The corpus callosum appears to be concerned, inter alia, with unification of the visual field across the mid-line. Berlucchi (1972, 1981) reviews neurophysiological evidence which demonstrates that visual cortical cells with callosal connections have receptive field areas which are, generally speaking, close to or superimposed upon the vertical meridian of the visual field. At the behavioural level, Mitchell and Blakemore (1969) reported that split-brain patients could not tell whether a stimulus was shown immediately in front of or behind a fixation point. They could, however, easily say whether the stimulus had appeared closer in or further out than the plane of fixation if the stimulus appeared to one side of fixation.

Although the callosum serves to bring together pattern information from the two half fields of vision, certain aspects of visual experience remain undivided after forebrain commissurotomy. The onset or offset of a light, gross bright differences, and movement in the periphery of either visual field, can be detected by each half of the brain. Furthermore, stimuli produced by simultaneously shrinking and expanding point sources of light, one on each side of fixation, are perceived as being part of a single, coupled-like motion. Lines inclined to each other at an oblique angle, or aligned horizontally or vertically, and moved synchronously in the two visual fields also give rise to a unified percept (Trevarthen and Sperry, 1973; Trevarthen, 1974a). While the mechanisms for cross-field visual identification of subjects are compromised by callosal section, the surgery does not eliminate the mechanisms for the allocation of visual attention to one visual half-field as a result of prior information projected to the other (Holtzman, Sidtis, Volpe, Wilson and Gazzaniga, 1981). Those aspects of visual experience which are not dependent upon the integrity of the callosum are presumably mediated largely by means of the superior colliculus of the mid-brain and/or the posterior commissure.

**Auditory Functions**

Gazzaniga (1970) states that localisation of a sound source, an ability dependent upon a comparison of the phase, intensity or time of arrival of the sound at the left and right ears, is not impaired in split-brain patients. This implies that the integration of certain kinds of information from the two ears occurs at a sub-callosal level. Data from patients who have had an entire hemisphere removed by surgery support this conclusion (Berlin, 1977).

In Man, each ear is connected to the cerebral hemisphere on the opposite side by the contralateral auditory pathway and to the hemisphere on the same side by the ipsilateral pathway (see Fig. 6, Chapter 4, page 84). When information is fed directly to either the left or right ear split-brain patients give an accurate verbal report of the input to either ear. Assuming left hemisphere speech output, the fact that there is no difficulty in reporting left ear input implies that the ipsilateral pathway from left ear to left hemisphere is fully functional. However, Milner, Taylor and Sperry (1968) and Sparks and Geschwind (1968) observed that under dichotic listening conditions, in which competing information is fed simultaneously to the left and right ears, their patients showed virtually no recall of left ear input. From this it can be argued that the contralateral pathway from the right ear to the left (speaking) hemisphere in some way inhibits or suppresses the information carried by the ipsilateral pathway from left ear to left hemisphere. This idea is consistent with a variety of physiological evidence and is discussed further in Chapter 4.

Given that there is no difference in recall between left and right ears of split-brain patients with monaural stimulus presentation, Milner et al. (1968) concluded from their results that the callosal route from the right to the left hemisphere is more important in mediating dichotic left ear performance of normal subjects than is the ipsilateral pathway from the left ear to the left side of the brain. Comparison of two patients who had undergone complete commissurotomy with two partial split-brain patients, in whom only a part of the corpus callosum was divided, suggested that the crucial region of the callosum involved in inter-hemispheric transfer of auditory information lies between the most posterior section (the splenium) and the anterior third of the callosum (Springer and Gazzaniga, 1975).

Since the early work on dichotic listening in split-brain subjects was carried out it has become clear that left ear responses are not always extinguished. Springer, Sidtis, Wilson and Gazzaniga (1978) compared the performance of 5 patients with complete commissurotomy (in one case sparing the anterior commissure) on dichotic recognition of CV (conso-nant-vowel) syllables and digits. For all patients, recall of left ear digits (using a written response) was generally high (over 80 per cent for four out of the five patients) and superior to recall of the syllables, of which approximately 20 per cent were correctly reported from the left ear in all but one patient. This patient recognised only three per cent. Springer et al. suggested that since

> ... the acoustic structure of the CV syllable pairs permits a considerably greater degree of spectral-temporal overlap within a stimulus pair than does the digit stimuli ... the degree of 'suppression' of the left ear input may well be a function of the extent to which that stimulus competes, cortically or subcortically, on an instant by instant basis with a right ear stimulus similar to it both spectrally and temporally. (p 309)

Greater familiarity of digits may also have played a role. (Since the CV syllables comprised a smaller set of stimuli than did the digits, and the stimuli used were shown on a list in front of the subject during testing, it cannot be argued that digits comprised a smaller set of items to be held in short term memory.)

The conclusion to be drawn from the above experiment is that during dichotic listening at least some information must be transmitted by way of the ipsilateral left ear–left hemisphere route.

## Somesthesis

Turning now to the sense of touch (somesthesis), laterality effects have been found to be subject to considerable individual variation. Gazzaniga, Bogen and Sperry (1963) reported that, if visual cues were excluded, the first patient studied was unable to point with one hand to where he had been touched on the opposite side of the body. This did not apply to the head or neck region for which cross-localisation was consistently good. However, the second case of the series was able to cross-localise on the torso and proximal extremities and this was interpreted as showing a greater degree of bilateral cerebral representation of each cutaneous half-field in this patient. In general, it appears that for the head and neck region all patients can cross-localise, for the torso and limbs the situation tends to be intermediate, while poorest transfer is seen for the hands and feet. Although the mere presence or absence of tactile stimulation may sometimes be signalled by either side, more complex forms of discrimination, such as between continuous cross-wise or lengthwise movement across the skin, fail to transfer. Nor can split-brain patients usually perform inter-manual comparisons of objects held in each hand (Sperry, Gazzaniga and Bogen, 1969).

## Olfaction

The sense of smell is represented independently in the hemisphere ipsilateral to each nostril (Gordon and Sperry, 1969). This was shown by plugging each nostril in turn and presenting different olfactory stimuli to be recognised subsequently by the same or the opposite nostril.

## Motor Functions

Studies of motor functions in the first nine patients of the series were summarised by Gazzaniga, Bogen and Sperry (1967). All patients initially showed left sided apraxia, that is, an inability to carry out actions with the left arm in accordance with verbal commands. However, the required movements could generally be imitated if first demonstrated by the examiner. Although an improvement in the selective dyspraxia of the left arm was observed during the early post-operative period Gazzaniga et al. stated that 'the condition probably persists chronically in all patients to some degree'. Subsequent observation, however, has revealed that though this applies to several patients it does not apply to them all (Bogen and Vogel, 1974).

Initially at least, the patients were unable to draw with the right hand an object seen in the left visual field, although they could do so with the left hand. Free drawing was also much better with the left than with the right hand (Bogen, 1969a). In addition, these patients showed some difficulty in performing Block Design tests with the right hand. These impairments suggest that the callosum is traversed during visuo-spatial or visuo-constructive performance by the right hand in neurologically intact subjects (Ettlinger and Blakemore, 1969).

Bi-manual coordination of split-brain patients during normal, everyday activities was generally good. For example, the first patient of the Bogen series was able to put on a tie, fold towels and tie shoelaces although occasionally antagonistic actions between the two hands were observed (Gazzaniga, Bogen and Sperry, 1962). Later patients of the series were also able to perform tasks such as catching a ball but were poor at carrying out two different acts concurrently, one with each hand (Gazzaniga, Bogen and Sperry, 1967). This, incidentally, was reported to be the most consistent symptom in the patients described by Akelaitis (Akelaitis, 1945; Smith and Akelaitis, 1942).

On tests requiring inter-dependent regulation of speed and timing between right and left hands split-brain patients are severely impaired (Preilowski, 1972; 1975). This contrasts with their good coordination with regard to highly over-learned activities and suggests a shift in nervous control to lower centres for long-established bi-manual actions (Sperry, 1974). It must be remarked, however, that the good bi-manual coordination shown for everyday skills is dependent upon the patient being able to see his hands. With vision excluded some difficulty in coordinating left and right hands persists in all patients with total commissurotomy even in the long term (Bogen and Vogel, 1974). It would be interesting to have a video-record of the patients' bi-manual performance under visual control. A careful analysis may reveal deficits in timing which might otherwise go unnoticed. Since it appears that bi-manual coordination deteriorates when the patient's attention wanders it might also be instructive to attempt to distinguish experimentally between visual and attentional factors.

### Comparison with Patients of the Akelaitis Series

In view of the dramatic disconnection symptoms disclosed by the careful studies of Sperry and his collaborators it is surprising that Akelaitis observed little effect of commissurotomy on the performance of those patients who underwent the operation in the 1940s. It has been suggested that the apparent lack of effects among the patients studied by Akelaitis may have been due to less complete section of the callosum than in the series of patients operated on by Bogen, Vogel and Wilson. Another possible reason is that the surgery was usually performed in two stages in the Akelaitis series but more commonly in a single operation in the more recent cases (Hurwitz, 1971). However, in a follow-up study of one of the patients from the Akelaitis series it was found that crossed visuo-tactile matching and inter-manual transfer of learning were both grossly defective, even 27 years after the operation (Goldstein and Joynt, 1969). Such enduring effects suggest that lack of sophistication in testing techniques may have played some part in masking certain of the basic split-brain phenomena in the patients studied by Akelaitis. Certainly, the patients become remarkably adept at cueing in one hemisphere to what is going on in the other (Gazzaniga, 1969; 1970). For example, by using his left hand to trace out on the back of the right hand a word presented to the left visual field it is possible for a split-brain patient to give the impression that his right hemisphere can read and/or speak since the tactile information from the right hand can be verbalised by the left hemisphere.

## Comparison with Callosal Agenesis

The effects of split-brain surgery may be compared with the condition known as callosal agenesis in which the callosum is absent from birth (for reviews see Selnes, 1974; Chiarello, 1980). In general, acallosal subjects do not show the dramatic disconnection effects seen after commissurotomy. For example, they do not show left ear extinction under dichotic listening conditions (Netley, 1977; Dennis, 1981; Lasonde, Lortie, Ptito and Geoffroy, 1981), and generally they are able to match visual stimuli presented to left and right visual fields (Chiarello, 1980). Nonetheless, defects have occasionally been reported. Lynn, Buchanan, Fenichel and Freeman (1981) mention that their patient sometimes misnamed objects held in the left hand. Gott and Saul (1978) reported that one acallosal patient could not cross-integrate complex visual information, but no control data are reported. Milner and Dunne (1977) found that their patient, unlike a normal control group of subjects, did not recognise that a tachistoscopically exposed picture made up of two halves of different faces, presented so that the join coincided with the vertical midline of the visual field, was in any way unusual. (Interestingly, the normal subjects did not notice the asymmetrical nature of the so-called chimeric stimulus if the line joining the two half-faces of the composite figure was obscured by a white strip in the final photograph shown to the subjects.)

Impairment of perceptuo-motor coordination among acallosals and in bilateral transfer of learning and tactile cross-localisation have been found by a number of workers (Russell and Reitan, 1955; Ettlinger, Blakemore and Wilson, 1972, 1974; Ferris and Dorsen, 1975; Reynolds and Jeeves, 1977). In two cases these have been found to persist during a 15 year period (Jeeves, 1979). Similar results have not been reported by others (Kinsbourne and Fisher, 1971; Ettlinger, Blakemore, Milner and Wilson, 1972). It seems probable that the discrepant findings among acallosals relate primarily to differences in the degree of difficulty of the experimental tasks employed (Jeeves, 1979) and to differences in cerebral pathology which almost always accompany the condition of agenesis. (Were it not for such associated problems the agenesis would hardly ever come to light since there would be no call for radiological evaluation.) Differences between patients in level of intellect may also contribute to the divergence in findings. (For an excellent review of behavioural studies of callosal agenesis see Milner and Jeeves, 1979.)

The absence of dramatic split-brain symptoms among acallosals can be attributed either to the use of non-callosal commissural systems and/or to the use of ipsilateral sensory systems, perhaps in conjunction with bilateralisation of function in the acallosal brain. The anterior commissure, for example, has been shown in one study of commissurotomised patients to be capable of supporting a considerable degree of inter-hemispheric transfer of verbal, pictorial and auditory information in some individuals (Risse, Le Doux, Springer, Wilson and Gazzaniga, 1978). This structure is often enlarged in callosal agenesis (Sheremata, Deonna and Romanul, 1973). (Embryological evidence suggests that the anterior commissure may regress after the appearance of the callosum (Harner, 1977) and thus its enlargement in callosal agenesis may reflect an earlier stage of brain development.) However, there appears to be considerable individual variation in the extent to which the anterior commissure can mediate inter-hemispheric transfer of information in different sensory modalities. McKeever, Sullivan, Ferguson and Rayport (1981) failed to find evidence of inter-hemisphere transfer in three split-brain patients in whom the anterior commissure was spared in contrast to the findings of Risse et al. who report good transfer in four other patients.

With regard to the bilateralisation of function hypothesis, the presence of language in the right hemisphere would account for the lack of left ear extinction under dichotic listening conditions and for the ability to name stimuli presented to the left hand or in the left visual field. Indeed, it has been found (Gott and Saul, 1978) that at least one acallosal subject had bilateral speech functions as determined by the Wada amytal test (see Chapter 5). However, this patient was not a right hander so the probability of bilateral speech is higher than for a fully dextral subject anyway. Dichotic listening scores for acallosals do tend to show, on the whole, somewhat reduced ear differences in comparison with normals (Chiarello, 1980) which is consistent with the view that language is more bilateralised in acallosals. Alternatively, increased reliance on ipsilateral pathways, combined with interhemispheric transfer by means of non-callosal commissures, would also produce decreased ear asymmetry.

Although acallosal patients do not show the pattern of performance characteristic of split-brain patients, they do not always perform at the same level as normal subjects. Chiarello (1980) in her review of the published cases concludes that acallosals sometimes show deficits on spatiomotor tasks, even those that presumably can be accomplished within a single hemisphere. She argues that associated brain damage cannot, on its own, explain this impairment. More recently Dennis (1981) examined the language abilities of a single acallosal patient and concluded that she showed some impairment in syntactic and pragmatic functions of language in comparison with normals.

It can be speculated that the presence of an intact corpus callosum allows for the optimum level of both linguistic and spatial performance from the same brain. Absence of the callosum may necessitate greater bilateralisation of function than usual. This may be inimical to the (hemispherically segregated) neural organisation best suited to serve both linguistic and visuospatial functions (Levy, 1969). The possibility can therefore be entertained that under normal circumstances the callosum serves to establish lateralisation of function. It may do this by one hemisphere exerting an inhibitory influence on its partner with respect to particular, specialised functions.

Dennis (1976) argued, on the basis of a comparison between acallosal and split-brain performance, that during ontogeny the callosum inhibits the development of ipsilateral sensory and motor pathways from hand to brain with regard to topographic localisation. It may be predicted, therefore, that such inhibition would be reduced by callosal section. In line with this general idea, Dimond, Scammell, Brouwers and Weeks (1977) found that their patient who had undergone section of the trunk of the callosum showed a post-operative increase in the use of his left hand. He also showed a switch from a pre-operative advantage of the right ear on a dichotic listening task to a post-operative left ear superiority. Dimond et al. suggested that the destruction of part of the callosum disrupted an inhibitory function normally mediated through this region.

The findings from callosal agenesis, though they do not mirror the results of the split-brain studies in any direct fashion, can be interpreted to support the conclusion that the corpus callosum is crucially involved in the integrated action of left and right hemispheres. Future studies of differences between longterm post-operative performance of split-brain patients and acallosals may serve to illustrate the degree of compensation possible for absence of the corpus callosum in the adult and immature brain respectively.

## LATERAL SPECIALISATION OF FUNCTION IN THE BISECTED BRAIN

It is quite clear that both halves of the brain can initiate motor responses to stimuli presented within their own half-field of vision; the response of the contralateral hand is invariably performed without impairment when stimuli are flashed to one or other hemisphere. Hand postures, for example, can be copied successfully from pictures and objects are manipulated with normal facility. In contrast, all patients except one in the Bogen series (Nebes and Sperry, 1971) have shown severe defects in ipsilateral manual control when asked to respond to a stimulus presented on the same side of the visual field as the responding hand. This was particularly true for the right hand/right hemisphere combination and may in part reflect the difficulty of controlling this hand without interference from the 'constant prevalence of dominant activity in the left hemisphere' (Gazzaniga, Bogen and Sperry, 1967). There seems, however, to be considerable long term improvement shown by a number of patients in the degree of ipsilateral manual control that they can achieve although many actions remain slower than normal even years after surgery (Zaidel and Sperry, 1977).

Split-brain patients readily carry out tests of intra-modal and inter-modal transfer involving visual and tactile stimuli, provided that both stimuli are presented to a single hemisphere, either the left or the right. In addition, matches can sometimes be made if a visual stimulus is directed to one hemisphere for an object to be retrieved by the ipsilateral hand. This would seem to be accomplished by the ipsilateral sensory system conveying sufficient information about specific features of the object, such as the presence or absence of edges, its length or weight, for a correct choice to be made from among the limited number of alternatives available.

## Verbal Functions

Complete commissurotomy is typically followed by a period of mutism from which the patient may take some time to recover. The reasons for this mutism are unclear (Bogen, 1977). Speech eventually returns although some patients have complained that they cannot sing as well as before (Bogen and Vogel, 1974). In the weeks after the operation patients can describe or name stimuli and read words or letters presented in the right visual field but cannot do so if the same stimuli are presented to the left visual field. Objects placed in the right hand can be readily named but an adequate verbal commentary cannot be given of tactile stimulation of the left side (except the head and neck region) or of objects placed in the left hand (Gazzaniga, Bogen and Sperry, 1965). Indeed, this left-hand anomia is one of the most 'ubiquitous and persistent' deficits observed during long-term follow-up examination (Bogen and Vogel, 1974). Writing with the right hand is carried out as efficiently as before the operation but the left hand is usually agraphic even if pre-operatively the patient could write with this hand (Bogen, 1969a).

Despite the lack of speech shown by the right hemisphere, this half of the brain is not without some verbal capacity. To recall a phrase of Hughlings Jackson (1874), 'The speechless man is not wordless'. If, instead of demanding a vocal response, a patient is required to retrieve with his left hand an object designated by a word flashed in the left field of vision, he may sometimes succeed, though denying verbally that he knows what object to choose. When two words are presented, one in each visual field, he may retrieve with his left hand the object designated by the word on the left of fixation while saying that it is the object whose name appears to the right of fixation (Sperry and Gazzaniga, 1967). He may also respond appropriately with his left hand to oral commands from the examiner, and retrieve vocally identified objects (Gazzaniga and Sperry, 1967). Thus there is some capacity for auditory-verbal as well as visuo-verbal comprehension in the right hemisphere. This implies that it is the ability to express itself verbally that the right hemisphere lacks. Nonetheless, Butler and Norsell (1968) reported that one patient was able to name symbols presented in the left visual field, provided that the exposure duration was sufficiently prolonged. Since precautions were taken to ensure that central fixation was maintained these authors wondered whether the right hemisphere could, in fact, express itself in words if given sufficient time to do so. However, the possibility that there was some information-transfer between the hemispheres as a result of sophisticated cross-cueing strategies remains open.

*Linguistic Functions of the Right Hemisphere*
Although the right hemisphere seemed to understand some words, the initial findings suggested that such words belong to a single grammatical class, namely nouns. The left hand could retrieve objects but could not perform actions to visually presented commands nor reliably pick out pictures corresponding to such actions (Gazzaniga and Sperry, 1967). More recently, Iwata, Sugishita, Toyokura, Yamada and Yoshioka (1974) studied three (Japanese) patients with section of the splenium of the callosum for the removal of tumours of the pineal gland and found no evidence for any right hemisphere comprehension of even simple nouns. Thus these patients, in whom childhood pathology may be presumed absent, did not confirm the earlier findings. It is possible that the American series of patients, all of whom sustained longstanding epilepsy prior to surgery, show some degree of neural re-organisation of function as a consequence of early brain damage. Such neural re-organisation is generally held to occur to a greater extent in the immature compared with the adult brain. However, even in the adult some re-organisation seems possible. Three years after operation the Japanese patients could understand some words written in Kanji (an ideographic script) but not words written in Kana (a syllabic script) when tested using their right hemisphere. This contrasted with their total failure to read anything with their right hemisphere immediately after the surgery (Sugishita, Iwata, Toyokura, Yoshioka and Yamada, 1978).

Initial studies of language comprehension in the right hemisphere suggested that the right half of the brain was unable to understand printed verbs (Gazzaniga, Bogen and Sperry, 1967; Gazzaniga and Hillyard, 1971) in so far as the subjects did not carry out the commands specified by the verbs, such as to tap, wave or smile. Levy (1974a) reports that she tested the ability of the right hemisphere to execute printed instructions using three different response methods: (i) performance of the specified action, (ii) pointing to a picture of the appropriate action and (iii) tactually retrieving an object associated with the action. She found that the patients could not carry out the action, were sometimes able to point to the correct picture but were often successful at retrieval of an associated object. Levy therefore argued that during tactual retrieval of the object by the left hand the sensory input was restricted almost entirely to the right hemisphere, which could thereby indicate its understanding without interference from the left hemisphere. With a pointing or 'action' response the left hemisphere, by utilising ipsilateral pathways to the left hand, tended to usurp right hemisphere control of responding.

The apparent inability of the right hemisphere to respond to verbs presented in the left visual field (although some patients could pick out from a multiple-choice array an object associated with the verb) stands in marked contrast to the understanding demonstrated by the ability of some split-brain patients to locate with the left hand an object described aloud by an experimenter, even if the latter only mentions the use to which the object is usually put.

In formal testing of the auditory vocabulary of the isolated right hemisphere Zaidel (1976) found no difference between comprehension of nouns and verbs. In general, the auditory single word vocabulary of the right hemisphere substantially exceeded its visual vocabulary in two representative commissurotomised patients and was only slightly inferior to the level observed for the left hemisphere.

According to Zaidel (1978) the right hemisphere analyses the meaning of a word in terms of its 'auditory gestalt' rather than by analysing its phonetic components. One line of evidence for this view is that discrimination of a word against background noise interferes far more with the ability of the right hemisphere to choose a matching picture than with that of the left hemisphere. Zaidel argues, '... the noise, lacking specific phonetic information, interferes with auditory analysis by the right hemisphere which depends on general acoustic cues, much more than it interferes with phonetic analysis by the left hemisphere.'

The fact that the right hemisphere has some ability to generate an auditory sound image (Zaidel, 1978) contrasts with the finding of Levy and Trevarthen (1977) that the right hemisphere is unable to match pictures of objects on the basis of whether the names of the objects rhyme (e.g. key/bee). This suggests that the right hemisphere does not spontaneously convert a visual stimulus to an acoustic representation. Zaidel and Peters (1981) have recently shown that the right hemisphere is unable to indicate whether a written word rhymes with the name of an object shown on a picture even in those instances where the patient is able to read the word using his right hemisphere. This suggests that right hemisphere reading is undertaken by means of purely visual analysis rather than by means of phonological decoding (see also Chapter 5).

Methodological limitations inherent in the use of tachistoscopic procedures have, until recently, restricted investigation of the right hemisphere's linguistic capacity to the level of the single word. However, Zaidel has devised a contact lens technique whereby the image of a stimulus always appears on the same area of retina whatever the position of the eye. This allows the subject to make free scanning movements of his eyes while still restricting visual input to a single hemisphere thus freeing an experimenter from the need to use tachistoscopic procedures which severely restrict the amount of information which can be presented at any given moment.

Using this technique Zaidel (1978) presented various phrases to his two split-brain subjects and tested their comprehension by means of a multi-choice array of pictures, where the meaning of the pictures was different from the stimulus by one critical word. Zaidel found that beyond about three words the right hemisphere's performance deteriorates. This he attributes to a severely limited short term verbal memory in the right hemisphere. He suggests that short term verbal memory is phonetically based or dependent upon an articulatory system, a view for which there is some independent experimental evidence (Baddeley and Hitch, 1974; Ellis and Hennelly, 1980). In Zaidel's view it is this system that the right hemisphere lacks. Again the auditory ability of the right hemisphere to understand phrases exceeds its capacity to follow phrases presented visually. Thus while phonetic decoding may be deficient in the right half of the brain, there is once more evidence of the capacity to abstract an acoustic gestalt from the auditory input.

The contact lens technique permitted examination of the sensitivity of the right hemisphere to nuances of meaning specified by contrasting grammatical constructions. In terms of syntactic competence, Zaidel (1977) found that the right hemisphere performs at approximately the level of a normal five year old child. However, Zaidel is at pains to point out that the right half of the brain is not simply a minor version of the left hemisphere, a case of arrested development as it were. The psycholinguistic performance of the right hemisphere shows a unique profile which does not correspond to any stage of first language acquisition nor to any specific pattern of language breakdown as in different categories of aphasia (Zaidel, 1978).

To consider now executive rather than receptive aspects of language, Levy, Nebes and Sperry (1971) found that using plastic letters one patient could spell simple words with his left hand, yet when asked to say aloud the word he had just arranged he was unable to do so. It thus appeared that the right hemisphere, at least in this patient, could effect some language expression, though not vocally.

In a second type of test, the examiner made simple words with plastic letters which the patient had to feel with his left hand. Although unable to say the word aloud, the patient could write it out of sight with his left hand. After doing so, he could often pronounce the word. It thus seems that bilateral kinaestheic feedback allowed the left hemisphere to discriminate what the left hand had written.

In a third type of test the patient was asked to write down the names of various objects placed in his left hand. It was observed that only the first two or three letters of each word were correct and that the patient gave in vocal response words which resembled the middle and end letters of his written response. Levy et al. therefore argued that the left hemisphere wrested control of the (left) writing hand after the initial few letters, and suggested that 'the minor hemisphere could probably talk more than it does, were it not for the grip which the major hemisphere maintains over the motor channels for speech'. This is an interesting suggestion in as much as it implies that occupying the left hemisphere on a task of its own may allow the right hemisphere a greater chance to express itself. Some such mechanism is suggested by the results of an experiment by Gordon (1980a).

Pairs of commands, to which the subject had to respond with either hand, were presented dichotically to left and right ears of split-brain patients. Normally, it will be remembered, split-brain patients show suppression of the left ear input in this situation, that is they recall little from the left ear (but see above – Springer et al. 1978). However, using manual output Gordon observed that patients were able to respond to a fair proportion of left ear commands with their left hand at the same time as the right hand carried out the instructions given to the right ear. The proportion of left ear commands responded to increased when input to the right ear consisted of a story which the subject had to shadow (i.e. repeat verbatim as it was being told). Thus with the left hemisphere occupied on the verbal shadowing task the right hemisphere gave evidence of being able to express its understanding through the appropritate motor action.

The linguistic competence of the right hemisphere seems to be considerably greater than might have been expected on the basis of studies of aphasic patients with restricted lesions of the left cerebral hemisphere. It is also an interesting fact that a restricted lesion of the left hemisphere may produce greater impairment of language than total left hemispherectomy (Smith, 1974). Attempts to assess the linguistic competence of the right hemisphere in normal subjects by means of reaction time studies, in which subjects had to indicate whether a word presented in one or other visual field was the same as a word heard previously, have indicated little right hemisphere ability (Moscovitch, 1972). On the basis of these observations Moscovitch (1973) has argued that the left hemisphere normally inhibits the comprehension and expression of speech by the right hemisphere. Inhibition is greatest in the intact brain, reduced with a unilateral left-sided lesion, reduced further by commissurotomy and eliminated altogether by left hemispherectomy (surgical removal of one hemisphere). This model of right hemisphere language behaviour stands in contrast to the traditional view that language functions are re-organised in the right hemisphere, following disease or surgery to the left, although the two models are not mutually exclusive. Certainly, it is noteworthy that the adult split-brain patients who have shown most evidence of right hemisphere expressive language are those who sustained cerebral birth injuries and underwent commissurotomy during their early teens.

One patient, known by his initials P.S., was able to spell certain words with his left hand (Gazzaniga, Le Doux and Wilson, 1977) and seemed to comprehend a range of language related stimuli presented to his right hemisphere in the period immediately following surgery. Some two years after the operation, at the age of 15 years, P.S. could verbally name stimuli presented in his left visual field, something he could not do during the early post-operative period. Since this patient cannot perform a visual same-different match across the vertical mid-line, his performance on naming tasks suggests the development of the ability to control vocalisation from his right hemisphere rather than leakage of visual information from his right hemisphere to his left (Gazzaniga, Volpe, Smylie, Wilson and Le Doux, 1979). This patient is unique among the split-brain patients, in this respect as well as in others, and has, in fact, had an entire monograph devoted very largely to his case (Gazzaniga and Le Doux, 1978).

The development of expressive language in the right hemisphere of patient P.S. is consistent with reports of the development of right hemi-sphere speech following aphasia-producing lesions of the left side of the brain (Kinsbourne, 1971; Pettit and Nolls, 1977) and even after total left hemispherectomy (Smith, 1974; Burkland and Smith, 1977). Recently Gazzaniga, Sidtis, Volpe, Holtzman and Wilson (1982) obtained evidence from P.S. which shows that even in the absence of the callosum some linguistic information crosses the mid-line in this patient. After learning with one hemisphere to associate a particular response word with a specified stimulus item P.S. was able to give the same response when the stimulus was presented to the opposite hemisphere. Since the stimulus was never *named* directly (and experiments showed that visual information did not transfer) this suggests that information of a linguistic nature was available to the naive or untrained hemisphere. As P.S. was unable to attach an associate to the *correct* member of a homophone pair using his untrained hemisphere it seems that the information which transfers is of a phonological rather than semantic nature.

It would be most interesting if someone were to test the isolated right hemisphere for the generative aspects of language. A number of sugges-tions have appeared over the years, to the effect that the right half of the brain may play some role in creative language functions (e.g. Critchley, 1962). Dimond, Scammell, Brouwers and Weeks (1977) commented that their patient who sustained section of the centre region of the corpus callosum showed an improverishment in speech behaviour in comparison with his language prior to operation. Pre-operatively, this patient could lucidly explain technical matters connected with his job but several months after the operation (to remove an angioma) the patient's speech was relatively simple and unsophisticated. Dimond et al. felt that the reduction in the quality of speech was specifically due to the callosal section. However it should be kept in mind that the patient's left hemisphere was retracted in the frontal region during operation in order for the surgeon to visualise the callosum. Milner (1974) has shown that lesions of the left frontal lobe produce impairment in the fluent production of words.

If the deficit observed by Dimond et al. in what might be termed 'meta-linguistic' function was specific to partial disconnection of the two cerebral hemispheres, why has a similar deficit not been reported for other split-brain patients? One possible explanation is that the American East and West Coast patients are all epileptics and have been so for a number of years. The opportunity for re-organisation of function within their brain has therefore been considerable. This does not explain why Jeeves, Simpson and Geffen (1979) did not report any 'meta-linguistic' defect in their three patients who sustained section of the trunk of the callosum (as did the patient studied by Dimond et al.) during removal of an intraventricular tumour. Be this as it may, it is still possible that somewhat elusive 'meta-linguistic' functions may derive in part from the right hemisphere of the intact brain. Wapner, Hamby and Gardner (1981) studied the language behaviour of patients with lesions restricted to the right half of the brain and came to the conclusion that such patients showed evidence of 'higher-level' linguistic difficulties. Using neurologically intact subjects Dimond and Beaumont (1974) report that when words were projected to the left visual field free associations to the stimuli were more unusual (in comparison with established norms) than with right field stimulation. This seems an area that will repay further careful study.

### Non-Verbal Functions in Split-Brain Patients

The left hand of split-brain patients is superior to the right hand in carrying out constructional tasks such as the Block Design sub-test of the Weschler Adult Intelligence Scale, the right hand being 'barely able to do the simplext problems' (Gazzaniga, Bogen and Sperry, 1965). Drawing and copying has also been observed to be consistently poor for the right compared with the left hand. This was the case even when pre-operative performance was better with the right hand and applied particularly 'in the copying of designs suggesting a third dimension' (Bogen and Gazzaniga, 1965). Such right hand 'dyscopia' (Bogen, 1969a) cannot be attributed to poor motor control since nearly all patients are able to *write* normally with this hand following their operation. Furthermore, a superiority of the right hemisphere may be observed on visuo-spatial matching tasks in which the executive component of performance is minimised.

Nebes (1971a) required commissurotomised subjects to match one of a number of arcs to the full circle from which it was derived. Matching was either by touch alone, by vision alone or by comparing the arcs felt in one hand with full circles seen in free vision, and vice-versa. The performance of the left hand was much more accurate than that of the right hand under all conditions even though there was no difference between the hands when either arcs or full circles were matched together. In a further experiment subjects were presented with fragmented stimulus figures in one or other visual field and were asked to retrieve with one hand a solid figure corresponding to what the fragmented stimulus would look like if it were not fragmented (Nebes, 1972). Again the right hemisphere was superior on this task despite the fact that there was no difficulty for either hemisphere when the visual stimuli were presented as unified figures. In a third experiment, three split-brain patients saw an array of dots in one or other visual field. On half the number of trials the array consisted of more dots along the horizontal than the vertical dimension and on the other half of the trials the reverse was the case. Normal control subjects perceived the dots as forming lines parallel to the axis containing the larger number of dots. Using this as criterion, the split-brain patients were found to be more 'accurate' with stimuli presented in the left compared with the right visual field (Nebes, 1973).

The experiments by Nebes and others (Milner and Taylor, 1972; Kumar, 1977) point to a superiority of the right over the left hemisphere in visual and tactile pattern perception and memory. However, the degree of difference between left and right hemispheres in this respect may be less than was at one time thought. Le Doux, Wilson and Gazzagnia (1977b) repeated with the patient P.S. the Nebes (1972) fragmented figures test and the test of tactile matching carried out by Milner and Taylor (1972) but, as well as requiring the patient to manipulate items by hand, Le Doux et al. also designed purely visual versions of the tests. The tasks became matching tasks in which stimulus and response cards had to be matched visually. The dramatic difference between the performance of left and right hemispheres was not seen in these versions of the test, even though the initial results were replicated with P.S. when the original form of the tests was employed. It is thus possible that much of the disparity between the two sides of the brain, previously found when the patients used their hands in a manipulative response, may be due not so much to visuo-spatial functions per se but to a manipulo-spatial advantage for the right hemisphere.

Levy-Agresti and Sperry (1968) devised a task which required forms felt by one hand to be matched with drawings of unfolded shapes. These shapes were such that if they were folded in three dimensions they would correspond to the forms which the subjects were exploring tactually. Split-brain patients were more accurate matching forms felt by the left hand with shapes in the left visual field than they were using the right hand/right field combination. Moreover, it seemed that the task was accomplished in different ways depending on which hemisphere was stimulated. The left hemisphere did best on those items which were amenable to verbal analysis, while the right hemisphere excelled on those items which were comparatively resistant to verbal description but could be readily visualised. Levy (1969; 1974a,b) has outlined a theory in which she argues that the neural organisation underlying language is in some way incompatible with the type of neural organisation required for visuo-spatial, holistic perception. In her opinion, 'cerebral asymmetry of function in an animal with language may be the only design for a brain which optimises both linguistic and Gestalt perceptual functions' (Levy, 1974b).

## GENERAL OBSERVATIONS ON THE EFFECTS OF COMISSUROTOMY

The results of the studies reviewed in this chapter demonstrate that the right as well as the left hemisphere has its own sphere of mental activity. In one sense, splitting of the brain appears to split the mind (Sperry, 1968). The question of whether the property of consciousness belongs only to one hemisphere, the left, as argued by Eccles (1973), or whether the brain has a double-consciousness, as suggested by Pucetti (1981), is not one that will be taken up here. Suffice to say that using his right hemisphere at least one patient is able to recognise photographs of himself and of other people and objects important in this patient's everyday life (Sperry, Zaidel and Zaidel, 1979). The right hemisphere can also express emotion appropriate to a situation (Le Doux, Wilson and Gazzaniga, 1977a; 1979). Nonetheless, it is a remarkable fact that to casual observation split-brain patients appear almost entirely normal during their everyday activities. With rare and transient exceptions, their mental life appears totally unified, and only careful experiment reveals the extent to which division of the corpus callosum results in two independently-functioning cerebral hemispheres.

Although split-brain patients appear on the surface to suffer no obvious gross cognitive impairment as a result of their operation the surgery may not be without some cost to normal mental function. Zaidel and Sperry (1974) tested ten patients with complete commissurotomy and two patients with partial section of the callosum and concluded that all complete patients were substantially impaired on tests of memory, relative to their general intellectual abilities. Unfortunately Zaidel and Sperry were unable to test their patients pre-operatively as well as post-operatively and it is not known how good their memory was prior to the operation. After extensive pre- and post-operative testing of a single patient – P.S. – Le Doux, Risse, Springer, Wilson and Gazzaniga (1977) found a post-operative improvement in all memory tasks and argued that commissurotomy per se was not harmful to cognitive processes generally. However, the improvement found for P.S. may reflect the reduction in fits as a result of the surgery; it does not necessarily mean that commissurotomy is without any cognitive cost.

Campbell, Bogen and Smith (1981) report that all of ten complete and two partial split-brain patients showed deficits on Benton's Visual Retention Test. Pre-operative scores were available for two patients (one complete, one partial commissurotomy). Both showed post-operative declines. Together with the low scores of the remaining patients this implies that maintenance of inter-hemispheric connections is necessary for adequate performance on the Visual Retention Test, perhaps because both verbal and non-verbal strategies are involved when performance is at an optimal level. A similar argument would help to explain the fact that on tests of word paired-associate learning the patients studied by Zaidel and Sperry (1974) all showed marked deficits in comparison with a small group of matched control subjects. This might be attributable to disconnection of the verbal hemisphere from imagery processes in the right hemisphere, since imagery is known to facilitate paired-associate learning. That image mediated verbal learning involves the right half of the brain is suggested by findings with both brain damaged and normal subjects. Jones-Gotman and Milner (1978) found that right sided temporal lobe damage impaired learning of words rated high in imagery, but not learning of abstract words. Similarly, Whitehouse (1981) reported that left sided anterior damage impairs verbal coding, leaving imaginal coding intact, while the converse is true with anterior right sided damage. In addition, several experimenters report that normal subjects show greater left visual field recognition of concrete or highly imageable words in comparison with abstract words (Ellis and Shepherd, 1974; Hines, 1976; 1977; Hatta, 1977; Day, 1977; 1979; Marcel and Patterson, 1978). However, an alternative explanation of these latter findings is that age-of-acquisition of words is the relevant variable (Beaton, Sykes, King and Jones, 1982).

In addition to the findings discussed above there are reports of memory defects following callosal surgery for non-epileptic conditions. Dimond, Scammell, Brouwers and Weeks (1977) commented that their patient with section of the centre region (trunk) of the callosum showed a persistent lack of memory for things he had done or said. Since it concerned his own behaviour, Dimond et al. referred to this memory impairment as 'auto-pragmatic amnesia'. Jeeves, Simpson and Geffen (1979) also mention a defect of short-term memory in a patient with a similar but smaller callosal section. However, bilateral damage to the fornix cannot be ruled out as a possible contributory factor in this case. Certainly another patient studied by the same group gave little evidence of any memory impairment, and in the case of the latter patient damage to the fornix was minimal (Geffen, Walsh, Simpson and Jeeves, 1980).

If there is a memory defect following commissurotomy does the problem lie in the initial acquisition stage or in the retention or retrieval of information? Huppert (1981) attempted to answer this question using a technique she developed for studying memory functions in patients with Korsakoff's syndrome. By equalising initial acquisition of information between groups of subjects it is possible to examine forgetting in the two groups without the confounding influence of any acquisition defect in either of them. Huppert equalised initial learning by manipulating the time required to view stimulus material such that it was recalled equally by each group in a test of immediate retention. In comparison with Korsakoff patients and normal control subjects three split-brain patients showed no quicker forgetting but two of the three patients required a longer period than the controls to reach equivalent levels of initial learning. These findings therefore suggest that if there is a memory defect in split-brain patients it lies in the initial acquisition of information. However, the single split-brain patient who performed as well as the control subjects at the initial stage had a post-operative IQ of 106 compared with IQs of 77 and 90 for the other two split-brain subjects. It is possible that this – rather than the callosal surgery – is responsible for the latter's acquisition defect.

Anecdotal evidence of an impairment among split-brain patients in the capacity to sustain attention has appeared in the literature (Sperry, 1974; Zangwill, 1974). In an attempt to investigate this experimentally Dimond (1976; 1979a,b) required six total and two partial commissurotomised patients to carry out traditional vigilance tasks in which the subject had to watch for and respond to infrequently occurring visual, auditory or tactual signals. Dimond found that the usual vigilance decrement set in much earlier for the total than for the partial split-brain patients, the latter performing in a manner similar to that expected of normals. When subjects failed to detect a signal, the signal was repeated every 1.5 seconds until the subject did respond. Dimond found that the number of additional signals required to evoke a response was much higher for the left than for the right hemisphere, increasingly so as the test session progressed. There is a close parallel here with the performance of normal subjects studied by Dimond and Beaumont (1973). Visual signals were presented only to one hemisphere throughout an experimental session. Although the detection rate for the left hemisphere was initially high, this declined fairly rapidly as the session wore on. Initial detection rate for the right hemisphere was not so high but showed comparatively little decline with time. This was not found to be the case when signals were presented randomly to either the left or right hemisphere throughout a session (Dimond and Beaumont, 1971a).

The implication to be drawn from Dimond's studies with normals and with split-brain patients is that the two halves of the brain maintain some sort of balance of arousal between each other by way of the corpus callosum (see also Bremer, 1966) and that the system can become 'de-coupled' by callosal section or the continued stimulation of only one hemisphere.

Dimond's results with commissurotomised patients were replicated by Ellenberg and Sperry (1979), in so far as a rapid decline in vigilance was observed when tactual signals were presented for response. However, if the task was a little more demanding, namely a simple sorting task, then no decline in the level of performance was observed throughout a testing session lasting one hour. Thus a detection task appears to entail a loss of sustained attention while a simple discrimination task does not. The greater cognitive complexity of the latter seems to be sufficient to maintain a uniform level of performance.

The word 'attention' refers to a slippery concept and hence has a number of different but related meanings (Kahneman, 1973). The experiments just described refer to the ability of the split-brain patient to 'pay attention' in the sense of concentrating for a protracted period of time. The question of how attention, in the sense of selecting how the resources of a limited capacity processing system are distributed between competing demands, is deferred until Chapter 13.

## META-CONTROL IN THE SPLIT-BRAIN

The results of the research discussed in this chapter demonstrate that each hemisphere is to some extent a complete brain in itself, able to perceive, learn and organise responses. What then determines which half of the brain will assume control?

When only one half of a vertically symmetrical figure is presented tachistoscopically to a split-brain patient, such that the vertical axis coincides with the mid-line of his visual field, the figure is perceived as being bilaterally symmetrical rather than as a mere half-figure. Exploiting this phenomenon of perceptual completion, Levy, Trevarthen and Sperry (1972) presented split-brain subjects with stimuli made up by combining one half of each of two different figures, such as two different faces. The composite stimulus, known as a chimeric stimulus, consisted of separate half-figures on each side of a vertical axis which was made to coincide with the visual mid-line. The task for each subject was to match the tachistoscopically presented stimulus with one of a number of alternative symmetrical figures subsequently presented in free vision. Prior to the experiment the subjects learned to attach names to each of these symmetrical figures.

The results showed that when a verbal response was required the subjects named the item corresponding to that half of the stimulus which was presented to the left hemisphere. If a manual response was called for, the subjects consistently chose the figure corresponding to the stimulus half presented to the right hemisphere, regardless of which hand was used for responding. The right hemisphere, therefore, seemed to predominate, provided that a verbal response was not required; with verbal responding the left hemisphere took control. As Levy et al. point out, it is of special interest that

> ... the minor right hemisphere dominated the voluntary motor response in most of the non-verbal tests where the sensory input to the right and left hemispheres was essentially similar, even when the major hemisphere was favoured by having the subjects use their right hands. (p 74)

It was noticed, however, that such right hemisphere responses were sometimes inferior to those obtained when the left hemisphere was forced to respond by, for example, requiring the patient to provide a verbal description of the stimulus pattern. This suggested to Levy and Trevarthen (1976) that hemispheric ability is not the same thing as hemispheric dominance. The general strategy in split-brain research has been to compare the performance of left and right half-brains independently which, as Levy and Trevarthen point out, precludes assessment of the *disposition* of one hemisphere to take control of behaviour. The split-brain findings inform us of the capacity, that is, the upper limits of performance, of each separated hemisphere. They do not normally tell us which hemisphere will attempt to control cognitive operations, a function which Levy and Trevarthen refer to as meta-control.

In an attempt to study meta-control, Levy and Trevarthen again employed chimeric stimuli. Four split-brain patients were asked to decide which stimulus from an array on a card in front of them 'went with' the 'one' they had just seen flashed on the screen. By looking at the match chosen by the patient, the experimenters were able to say whether the left or the right hemisphere was in control. Previous work had suggested that when instructed to choose a matching item that 'went with' a previously exposed (non-chimeric) item, split-brain patients using their right hemisphere chose items that showed visuo-structural similarities to the tachisto-scopically exposed stimulus (an 'appearance' match). Presenting a stimulus to their left hemisphere, on the other hand, led to the patients choosing matching items that showed functional-conceptual similarities (a 'function match'). For example, if a picture of a hat were flashed on the screen, patients using their right hemisphere tended to choose cake on a plate (which *looked* similar in outline to the hat) whereas with their left hemisphere they would tend to pick out spectacles which, like a hat, are worn by people, but the two items look totally dissimilar (see also McCarthy and Donchin, 1978; Landis, Assal and Perret, 1979). Thus if one half of a chimeric stimulus presented to a split-brain subject's left hemis-phere was, for example, (half of) a hat and the other half presented to the right hemisphere was (half of) a pair of scissors, a patient responding with a pair of spectacles as the matching item was assumed to have made a function match with his left hemisphere in control. Conversely, if the patient chose the cake, he was said to have made an 'appearance' match with his left hemisphere.

Levy and Trevarthen were able to show that with instructions to choose an item that 'looked like' the one just seen, patients usually picked out items that were similar in appearance to the half of the chimera presented to the right hemisphere, while instructions to choose an item 'that you would use with' the one just seen generally led to the item presented to the left hemisphere being chosen as the basis for comparison. However, three out of four patients not uncommonly showed a dissociation between the kind of matching performed and the hemisphere performing the match. That is, with 'appearance' instructions matches sometimes were given to the half of the chimeric stimulus presented in the right field of vision or with function instructions matches were made to the half chimera presented to the left visual field. Their findings led Levy and Trevarthen to argue that there can. be

> ... little doubt that meta-control systems exist, that these systems tend to activate that hemisphere which is appropriate for some task, and that they do so independently of whether the then activated hemisphere actually utilises its specialisations. On some occasions, however, the meta-control systems can fail to arouse the appropriately specialised hemisphere, in spite of the fact that the other one must then proceed to perform in a cognitively inappropriate mode. It is clearly the case that, in general, dominance of one hemisphere over behaviour cannot be due to a skill or speed contest between the two halves of the brain, but must depend upon the *expectations* as to the cognitive requirements, irrespective

of whether those cognitive specialities are actually utilised. It would appear that hemispheric activation does not depend upon the hemisphere's real aptitude, but on what it *thinks* it can do. (p 310)

The problem of meta-control cannot be confined to the split-brain but must arise for the intact brain also.

## SUMMARY

From evidence reviewed in this chapter, it is quite clear that each half of the brain can, when required to do so, independently perceive, remember and respond to stimuli presented to it. Such a remarkable degree of independence is probably not possible in the non-commissurotomised individual in whom the corpus callosum conducts information from one side of the brain to the other.

It is also clear that the left and right hemispheres of the split-brain patient differ to some extent in the functions that they each carry out. However, every patient who has undergone total commissurotomy has previously suffered from long-standing epilepsy. This may have caused some cerebral re-organisation of function. Results obtained from split-brain patients, therefore, may not apply to normal individuals with an intact corpus callosum. The question therefore arises as to whether hemispheric asymmetry of function can be demonstrated in neurologically normal subjects. This issue is explored in the chapter which follows.

Dramatic disconnection effects are not seen in cases of congenital absence of the corpus callosum although careful study may reveal subtle deficits in some patients. Increased reliance on ipsilateral sensory pathways and/or greater bilateralisation of function may mask more serious defects of mid-line transfer. Comparison of callosal agenesis with the long-term effects of commissurotomy may illustrate the degree of compensation possible for absence of the callosum in children and adults respectively.

# 4

# Hemispheric asymmetry in normal subjects

## methods, findings and issues

The previous chapters discussed hemispheric asymmetry as revealed by brain bisection and the study of patients with unilateral cerebral lesions. However, the split-brain findings may not apply to normal subjects, and attempts to compare the relative effects of damage to left and right sides of the brain are notoriously fraught with problems. Chief among these is that of ensuring that left and right hemisphere damaged groups are appropriately matched in extent and type of pathology as well as in locus of lesion. As the effects of right sided damage are likely to be less immediately troublesome than comparable damage on the left, which frequently causes some degree of aphasia, naturally occurring left hemisphere lesions may well lead the patient to seek medical advice earlier in the disease process than would equal damage at the right side of the brain. This possibility might easily introduce a bias into the composition of left and right brain damaged groups. In the absence of post-mortem verification of the site and size of the cerebral lesion it is difficult, on purely clinical and radiological grounds, to be certain that the two groups are evenly matched in extent or severity of damage. With the advent of computerised axial tomography (CAT scan) this is likely to be less of a problem in the future, but in the past researchers have had to be content with less sophisticated methods.

One group of Italian workers attempted to deal with the problem of ensuring equivalent left and right hemisphere damage in their two groups of patients by assessing simple reaction time to a visually presented stimulus. Reaction time has been found to be extended in cases of brain damage (Costa, 1962) and it may reasonably be thought that latency to respond is related to the degree of damage. By entering reaction time data in an analysis of covariance, a statistical technique which enables the influence of one or more extraneous variables to be taken into account when assessing the effect of the principle variable of interest, De Renzi and his colleagues (Arrigoni and De Renzi, 1964; De Renzi and Fagglioni, 1965) hoped to control for possible differences in extent of damage between left and right hemisphere groups. A problem with this manoeuvre, however, is that certain evidence implicates the right hemisphere in control of simple reaction time (Benson and Barton, 1970; Howes and Boller, 1975; Nakamura and Taniguchi, 1977) and thus equivalent response times for left and right hemisphere groups may not in fact reflect equivalent damage at the two sides of the brain.

A second problem which besets the researcher investigating the effects of unilateral cerebral lesions is the possibility that the same functions may be organised in different ways in left and right hemispheres (Semmes, 1968). If this is so, identically placed lesions at the two sides will not necessarily have equivalent effects, even though each intact hemisphere may be equally as capable as its fellow of subserving a given function. Furthermore, it is usually impossible to say whether the effects of a lesion in one hemisphere, in terms of loss or impairment of a particular function, follow from the destruction of the true neural locus of that function, or instead reflect the influence of abnormal tissue on other brain areas that actually subserve the function in question. To localise a region of dysfunction, therefore, is not to localise a function, as pointed out by Hughlings Jackson in the nineteenth century.

A final difficulty is that there are almost always generalised, as opposed to specific, effects of localised cerebral lesions. To be sure that the locus of a given lesion is responsible for loss or impairment of a particular capacity, it is helpful to invoke the principle of doubled dissociation proposed by Teuber (1955). If lesion A affects function X but not function Y while lesion B affects function Y but not function X, it is reasonable to assume that the effects observed are specific to the loci of lesions A and B rather than to the general effects of brain damage.

Because of the sort of problems outlined above (for a more detailed discussion the reader is referred to Mayer (1960) and Piercy (1964) it is desirable that the results of studies carried out on patients with brain damage be confirmed, and if possible extended, among normal subjects. Within the context of hemispheric asymmetry there are several means whereby this can be achieved, and each of these will be considered in turn. No attempt is made here to provide a comprehensive survey of the literature; the intention rather is to introduce the basic experimental techniques and to summarise some of the principal findings and theoretical issues. Reference will be made to fuller reviews at appropriate points in the text.

## TACHISTOSCOPIC VISUAL HALF-FIELD STUDIES

Directing visual input to a single hemisphere of the brain by flashing a stimulus to one side of a central fixation point was explained in the previous chapter. This technique can be employed with normal subjects, as well as with commissurotomised patients, although the presence of intact mid-line commissures in normals means that visual information presumably does not remain lateralised to one hemisphere as it does in split-brain patients.

## Ocular dominance

The term 'ocular dominance' (or 'eyedness') has been used to refer to a number of different phenomena (Money, 1972). In the present context it has usually been used to mean acuity dominance, that is, superior acuity in one eye, or sighting dominance, the tendency to use one particular eye in preference to the other during monocular viewing. The sighting dominant eye may be, but is not necessarily, the eye having the greater acuity.

Under conditions of binocular vision the two eyes do not contribute equally to providing a stable binocular percept. When the image presented to each eye is not the same, a state of rivalry may exist which is resolved in favour of each eye alternately (see Walker, 1978). The eye providing the more stable percept is then, by definition, the dominant eye. On this definition, the dominant eye may or may not be the same as the eye chosen preferentially on a monocular sighting test.

The importance of ocular dominance within the tachistoscopic laterality paradigm is twofold. Firstly, evidence suggests that there may be a superiority of the crossed compared with the uncrossed optic pathways (Sampson, 1969; Maddess, 1975; Osaka, 1978; Hubel and Wiesel, 1959). If this is so, and if there is a sighting or acuity difference between a subject's left and right eyes, then either the left or the right nasal pathway may be favoured accordingly. Hence there may be a bias in recognition towards either the right or left cerebral hemisphere respectively. Kershner and Jeng (1972) found right eye sighting dominant subjects to be more accurate than left dominant subjects on a tachistoscopic verbal task, while the reverse was the case on a non-verbal task. This suggests some interaction between hemispheric specialisation and a functional superiority of the crossed optic pathways (see also Polich, 1978). It is worth mentioning, however, that it is not always the nasal pathways that have been observed to be superior (Neill, Sampson and Gribben, 1971) and that naso-temporal differences have sometimes been observed for one eye but not for the other (Overton and Weiner, 1966; Markowitz and Weitzman, 1969; Beaton, 1977; 1979a).

Whatever the direction of naso-temporal differences their mere existence dictates caution in the design and interpretation of tachistoscopic laterality experiments. It is normally assumed that because each eye projects to both hemispheres any imbalance between the two eyes is adequately controlled. Yet if a naso-temporal difference exists for one eye but not for the other, the set-up is not balanced with respect to left and right visual fields. Practically speaking, this may not be important except at very brief exposure durations, but it is as well to be aware of the problem.

The second reason for the importance of ocular dominance with regard to visual laterality research lies in the possibility of a relationship between ocular dominance and cerebral lateralisation. Since some forms of eyedness are apparently under genetic control (Merrell, 1957) this is not an unreasonable suggestion. In fact, a relationship between visual laterality effects and eye dominance has been found in some studies (Hayashi and Bryden, 1967; Zurif and Bryden, 1969; Bryden, 1973).

Attempts have been made to determine whether there is a correlation between manual preference and eye dominance, but to date the results are conflicting. Some authors have reported that right eye sighting dominance is more common than a dominant left eye among dextrals while the association between the side of the preferred eye and preferred hand is less predictable among sinistrals (Friedlander, 1971; Porac and Coren, 1975). Other researchers have found no significant relationship between the dominant hand and eye (Gronwall and Sampson, 1971; Coren and Kaplan, 1973; Porac and Coren, 1976; 1979b). It is possible that a failure to distinguish between the two sexes can explain the lack of agreement. Gur and Gur (1977) found handedness and *sighting* dominance to be associated for males but not females and *acuity* dominance with handedness for females but not males. If there is a relationship between eyedness and handedness then there may also be a relationship between eyedness and braineness. If so, groups of subjects composed of individuals with opposite eye dominance may not be homogeneous with respect to cerebral laterality.

The dependent variable in tachistoscopic visual field studies is usually accuracy of recognition or recall and/or simple or discriminative reaction time. Simple reaction time refers to a response, such as a key press, indicating merely that the stimulus has been detected, whereas discriminative reaction time refers to a response where some kind of discrimination is called for between two stimuli presented either simultaneously or successively. For example, subjects are often required to decide whether two stimuli are the same or different and then to respond on one key for 'same' matches and on a second key for 'different' matches; or else to respond to instances of one kind of match and withhold responding to instances of the other kind (a go/no-go discrimination).

In the short review which follows no particular effort has been made to distinguish between the results obtained with different classes of response measure (accuracy or reaction time) as the findings, by and large, are consistent with each other. Speed versus accuracy trade-offs, for example, have rarely been reported. There are, however, suggestions that reaction time may be more sensitive to visual field differences than accuracy scores (White, 1972).

In this chapter (and those which follow) the terms left or right 'visual field', 'hemifield' and 'half-field' are used interchangeably.

## Tachistoscopic studies using verbal stimuli

The upsurge of interest in tachistoscopic laterality research which has taken place during the last two decades or so can be traced in part to the split-brain investigations and in part to an experiment conducted by Mishkin and Forgays (1952). Although these investigators were not specifically concerned with the notion of cerebral asymmetry of function, their finding that bilingual subjects recognised English words more accurately from the right of fixation but Yiddish words (which are read from right-to-left) more accurately from the left of fixation (see also Orbach, 1967) sparked off a number of experiments designed to uncover the relationship between visual field asymmetry and a range of procedural and subject variables. A crucial distinction turned out to be whether the stimuli were presented only to one side of fixation at a time or simultaneously to both sides. Heron (1957) found that unilaterally presented words were better recognised from the right visual field but that with bilateral presentation there was an advantage of words to the left of fixation. He therefore suggested that:

> faced with a line of print there is one tendency to fixate near the beginning of the line and another to move the eye along it from left to right. When alphabetic material is exposed in the right field alone, the two tendencies will be acting together. When, however, it is exposed in the left field alone, the tendency to move the eyes at the beginning of the line (presumably the dominant one) would be in conflict with the tendency to move the eyes from left to right. (pp. 146-7)

Heron was then able to explain his results by supposing that the neural traces corresponding to the tachistoscopically exposed stimuli are processed in the same way. However, with bilateral stimulus presentation a tendency to report, rather than process, the left-most stimuli first provides an equally plausible explanation (Ayres, 1966; Dick and Mewhort, '1967; Coltheart and Arthur, 1971; Wilkins and Stewart, 1974).

Support for the scanning hypothesis of lateral differences in tachistoscopic recognition came in the form of findings which showed the direction of visual field asymmetry to be related to whether the stimuli are symmetrically shaped (Bryden, 1968) or are presented in normal or in mirror image orientation (Harcum and Filion, 1963).

An alternative to the scanning hypothesis of laterality differences in tachistoscopic recognition is that, with unilateral stimulus presentation at least, words (Terrace, 1959) and letters (Bryden, 1966) and material for which a verbal label is readily available (Wyke and Ettlinger, 1961; Bryden and Rainey, 1963) are more accurately recognised in the right visual field as a consequence of the more direct neural pathway from the right than from the left side of fixation to language areas of the left cerebral hemisphere. This cerebral dominance hypothesis is able to explain the results of an experiment by Barton, Goodglass and Shai (1965) who presented three-letter words in a vertical orientation so as to minimise putative scanning mechanisms. These investigators found their unilaterally presented stimuli to be more accurately recalled from the right visual field both by American subjects viewing English words and by Israeli subjects seeing Yiddish words.

The early experiments have been reviewed by White (1969a, 1972), McKeever (1974), Pirozollo (1977) and, more recently, by Bradshaw, Nettleton and Taylor (1981a). The latter authors argued that these early studies were marred by methodological deficiencies. Nonetheless, Bradshaw et al. conclude on the basis of recent work on tachistoscopic word recognition that, at least with single-syllable words exposed one at a time to left or right visual hemifield, artefacts due to directional scanning contribute little if anything to hemified asymmetry. With non-word letter strings or multi-syllabic words the position is less clear.

Kinsbourne (1970) proposed a modification of the cerebral dominance model to explain tachistoscopic hemifield asymmetries. He suggested that when subjects know to expect verbal stimuli, attentional mechanisms prime the left hemisphere for reception of this material and it is this attentional bias, rather than any inherent perceptual advantage, which gives rise to the right hemifield superiority in word and letter recognition typically obtained with unilateral stimulus presentation. Kinsbourne initially based his suggestion on the result of an experiment he carried out using a square with a small gap in one of its sides. Normally subjects were able to detect the presence of this gap equally well whether the square was presented to left or right of the fixation point. Yet if the subjects were required to repeat the alphabet to themselves during the task this introduced a bias in gap detection favouring the right visual field. Although there is evidence to support Kinsbourne's hypothesis (Kinsbourne, 1975a; Cohen, 1975a) it cannot account for the results of those studies in which opposite field superiorities have been obtained for verbal and non-verbal stimuli presented randomly or simultaneously to each half field of vision (Terrace, 1959; Berlucchi, Brizzolara, Marzi, Rizzolatti and Umiltà, 1974; Hines, 1975; Klein, Moscovitch and Vigna, 1976; Pirozollo and Rayner, 1977). Nor can it explain the shift from an initial RVF superiority on a visual reaction time task to a LVF superiority when a concurrent memory list of nouns is superimposed on the perceptual task (Hellige, Cox and Litvac, 1979).

One of the difficulties of traditional tachistoscopic procedures is that of relying on subjects to maintain their fixation during stimulus presentation. McKeever and Huling (1970a) tried to get over this problem by presenting a digit at fixation point at the same time as a stimulus appeared in one or other visual field. The aim was to ensure that subjects had been fixating correctly by requiring them to report this digit immediately prior to recalling the stimulus. Trials on which the digit was not correctly identified were discarded from data analysis and substituted later in the series. Using this technique these workers reported a right field advantage both with unilateral and with bilateral word presentation and attributed their results to 'greater transmission fidelity and/or lesser transmission time' of the shorter pathway to the left, verbal, hemisphere (McKeever and

Huling, 1971a). Since a right visual field advantage was still found when the word in the right field was exposed 20 msecs. later than the word in the left field it was subsequently concluded that temporal factors were not importantly involved (McKeever and Huling, 1971b).

The atypical but consistent finding of McKeever and his collaborators that a right field advantage can be obtained with both unilateral and bilateral stimulus presentation may be attributable in part to their use of very short exposure durations, and in part to their use of a digit as a 'fixation forcer'. This technique has been widely adopted by researchers in recent years. However, it is not yet clear that the use of a fixation stimulus does not introduce as many difficulties as it seeks to eliminate. Hines (1972), for example, using exactly the same words as McKeever and Huling (1971a) replicated the latter's finding of a right field superiority in bilateral recognition when a digit was presented at fixation but obtained a significant left field advantage without the digit. McKeever, Suberi and Van Deventer (1972) replicated Hines's experiment but found that neither the magnitude nor direction of visual field asymmetry was affected by the presence of a central fixation digit. A similar finding was reported by MacKavey, Curcio and Rosen (1975). However, reversal from left to right visual field superiority was observed by Kaufer, Morais and Bertelson (1975) when they investigated the effect of presenting a Landolt's ring at fixation for subjects to report the position of a small gap in the ring prior to recalling the stimuli presented in the left and right visual fields. It is therefore possible that in some circumstances a stimulus at fixation serves as an anchor such that processing occurs first for material appearing to the right of fixation. Moreover, it has been suggested on the basis of experiments carried out with children that the nature of a fixation stimulus may affect visual hemifield asymmetry by inducing a bias towards either verbal or non-verbal processing, depending on the nature of the stimulus (Kershner, Thomae and Callaway, 1977; Carter and Kinsbourne, 1979). This claim however was not substantiated in a study with adults (Hines, 1978).

## Tachistoscopic studies using non-verbal stimuli

### Form recognition

A vast literature now exists reporting a right visual field (RVF) superiority in recognition or recall of words and letters. A large number of experiments have also been carried out to investigate laterality effects in tachistoscopic perception of non-verbal stimuli. One of the earliest was that of Terrace (1959) who introduced geometric forms into a series of unilaterally presented words in an attempt to prevent the development in his subjects of an exclusively verbal 'set'. Terrace found a slight though not significant advantage for the left hemifield for these forms in contrast to a right field superiority for words. Bryden and Rainey (1963) also failed to find a significant hemifield difference in the perception of geometric forms, as did

Lordahl, Kleinman, Levy, Massoth, Pessin, Storandt, Tucker and Vander-plas (1965) using random shapes as stimuli. However, Gross (1972) obtained an advantage for the left visual field in recall of partially shaded grid matrices and Fontenot (1973), Dee and Hannay (1973) and Dee and Fontenot (1973) found a left field superiority in recall of random shapes.

*Dot enumeration and localisation*
A non-verbal, visual task of a different kind was utilised by Kimura (1966). In this task, known to be sensitive to right hemisphere damage (Kimura, 1963a), subjects were required to report the number of dots that were presented randomly in one or other visual field. Kimura found an advantage for the left field in this task as also when geometric forms were substitued for the dots. Adams (1971) found no significant asymmetry in a dot enumeration task but it is possible that the relatively long exposure duration he employed was sufficient to eliminate any potential difference between the visual fields.

The left hemifield advantage reported by Kimura (1966) for the enumeration of dots was subsequently extended to include localisation of a single dot presented in any one of 25 different spatial positions (see also Bryden, 1976). This might help to explain McKeever and Huling's (1970b) finding that reproduction of dotted designs was more accurate for stimuli presented to the left of fixation whereas there was no asymmetry for solid line designs (Kimura and Durnford, 1974). On the other hand there is clinical evidence that the right hemisphere is better than the left in recognition of visually degraded stimuli even when these are letters and thus verbal in nature (Faglioni, Scotti and Spinnler, 1969).

*Perception of line orientation*
A superiority has been found for the left field discrimination of curvature (Longden, Ellis and Iversen, 1976) and in perception of the spatial orientation of a single line (Fontenot and Benton, 1972; Sasanuma and Kobayashi, 1978). In addition, Atkinson and Egeth (1973) found that responses were faster for the left visual field on a matching task in which subjects had to indicate whether or not two lines showed the same orientation. These findings are unlikely to be due to some superior 'tuning' for line detection in the right hemisphere,although Tei and Owen (1980) have argued that the right hemisphere is neurophysiologically more sensitive to orientation than is the left hemisphere.

The study carried out by Tei and Owen (1980) involved presentation of a grating in central vision while subjects fixated a central point. Immediately following offset of the adapting grating a second grating (of the same spatial frequency) was presented approximately 4 degrees to the left or right of the fixation point for 100 msec. This second grating had either the same or a different orientation as the adapting grating and the subjects' task was to indicate by means of a lever response whether the orientations were the same or different. The results showed that there were more errors made, and responses were slower, in the left visual field than in the right visual field.

In their analysis of error scores Tei and Owen combine data from trials in which the adaptation and test gratings were the same with data from trials on which they were different. It is thus not possible to tell whether the significance of the laterality effect observed was due to 'same' and/or 'different' trials. Inspection of Tei and Owen's Tables 1 and 3 reveals that the laterality effect was greater for 'same' trials but it is theoretically important to know whether there was any significant interaction between stimulus type and visual hemifield. A significant difference between the LVF and RVF on 'same' trials might imply a faster rate and/or greater magnitude of adaptation in one visual hemifield. A significant laterality effect for 'different' trials would suggest broader tuning of orientation detectors in one cerebral hemisphere than in the other. These two possibilities are not mutually exclusive.

Tei and Owen provide no breakdown of their data in terms of the angular separation between adapting and test stimuli on their 'different' trials so it is not possible to see whether adaptation is effective over a greater range of orientations in the LVF compared to the RVF. Such a finding might be seen as inconsistent with data from brain damaged patients (Warrington and Rabin, 1970; De Renzi, Faglioni and Scotti, 1971) which suggest that the right cerebral hemisphere is more *accurate* than the left hemisphere in judging the orientation of a line. The finding that adaptation is more profound in the LVF than the RVF, on the other hand, would be consistent with the results of Meyer (1976), who found a stronger McCollough effect in the LVF, but would not support the findings of Beaton and Blakemore (1981) who observed no hemispheric difference in extent of adaptation or orientation selectivity in either of their two well-practised subjects. The possibility of a left-right difference in *rate* of adaptation, however, remains to be explored.

The frequent finding of a left visual hemifield advantage in the perception or discrimination of line orientation may reflect an overall right hemisphere superiority for tasks which stress the spatial arrangement of elements in a stimulus display (Gross, 1972; Kimura and Durnford, 1974; Robertshaw and Sheldon, 1976; Pitblado, 1979a). This superiority in the visuo-spatial domain has been found in some studies to extend to perception of the third dimension.

*Stereopsis*
Stereoscopic depth perception was studied by Durnford and Kimura (1971) by asking subjects to judge whether two rods, one presented at the point of fixation and one presented in the left or right visual field, were horizontally aligned. With monocular viewing no field difference was found but with binocular viewing a left field superiority was obtained. Binocular viewing of random dot stereograms also yielded higher scores for the left field in identification of the fused forms. Durnford and Kimura

therefore concluded that the right hemisphere is specialised for the perception of depth cued by retinal disparity. Dimond, Bureš, Farrington and Brouwers (1975), using a different technique, also claimed a right hemisphere dominance for stereoscopic depth perception. This supports certain findings from brain damaged patients (Carmon and Bechtoldt, 1969; Benton and Hécaen, 1970; Hamsher, 1978) but the issue of whether there is a hemispheric dominance for stereopsis is unresolved (Julesz, Breitmeyer and Kropfl, 1976; Danta, Hilton and O'Boyle, 1978; Pitblado, 1979b).

*Facial recognition*
Prosopagnosia is a rare clinical condition in which visual recognition of faces is impaired as a result of brain damage. The few reported cases that have come to autopsy have generally shown bilateral cerebral lesions but the right sided damage has always been in the region of the occipito-temporal junction whereas the lesions on the left have not been uniformly located (Benson, Segarra and Albert, 1974; Meadows, 1974; Whiteley and Warrington, 1977). Since prosopagnosia has been associated with right hemisphere damage on purely clinical grounds (Hécaen and Angelergues, 1962; Warrington and James, 1967; but see Cole and Perez-Cruet, 1964) it is reasonable to interpret the post-mortem findings as confirming this association. Facial recognition has therefore been investigated with neurologically intact subjects in the expectation of finding a tachistoscopic left visual field superiority.

Reaction times to photographs of faces in a same-different task were reported to be faster for left than for right hemifield presentation by Geffen, Bradshaw and Wallace (1971), Rizzolatti, Umiltà and Berlucchi (1971), Berlucchi, Brizzolara, Marzi, Rizzolatti and Umiltà (1974) and St John (1981). With accuracy of recognition as the dependent variable a left field advantage was found by Hilliard (1973), Ellis and Shepherd (1975), Jones (1979) and Leehey and Cahn (1979). Although a superiority of the left visual field has not always been reported, this has been the common finding when the faces to be responded to were highly discriminable and/or unfamiliar (Sergent and Bindra, 1981). When schematic drawings of faces have been used as stimuli visual hemifield asymmetry has been found to be a function of several factors, including the degree of similarity between the stimuli to be compared and the inter-stimulus interval employed (Patterson and Bradshaw, 1975).

The question of whether the recognition of faces is a specific ability (Yin, 1970; Tzvaras, Hécaen and Le Bras, 1971; Whiteley and Warrington, 1977) or is part of a more general visuo-spatial processing system is unsettled (Ellis, 1975; Hay and Young, 1983). The typical finding of a left hemifield superiority in facial recognition is, however, consistent with other findings reviewed in this chapter which point to a relative dominance of the right hemisphere in dealing with non-verbal events in general and visuo-spatial processes in particular. This is entirely consistent with a wealth of clinical evidence which indicates that the integrity of the right cerebral hemisphere is more important than that of the left for carrying out visuo-spatial tasks (Milner, 1965; Newcombe, 1969; Benton, 1969a; De Renzi, Faglioni and Scotti, 1971; Taylor and Warrington, 1973; Hécaen and Albert, 1978).

## Theoretical interpretations of laterality effects

It is generally assumed that in the absence of methodological artifacts visual laterality effects in normals arise either because one cerebral hemisphere is relatively inefficient at processing or retrieving the stimulus material presented and/or because one hemisphere cannot fully process the information and has to send it across the corpus callosum to the opposite hemisphere. Transmission across the callosum takes time and necessitates crossing at least one synaptic junction, during which the information is said to undergo some degree of transformation such that it arrives at the second hemisphere in a comparatively degraded state (McKeever and Huling, 1971a; Gross, 1972; Gibson, Dimond and Gazzaniga, 1972). Thus either the comparative inefficiency of one hemisphere in dealing with the information presented and/or a reduction in stimulus fidelity consequent upon hemispheric transfer are held to account for the advantage obtained for a particular half of the visual field.

Attempts have been made to decide between the above two explanations. Such attempts have usually been based on the notion that the response to a stimulus presented in, say, the left visual field, should be quicker from the left hand than from the right. Although there is evidence that both hemispheres can control movements of the upper arm (Sperry, Gazzaniga and Bogen, 1969; Van Der Staak, 1975) the extremities appear to be under exclusively contralateral control (Brinkman and Kuypers, 1972; 1973). Thus in the former situation the stimulus is received and the response initiated by the same hemisphere (direct or uncrossed reaction). A right hand response to a left field presentation (indirect or crossed reaction), however, would mean that the hemisphere initiating the manual response is not the hemisphere receiving the stimulus and therefore some finite interval of time is required for a nervous impulse to cross from one half of the brain to the other in order to generate a response. Thus reaction time in this situation would be slower than when a response originates from the directly stimulated hemisphere.

Based on the argument outlined above, measures of inter-hemispheric transfer time have ranged from around 3 msec. (Poffenberger, 1912) to approximately 30 msec. (Filbey and Gazzaniga, 1969). One reason for this discrepancy in estimated values undoubtedly relates to the different kinds of stimulus material and task employed in the different experiments. Another related reason may be that it is not at all clear what it is that is transmitted during inter-hemispheric transfer. Rather naively, it has sometimes been suggested, either explicitly or implicitly, that it is some more or less faithful copy of the stimulus which is shunted from one side of the brain to the other. This assumption has been most conspicuous in the context of experiments comparing manual response times with vocal reaction time. It has been supposed that a vocal response must necessarily emanate from the left hemisphere and that a stimulus presented in the left half field of vision must therefore be transferred from the right to the left hemisphere. If this were so one would expect measures of inter-hemispheric transmission time to remain roughly constant for the same stimuli and subjects, yet this does not appear to be the case (Bashore, 1981).

A tactic for distinguishing an effect on reaction time of a hemispheric processing inefficiency from an effect due to inter-hemispheric transfer is suggested by the application of a memory search procedure (Sternberg, 1966). A subject is given a number of letters to hold in memory and is then presented with a probe letter. He has to decide whether the probe letter matches one of the letters in the memory set. It turns out that the time taken to respond increases linearly with the number of items in the memory set. A left hemisphere advantage for matching by name (Cohen, 1972) would be expected to lead to a RVF superiority on a memory search for letter names. If the right hemisphere is merely slower than the left on this task, then the more matches that have to be made the greater the relative right hemispheric disadvantage should be. The difference in reaction times between RVF and LVF should thus be an increasing function of the number of items held in memory. Conversely, if the right hemisphere cannot perform the task at all but has to send the probe stimulus to the left hemisphere for a comparison with memory items then the relative disadvantage in reaction time for the right hemisphere should be constant regardless of the size of the memory set. This experiment has been carried out by Klatzky (1970) and Klatzky and Atkinson (1971) with results favouring the inter-hemispheric transfer interpretation. There was also an effect due to the hand used to make the discriminative response.

A difficulty in interpreting laterality reaction time data according to a fixed anatomical model is that shorter response latencies for uncrossed as compared with crossed reactions would be expected simply on the grounds of stimulus-response (S-R) compatibility. It is well known that responses to a stimulus presented off-centre will be faster when made by the limb on the same side of the body as the stimulus appears, that is, when stimulus and response are said to be compatible (Fitts and Seeger, 1953).

It is possible to argue that the S-R compatibility effect is in some sense due to the nature of the anatomical pathways involved. However, compatibility effects are still observed when subjects cross their arms such that the right hand responds to stimuli presented in the left field of vision and the left hand responds to right field presentations. In such a situation it is sometimes possible using very simple stimuli to dissociate the effects of stimulus-response compatibility from those due to eye-hand connections (Berlucchi, Crea, Di Stefano and Tassinari, 1977) but the relevance of the different anatomical pathways in situations employing more complex stimuli is doubtful. Generally speaking, therefore, difficulties of interpretation in this area mean that very rarely does the tachistoscopic paradigm allow any firm conclusion to be drawn as to whether an observed inferiority of one half field of vision is due to time or information loss during callosal transfer and/or to less efficient processing by the contralateral hemisphere. Nonetheless any differences in some aspect of the responses made by subjects to stimuli presented in left and right visual fields may plausibly be attributed, in the absence of more compelling alternative explanations, to differences of one kind or another between left and right cerebral hemispheres, at least where a visual field effect clearly over-rides the effect of stimulus-response compatibility factors (e.g. Anzola, Bertolini, Buchtel and Rizzolatti, 1977) and/or procedures to counterbalance the effect of the responding hand are incorporated into the experimental design.

Although this review has so far concentrated on those findings which reveal most clearly opposite visual hemifield superiorities for verbal and non-verbal (especially visuo-spatial) stimuli it is impossible for one familiar with work in this field not to be struck by the lability of the laterality effects reported (Cohen, 1982). Failures to replicate results are common. If a superiority of one half-field derives from some fixed advantage of the contralateral cerebral hemisphere for the stimulus material in question then comparatively minor procedural differences (e.g. Hiscock and Bergstrom, 1982) should have little or no influence on the outcome of an experiment. As it is, the magnitude and direction of laterality effects vary not only between different subjects and different experiments but also within the same subject at different stages within an experiment (e.g. Hellige, 1976; Shefsky, Stenson and Miller, 1981). Structural models of hemisphere specialisation which posit that perceptual asymmetries arise because the brain structures that deal with a particular class of stimuli are lateralised exclusively or predominantly to one hemisphere rather than the other cannot cope with this variability. A possible resolution of this difficulty is offered by accounts such as those of Kinsbourne (1970, 1973, 1977) which stress dynamic as opposed to structural determinants of perceptual laterality effects. Thus Kinsbourne's attentional model views shifts in the magnitude or direction of a visual field difference as due to changes in the relative activation of one hemisphere as compared with the other. Such a theory predicts that an increase in advantage for a particular visual hemifield should be accompanied by an equivalent decrease in performance for the other hemifield, yet this does not appear to be the case (Hellige and Cox, 1976). Also, the magnitude of a particular visual field difference should be the same for equivalent stimuli presented in the same experimental situation. Hellige and Cox (1976), however, obtained a slightly larger hemifield superiority for random forms having 12 points than for forms having 16 points even though these two types of form were distributed randomly throughout a series of trials. Taken together with the finding that an initial RVF superiority on a visual reaction time task gave way to a LVF superiority when subjects were required to simultaneously hold in memory a list of nouns (Hellige, Cox and Litvac, 1979), this implies that dynamic shifts in attention are insufficient to explain all perceptual laterality effects.

Models of hemisphere specialisation have been critically reviewed recently by Cohen (1982). She distinguishes between those models that treat hemisphere specialisation as absolute, according to which a given function can *only* be performed by a particular hemisphere, and those that regard specialisation as relative. According to the relative specialisation models, both hemispheres can perform a given function but one is faster or more efficient than the other. These models therefore differ in the degree to which specialisation for a particular function is thought to exist. In addition, models differ as to the nature of the specialisation which they postulate to occur for the two hemispheres. In her review Cohen argues that there is little evidence in favour of the absolute specialisation models and points out that, as regards the nature of hemispheric specialisation, differences between the two halves of the brain have been conceptualised in terms of specialisation for different types of material, specialisation for different processes and specialisation for different stages of processing.

It was pointed out above that specialisation for different types of material cannot explain the lability of many perceptual laterality effects. Interest therefore centres on the notion that left and right cerebral hemispheres differ in the processes which they characteristically employ or in some other aspect of information processing.

### Information processing and visual field asymmetry

In recent years a growing sophistication in tachistoscopic half-field investigations has derived from a conceptual and methodological framework known as information processing theory. This term refers not to a single theory but rather to a set of assumptions underlying a particular approach to the study of perception and cognition. The basis of this approach is the belief that the response which a subject makes is not an immediate outcome of sensory stimulation but results from a number of processes which occur over time, stimulation of the sensory receptors being only the first stage in a series of events. Information processing theorists attempt to define these stages and the order in which they occur. Models of human information processing have been developed by a number of theorists of this persuasion and the interested reader is referred to the original publications for further details (Broadbent, 1958; 1971; Neisser, 1967; Sperling, 1963; 1967; Atkinson and Shiffrin, 1968; Turvey, 1973). Common to all models is the view that stimulation of sensory receptors sets up neural activity which outlasts the duration of the physical stimulus. This persistence is referred to as brief visual (iconic) or auditory (echoic) storage and is believed to last up to about one quarter of a second (Sperling, 1960; Coltheart, Lea and Thompson, 1974). During this time various encoding operations may be performed upon the stored information which enables some representation of the stimulus to be held in a short term memory store, after which it either drops out or is transferred into some other more durable store. Thus information processing models at their most simple distinguish between three distinct stages: registration, coding and retrieval.

*Iconic registration*
Cohen (1976) argued that visual laterality effects may be attributable to asymmetry at different stages of information processing. She presented three rows of two letters in either the left or right field and asked subjects either for a full report of all six letters or simply to recall a particular pair of letters (partial report condition) corresponding to a single column of the display. In the latter situation subjects were given a cue prior to the stimulus presentation (pre-cued condition) so that they knew beforehand whether to report the inner or outer column or the letters printed in a particular colour. In another condition of the experiment a 'masking' stimulus was presented 20 milliseconds (msec.) after the offset of the letter display (post-cued condition). The reasoning behind Cohen's experiment ran as follows. Firstly, a hemisphere difference in the rate at which information is encoded and/or read out from some sensory representation of the stimulus should be revealed by a differential effect of masking in left and right visual fields since the presumed effect of the mask is to prevent any further processing. Secondly, a difference in the ability of the two halves of the brain to utilise advance information in selecting the required stimulus feature for further processing should result in an asymmetrical effect of pre-cueing in the partial report condition as compared with the full report condition. Cohen's results showed that masking produced a significant decrement in the left visual field but not in the right visual field, thus implying a superiority in rate of encoding for the left hemisphere. The effect of pre-cueing was also greater in the right field, suggesting a hemispheric difference in the efficiency of selective sampling. The former result supports similar findings by other investigators (Oscar-Berman, Goodglass and Cherlow, 1973; Turvey, 1973; McKeever and Suberi, 1974; Ward and Ross, 1977) while the latter finding is consistent with the results of another experiment by Cohen in which she used different cues and a different task (Cohen, 1975b).

As well as the experimental conditions referred to above, Cohen (1976) employed conditions in which the cue to recall a particular column of letters was presented at various intervals after stimulus presentation. Contrary to most published studies, in the post-cued conditions Cohen found no advantage for partial as compared with full report at any of the delay intervals used in her experiment. This result she attributed to the very short duration of persistence of the icon, which meant that it had faded before the cue could be utilised to sample selectively. According to calculations carried out on her data, Cohen estimated the duration of the icon to be longer in the left (57 msec.) than the right (34 msec.) visual field.

More recently, Marzi, Stefano, Tassinari and Crea (1979), also using a post-cued partial report technique but without a mask, obtained a partial report advantage which decreased with increasing intervals between stimulus presentation and cue. This effect was symmetrical for the two visual fields, suggesting that there is *no* hemisphere difference in persistence of the icon.

Marzi et al. explained the discrepancy between their finding and Cohen's estimate of a longer lasting icon in the left visual field by arguing that the masking paradigm employed by Cohen does not allow measurement of icon persistence independently from encoding rate. Better performance for one hemifield could be due either to a longer duration of icon, thus allowing more information to be encoded before the icon fades, or to a faster encoding rate, allowing more information to be encoded before the arrival of the masking stimulus. Marzi et al. therefore saw Cohen's results as due to a confounding of these two aspects of information processing. In counter-argument it might be suggested that the Marzi et al. finding of an equivalent partial report advantage for left and right hemifields was due to their use of relatively long delay intervals (at least 300 msec.) between stimulus presentation and partial report cue. It is generally held that the icon has virtually faded after about quarter of a second from stimulus offset. With the extended delay intervals used by Marzi et al., therefore, any hemispheric difference in duration of the icon will be difficult to detect using the partial report technique. Cohen's estimate of the hemisphere difference in icon persistence was 23 msec. which is less than 10 per cent of the duration of the shortest delay interval used by Marzi et al.

Cohen's finding that masking produced a significant decrement in the left visual field but not in the right visual field is not supported by the results of Hellige and Webster (1979). These workers presented target letters followed at very short intervals by curved or straight line masking stimuli and found a LVF superiority. They also obtained a LVF superiority for simultaneous masking (both letters presented together but superimposed) and forward masking (mask presented prior to target letter). Given the very short inter-stimulus intervals employed (less than 20 msec.) and the nature of the masking stimuli this suggests the possibility that the LVF superiority arose from some right hemisphere advantage in discriminating the target stimulus from the mask. There is evidence that the right hemisphere is better than the left in making elementary sensory discriminations concerning brightness (Davidoff, 1975) colour (Davidoff, 1976; 1977; Pennal, 1977; Pirot, Pulton and Sutker, 1977) and velocity (Bertolini, Anzola, Buchtel and Rizzolatti, 1978). A right hemisphere advantage for this stage of processing is not necessarily incompatible with a left hemisphere advantage in rate of encoding.

Moscovitch, Scullion and Christie, (1976) also carried out an experiment within the framework of information processing theory. These authors presented pairs of faces to one or other visual field and required their subjects to match the faces either to each other or to previously presented faces. By varying the interval between presentation of target and test stimuli Moscovitch et al. found that a left field superiority in reaction time emerged only with an inter-stimulus interval of 100 msec. or longer. This was interpreted as indicating that at the lower inter-stimulus intervals employed (5 and 50 msec.) both hemispheres have access to a short-lived visual trace (i.e. the icon) but once the trace decays the right hemisphere has preferential access to, or operates more efficiently on, some more stable representation of the stimulus. If this was so, Moscovitch et al. argued, then the effect of a mask which eliminated the short term trace should force the subject to utilise the more 'highly processed' stable representation and thus reveal a right hemisphere advantage even with inter-stimulus intervals of less than 100 msec. Moreover, if the permanent representation is more 'highly processed', then in a task which can only be performed by referring to this higher-level information a right hemisphere advantage should be found even with a zero inter-stimulus interval, that is, with simultaneous presentation of the two stimuli to be matched. In order to force the subject to use higher-order processing Moscovitch et al. presented photographs and caricature pictures in a simultaneous same-different matching task. This yielded faster response times to stimuli presented in the left visual field although simultaneous matching of photographs alone, which could be matched on the basis of lower-level processes such as feature analysis, revealed no field advantage. Thus both predictions were confirmed.

The fact that no hemispheric difference was found with inter-stimulus intervals of less than 100 msec. in the first experiment of Moscovitch et al. was seen as indicating that the initial registration of a stimulus (the icon) is equivalent in left and right hemispheres. This supports the findings of Marzi et al. (1979) discussed above. That a LVF superiority in reaction time emerged at 100 msec., however, was taken by Moscovitch et al. to mean that the right hemisphere had some advantage with respect to a more stable representation of the stimulus. They argued that hemispheric differences only emerge at later stages of processing beyond immediate registration. Other authors have claimed that visual laterality effects occur when judgements have to be made from memory, that is, beyond the coding stage (Hardyck, Tzeng and Wang, 1977; 1978; Kirsner, 1980; Kirsner and Brown, 1981.) Although there are cases where a significant field difference has only been found when some form of retention interval has been imposed on the subject (e.g. Dee and Fontenot, 1973) it is unlikely that these are the only circumstances in which reliable effects occur (Schmuller, 1980). In an experiment by Berrini, Della Salla, Spinnler, Sterzi and Vallar (1982) subjects were presented with one stimulus in central vision and a comparison stimulus in one or other visual field. In the first condition of the experiment the laterally displaced stimulus was presented first followed by the central stimulus second. This produced the usual visual field asymmetry. Presenting the central stimulus first and the lateralised stimulus second produced no difference between the visual fields. The authors therefore saw the asymmetry as arising from the input rather then the recognition stages of processing.

*Hemisphere differences in coding*

The nature of stimulus coding has been studied by Posner and Mitchell (1967). Subjects were asked to respond 'same' or 'different' to pairs of letters. This could be on the basis of the shape of the letters (physical identity) or according to the name of the letters (nominal identity). When subjects were asked to perform a physical identity (PI) match (e.g. A-A or a-a) they correctly rejected instances of nominal identity (A-a) as quickly as pairs with different names (e.g. A-b). This suggests that physical matching was based entirely on a non-verbal mental representation of the stimuli. Conversely, nominal identity matches such as A-a were performed more slowly than physical identity matches (A-A) suggesting a separate representation for nominal as opposed to physical matches. These findings support the view of Paivio (1971, 1975) that verbal and non-verbal information is represented and processed in distinct but connected systems. His research has shown that the probability of recalling items is greatest for actual objects, not quite as great for pictures, reduced again for words representing concrete objects and least of all for words representing abstract concepts. The explanation usually offered for this phenomenon is that subjects implicitly attach a verbal label to pictures and objects and hence a verbal memory trace as well as a pictorial memory trace is established. A word describing a 'concrete' object also gives rise to a 'pictorial' trace but an 'abstract' word establishes only a verbal trace. The increased probability of recall of 'concrete' as compared with 'abstract' words is thus due to the availability of two codes as compared with one; if one system fails, the other may succeed.

Although the existence of a pictorial code has been disputed (Pylyshyn, 1973) the dual-coding hypothesis has gained wide acceptance. The verbal system may operate at either a semantic or acoustic level and the non-verbal system according to the kind of transformations (such as inversion, rotation, size reduction) that can be performed on visual images. Even if a non-verbal visual form of representation is not in the form of an implicit picture or image – and there must be non-verbal representations of a non-visual kind, for how else could we recognise the sound of a car door or the smell of supper? – there is a wealth of experimental evidence (Bower, 1970; Nelson and Brooks, 1973) to support the notion of two separate coding systems, verbal and non-verbal. (It is perhaps as well to make clear that these systems are not necessarily amenable to introspection.)

Both clinical (Jones-Gotman and Milner, 1978; Whitehouse, 1981) and experimental evidence link the verbal and non-verbal codes to left and right hemispheres respectively. Cohen (1972) presented the Posner task to the left or right visual field and found a RVF superiority for NI matches and a LVF superiority for PI matches, as did Geffen, Bradshaw and

Nettleton (1972). These results are important in showing that even with verbal stimuli a tachistoscopic LVF advantage may be obtained when the task can be performed non-verbally. Thus it is not the ostensible nature of the stimuli that is crucial in determining any difference between the visual fields as the kind of cognitive processing that the subject undertakes (but see Simion, Bagwara, Bisiacchi, Roncato and Umiltà, 1980). This is shown most dramatically by the results of studies in which, using exactly the same stimulus material, opposite field advantages have been obtained according to the task requirements (Klatzky, 1972; Seamon and Gazzaniga, 1973; Robertshaw and Sheldon, 1976; Niederbuhl and Springer, 1979).

That a stimulus can be processed either verbally or non-verbally helps to make sense of those otherwise anomalous findings in which verbal stimuli give rise to a LVF superiority (Gibson, Dimond and Gazzaniga, 1972; Schmit and Davis, 1974; Wilkins and Stewart, 1974; Hellige, 1976; Martin, 1978; Jonides, 1979; Niederbuhl and Springer, 1979) and non-verbal stimuli such as faces (Patterson and Bradshaw, 1975; Marzi and Berlucchi, 1977; Umiltà, Brizzolara, Tabossi and Fairweather, 1978), colours (Malone and Hannay, 1978) or pictures (Wyke and Ettlinger, 1961) produce an advantage for the right visual field. Even within the same series of experimental trials opposite field superiorities may be obtained for similar stimuli processed in different ways. For example, Umiltà, Rizzolatti, Marzi, Zamboni, Franzini, Camarda and Berlucchi (1974) required subjects to judge the orientation of a single line. Those lines which were oriented so as to be readily named (e.g. vertical, horizontal, diagonal) gave a RVF superiority in reaction time whereas the remaining orientations favoured the left visual field. (This may explain the results of White (1971) who presented lines only in horizontal, vertical and diagonal orientations and found a RVF superiority.)

The effect of providing subjects with a verbal label for random forms was investigated by Hannay, Dee, Burns and Masek (1981). A left field superiority when the forms were presented alone was converted to a RVF superiority when subjects had to use a label for each form. This implies that unwanted variance may be introduced into the data of those experiments where subjects are free to spontaneously attach their own labels to non-verbal stimuli.

A preference for using one code rather than the other could explain differences between individual subjects in the direction of hemifield asymmetry for different tasks (Kroll and Madden, 1978). However, there is a risk of becoming circular in accounting in this way for any particular visual field difference that is observed. What is required is independent evidence that a particular code is in fact being used.

Seamon and Gazzaniga (1973) presented subjects with two words referring to common objects (e.g. Hat–Duck). In one condition subjects were specifically instructed to engage in verbal rehearsal of these word pairs and in a second condition they were instructed to use visual imagery to relate the two objects. A picture was subsequently presented in one or other visual field and the subject's task was to indicate whether or not the probe picture had previously been presented among the memory set. It was found that, with instructions to verbally rehearse, reaction times were faster to left hemisphere probes, while under the relational imagery condition reaction times were faster to right hemisphere probes. It was argued that in the verbal rehearsal condition subjects had to name the picture probe before a match could be performed on a verbal basis, while in the imagery condition a match could be made directly. The statistically significant interaction between condition of retention and visual field of presentation was held to confirm the hemispheric locus of visual and verbal codes. These findings have been replicated by Metzger and Antes (1976), though found to be statistically significant only for 'same' trials,

*Modes of processing*

A subject's performance on an experimental task may depend not only on the way which he encodes the stimuli but also on the strategy he adopts in carrying out the task. Information processing theorists distinguish between serial and parallel processing. Serial processing refers to cognitive operations carried out successively whereas parallel processing is carried out simultaneously. Attempts have been made to map these two modes of processing on to the left and right hemispheres respectively.

Cohen (1973) presented subjects with from 2 to 4 letters in a same–different matching task. On 'same' trials all the letters 'were identical; on 'different' trials at least one letter was different from the others. She found that, for the RVF, manual reaction times for 'same' trials increased with the number of letters presented but, for the LVF, reaction time was more or less constant regardless of the number of letters presented. Since a process of comparing each letter in turn with the other letters of the stimulus set would lead to reaction time increasing with set size this result has been quoted as showing that the left hemisphere operates in a serial manner. Conversely, the constant reaction time found with left visual field presentations has been taken to imply that the right hemisphere processes information simultaneously, that is in parallel. However, Cohen also carried out the same experiment using non-letter characters from a typewriter and obtained no consistent effect of set size in either visual half field. When these characters were combined with letters in a third experiment there was again an increase in reaction time with set size for the RVF but only for letters, not for the typological characters. It thus seemed that serial processing was confined to verbal stimuli. However, this neat conclusion is upset by the finding of White and White (1975) that even with letters as stimuli there was no significant visual field-by-set size interaction when subjects performed nominal and physical identity matches. Furthermore, the notion that it is only the left hemisphere that operates in a serial manner is not supported by the results of Gross (1972), Niederbuhl and Springer (1979) and Polich (1980) all of whom found evidence of serial processing in both visual fields.

Although there is evidence that the left hemisphere is better at dealing with sequences of stimuli in different modalities (Carmon and Nachson, 1971; Swisher and Hirsch, 1972; Brookshire, 1975; Nachson and Carmon, 1975; Tallal and Newcombe, 1978; Sherwin and Efron, 1980; Mills and Rollman, 1980) this does not necessarily entail that the left hemisphere itself processes information sequentially. Nor should this be confused with evidence (see Chapter 5) that the left hemisphere has a special role to play in the organisation of sequential motor behaviour.

A second distinction that has been drawn is between analytic and holistic modes of processing (see Bever, 1975), sometimes referred to as analytic versus gestalt processing. It seems to have arisen from the report of Levy-Agresti and Sperry (1968) that in a visuo-tactile matching task the two hemispheres of split-brain patients solved the problems in characteristically different ways. The right hemisphere seemed able to grasp the shape of a three-dimensional form as a unified whole whereas the left hemisphere concentrated in turn on each of the edges and corners of the forms. The analytic versus holistic dichtomy is often confused with the serial versus parallel dichotomy but the two are conceptually distinct.

The analytic versus holistic dichotomy as it applies to laterality research has been more often invoked to explain results in a post hoc fashion than it has itself been subjected to experimental scrutiny. However, Martin (1979) predicted that if the shape of a letter was made up from smaller letters, then with instructions to respond on the basis of the overall configuration there should be a LVF advantage. With instructions to respond on the basis of the smaller letters there should be an advantage for the RVF. In the event, the global task (responding to the overall configuration) showed a non-significant LVF superiority and the local task (responding to constituent letters) gave a significant RVF superiority in reaction time.

Since the strategy which a subject adopts towards an experimental task is under his own control, unless otherwise constrained, any visual field asymmetry found in a particular experiment is as likely to reflect aspects of information processing as any intrinsic difference between the hemispheres for the task in question (e.g. Ross and Turkewitz, 1981). Even with subjects' strategies constrained, it will be necessary to know how those interact with hemispheric specialisation as a function of different cognitive tasks and experimental variables if a complete account of visual hemifield asymmetries is to be achieved.

Visual field differences have been related to a host of methodological, procedural and stimulus factors. These include the spatial or temporal intervals between the elements making up a display (Kimura, 1969; Hines and Satz, 1971), the distance of stimuli from fixation (Bryden, 1966; McKeever and Gill 1972c; Carmon and Nachson, 1973; Curcio, MacKavey and Rosen, 1974), the spatial frequency of the stimulus (Rao, Rourke and Whitman, 1981), the directional characteristics of words and letters (Harcum and Filion, 1963; Bryden, 1966; 1968; White, 1969b), the number of times a stimulus is presented (Hardyck, Tzeng and Wang, 1977; 1978;

Schmuller, 1980), exposure duration (Bryden, 1965; Gill and McKeever, 1974; Beaumont and Dimond, 1975), stimulus size (Pring, 1981; Pitblado, 1979b), typeface (Bryden and Allard, 1976), complexity (Fontenot, 1973) and discriminability (Patterson and Bradshaw, 1975). In addition to these stimulus characteristics, a large number of subject variables have been studied in relation to visual hemifield asymmetry. Many of these are covered in the chapters which follow.

## DICHOTIC LISTENING

One of the techniques now employed in many laboratories to assess differences in function between the two halves of the brain entails simultaneous presentation of competing information to the left and right ears. Pairs of stimuli are aligned on two tracks of a magnetic tape such that the onset and offset of one stimulus coincide exactly with the onset and offset of the second stimulus. Intensity of the two stimuli is also carefully balanced. The tape is played over a pair of stereo headphones to the subject who hears one member of the stimulus pair at one ear and the second member of the pair at the opposite ear. This procedure, referred to as dichotic presentation, has its origins in experiments on selective attention carried out by Broadbent (1954, 1958). However, the rapid development of interest in dichotic listening as a tool for the investigation of hemispheric asymmetry stems largely from the work of Kimura in the early sixties.

Kimura (1961a) showed that temporal lobectomy results in a significant deficit in recall of digits presented at the ear opposite the side of the surgery, but only under conditions of dichotic presentation and not when input is restricted to one ear alone. She argued that this was consistent with the view that the auditory pathways from each ear to the contralateral cerebral cortex are more effective than the ipsilateral pathways from each ear to the same side of the brain. Although the contralateral impairment observed with simulanteous digit presentation to the two ears was found after both left and right sided lobectomy, the overall efficiency of recall was much higher for the right temporal group of patients. Kimura saw this result as confirming that the left temporal lobe is more important than the right temporal lobe in the perception of spoken material.

### Models of dichotic listening

The fibres from the receiving station of the inner ear, the organ of Corti, synapse first at the cochlear nucleus. From there, some fibres continue to the superior olivary body on the same side of the brain stem as the stimulated ear but the majority cross over to the olivary body on the opposite side of the brain stem. Pathways from the two ears thus converge on the olivary body at each side (Figure 6). From here both ipsilateral (uncrossed) and contralateral (crossed) fibres pass successively upwards to

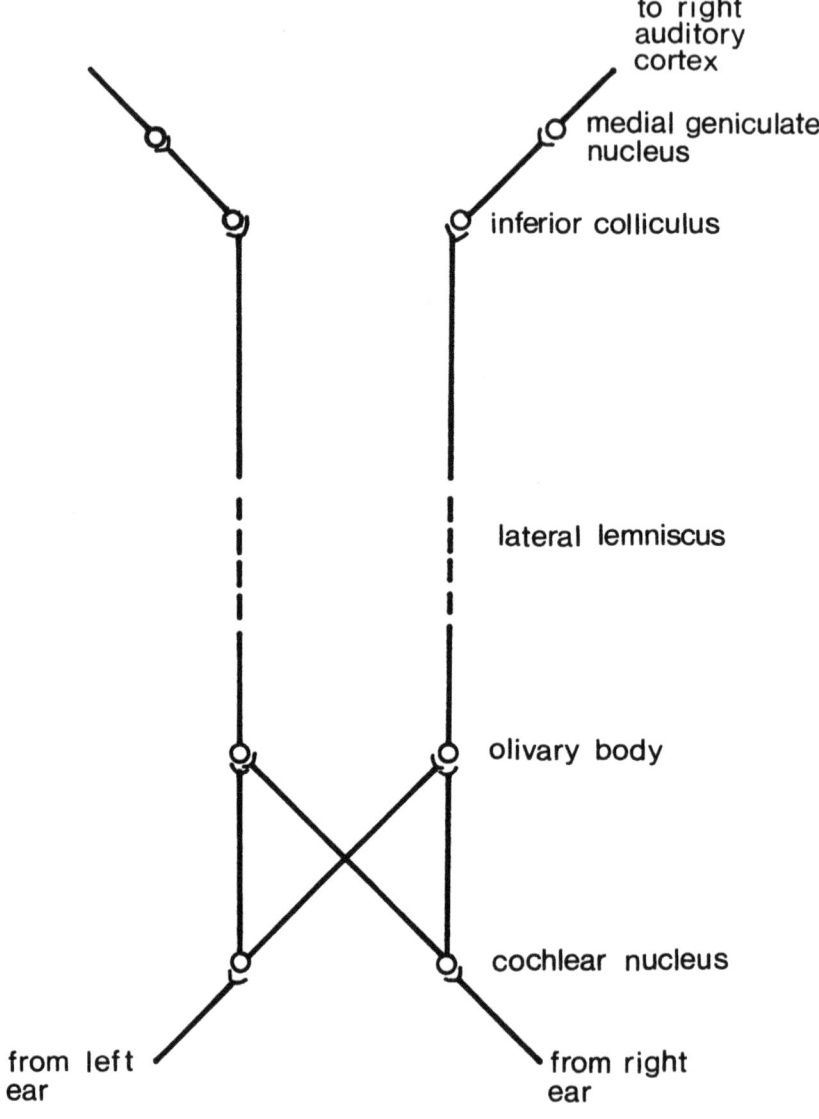

**Fig. 6** Diagram of auditory pathways

the lateral lemniscus, inferior colliculus and medial geniculate nucleus before arriving finally at the superior gyrus of the temporal lobe.

There is now substantial clinical (Bocca, Calearo, Cassinari and Miglia-vacca, 1955; Oxbury and Oxbury, 1969; Celesia, 1976) and experimental (Rosenzweig, 1951; Satz, Levy and Tyson, 1970; Majkowski, Bochenek, Bochenek, Knapic-Fijalkowska and Kopec, 1971; Andreassi, De Simone,

Friend and Grota, 1975; Mononen and Seitz, 1977) evidence to support the view that the crossed pathways are more important than the uncrossed pathways in transmitting auditory information. In particular, Milner, Taylor and Sperry (1968) and Sparks and Geschwind (1968) found that split-brain patients were able to repeat digits presented to either ear alone but showed almost total absence of report of digits presented to the left ear when different digits were simultaneously presented to the right ear. This suggests that impulses travelling along the uncrossed pathways suppress or inhibit impulses travelling to the cortex by way of uncrossed fibres, although recent work suggests that the extent of ipsilateral suppression depends upon the nature of the competing stimuli (Springer, Sidtis, Gazzaniga and Wilson, 1978). This suppression may take place at different levels, sub-cortical as well as cortical, within the auditory system (Berlin, Lowe-Bell, Cullen, Thompson and Loovis, 1973; Berlin, 1977). The effect of suppression, wherever it occurs, is that stimuli presented to the left ear are destined predominantly for the right hemisphere and stimuli heard at the right ear arrive mainly in the left hemisphere. The extinction of left ear responses in the dichotic listening performance of split-brain subjects can thus be explained by the fact that material in the right hemisphere cannot reach the speaking left hemisphere unless the anterior two thirds of the callosum (Springer and Gazzaniga, 1975; Hécaen, Gosnave, Vedrenne and Szikla, 1978) and/or, in some patients, the anterior commissure, (Risse, Le Doux, Springer, Wilson and Gazzaniga, 1978) is intact.

Kimura (1961b) presented neurosurgical patients with three pairs of digits, one member of each pair to the left ear and the other member to the right ear. Twelve of the patients had right hemisphere speech representation as determined by the Wada sodium amytal test. In this group of patients the mean number of items recalled from the left ear was higher than from the right ear. In the remaining 103 patients, with presumed left hemisphere speech, mean recall was higher from the right ear. Kimura also found that among normal right handed subjects there was a small but statistically significant advantage in recall of verbal material presented to the right ear. Dichotic listening thus appeared to offer a comparatively simple, non-invasive technique for determining the speech dominant hemisphere in neurologically intact subjects as well as in neurosurgical patients, since the side of speech dominance shows up as an advantage for digits presented to the contralateral ear. The right ear advantage in normals was attributed by Kimura to a combination of two factors. First, the crossed auditory pathways inhibit the uncrossed pathways such that stimuli presented to the right ear arrive intact at the left hemisphere and stimuli presented to the left ear arrive at the right hemisphere. Second, information is either degraded in passing from the right to the left hemisphere or else is less efficiently processed by the right half of the brain.

Following publication of Kimura's paper the dichotic listening technique was taken up enthusiastically as a means of assessing language laterality in normal subjects. Her finding of a mean right ear advantage

(REA) in the recall of dichotically presented material by right handers has been replicated by many investigators (see Studdert-Kennedy, 1975; Berlin and McNeill, 1976) and is now a firmly established phenomenon. However, Kimura's interpretation of this finding did not go unchallenged.

When digits are presented in pairs to left and right ears, there is an almost universal tendency (Broadbent, 1954) for subjects to report all of the items presented to one ear before reporting those items presented to the other ear. Inglis (1965) summarised data which showed that among individuals with memory defects only the number of items recalled from the second ear differed from the number recalled by normal control subjects, whereas recall from the initial ear was similar for both groups. Inglis argued that his result supported an interpretation of the usual right ear effect in terms of memory rather than sensory competition. He suggested that a tendency to report first the material entering the right ear might allow information from the left ear to decay in short term memory and thus give rise to the observed superiority of the right ear.

An order of report interpretation was considered an insufficient explanation by Bryden (1967), who continued to find a right ear advantage even when analysing only responses given from the ear reported first, but such an explanation continues to surface from time to time (e.g. Friedes, 1977).

Notwithstanding Bryden's (1967) results, it may happen that subjects primarily attend to, as opposed to recall, information presented to the right ear in the absence of constraints to do otherwise (Simon, 1967; Haydon and Spellacy, 1973; Levy and Bowers, 1974). Perhaps that is why sounds are recognised more accurately (Hublet, Morais and Bertelson, 1977) to the right of a subject and appear louder than sounds of equal intensity heard on the subject's left (Kellar, 1978; Wexler and Halwes, 1981). However, an attentional hypothesis cannot explain the finding of a right ear advantage when subjects are asked to attend to the input presented to the left ear (Bryden, 1969). Furthermore, Kallman (1978) found a right ear superiority for words and a trend towards a left ear advantage for music in a target detection task in which the two types of stimuli were randomly interspersed. An attentional model cannot account for the statistically significant ear-by-stimulus type interaction. Finally, the theory that attention is directed to the left or right side as a consequence of asymmetrical activation of the cerebral hemispheres (Kinsbourne, 1970; 1973; 1975a) predicts that in hemispherectomised patients attention should be directed almost exclusively to the side contralateral to the intact hemisphere. Nebes, Madden and Berg (1981) found no evidence to support this hypothesis. This is not to argue that prior activation of one hemisphere cannot introduce a bias towards the opposite ear (Nachson, 1973; Morais and Landercy, 1977) but it does suggest that the effect of activation may be to engage the processing mechanisms of a particular hemisphere rather than upset a hemispheric balance of attention.

The possibility that a dichotic right ear superiority is due to subjects being required to give a verbal response prompted Springer (1971, 1973) to present consonant-vowel syllables either dichotically or opposed by white noise (a mixture of all possible sound frequencies) and to use manual reaction time to a target syllable as her response measure. In both conditions the usual asymmetry emerged suggesting that factors associated with verbal output do not determine the advantage for the right ear. Thus although Kimura's original findings related ear asymmetry to the hemispheric side of speech *production* it is possible that differences between the ears reflect asymmetry between the hemispheres in speech *perception* rather than production.

Morais and Bertelson (1975) pointed out that the usual dichotic listening paradigm does not allow one to distinguish between an interpretation of the REA in terms of ear of stimulus presentation and an alternative explanation favouring input coming from the right side of space. That is, one explanation sees the right ear advantage to be the result of stimulating a particular ear, the other as the result of receiving sounds from a particular side of space. Morais and Bertelson therefore carried out an experiment in which the apparent spatial localisation of a sound source was achieved by manipulating either the time or the intensity difference of the same stimulus heard at the two ears. By stereophonic means a situation was created such that each ear heard the same two virtually simultaneous messages but, due to a very small difference in the time of arrival of a given stimulus at the two ears, one message appeared to come from the subject's left while the other appeared to come from the right. Thus ear of entry was eliminated as a potential source of dichotic advantage since both ears heard the same two messages. It turned out that Morais and Bertelson obtained a significant advantage for those stimuli which appeared to come from the right side of space. They argued that an interpretation in terms of the relative potency of crossed and uncrossed pathways could not account for this result. Kinsbourne's (1970) explanation of perceptual asymmetries as resulting from activation of the contralateral hemisphere could, however, handle their data. In support of the idea that ear differences in recall may have something to do with position in lateral space there is an intriguing but brief report in the literature that the magnitude of ear asymmetry can be modified by requiring subjects to wear prism lenses that displace the visual field to one side of the true position (Goldstein and Lackner, 1974).

The fact that a right ear advantage was obtained in the experiment by Morais and Bertelson does not mean that there is no difference in functional potency of crossed and uncrossed pathways, but implies that competition between the two ears is not the sole and necessary determinant of auditory laterality effects. This was in any case already known from the fact that statistically significant ear differences in reaction time or accuracy of report have been observed even with monaural stimulus presentation (for a bibliography see Henry, 1979). In general, it appears that the more complex the level of stimulus processing required, the more likely it is that a difference between the ears will be revealed with monaural presentation.

Few, if any, workers would deny that a superiority for the right ear has, other things being equal, something to do with the fact that the left hemisphere is more efficient at verbal tasks generally. Yet the nature of this 'something' will vary with different stimuli, tasks and subjects. Not all dichotic listening experiments tap the same psychological operations which may well differ in the extent of their underlying cerebral lateralisation.

Lateralised processing mechanisms may be engaged at either the acoustic, phonetic, syntactic or semantic levels of language. At the acoustic level Kimura and Folb (1968) found the usual right ear advantage in the perception of dichotically presented backward speech, even though such sounds are totally unintelligible. Zurif and Sait (1970) also obtained a difference between the ears in recognition of meaningless speech stimuli. In one condition of their experiment nonsense sequences followed a structural pattern of English in so far as replacing the nonsense stems by English stems would have resulted in a grammatically correct sequence. In the control condition this was not the case. A superiority of the right ear was obtained in the structured but not in the unstructured condition. This effect could not be attributed to different patterns of intonation in the two conditions (Zurif and Mendelsohn, 1972) which implies that the effect was mediated at a syntactic level. However, Harriman and Buxton (1979) presented semantically anomalous sentences spoken either in a monotone or with appropriate intonation, and obtained a REA only in the latter condition.

One of the simplest sets of stimuli used in dichotic listening experiments consists of natural or computer generated pairs of syllables, such as /pa/ and /da/. Shankweiler and Studdert-Kennedy (1967) used a device to generate consonant-vowel (cv) syllables from different consonants paired with the same vowel. Pairs of syllables carefully synchronised were then presented dichotically to subjects who had to report both syllables on each trial. A right ear advantage was found which was greater for those syllables differing in two articulatory features, voice and place, than for syllables differing in only one such feature ('voicing' refers to vibration of the vocal chords; 'place' refers to the place of articulation of the sound in the vocal tract). Using steady state vowels, which do not convey information by means of distinctive articulatory features, no laterality effect was obtained. Shankweiler and Studdert-Kennedy therefore concluded that the language processing system may be engaged at the level of the sound structure of language and involves an analysis-by-feature mechanism. Thus not only can the dichotic listening paradigm be employed as a tool for investigating brain asymmetry but it can also be used to probe the nature of speech perception.

Liberman, Cooper, Shankweiler and Studdert-Kennedy (1967) argued that the perception of speech involves analysis of an acoustic message into the articulatory movements which would be required to reproduce that message, the so-called motor theory of speech perception. Such a theory implies that decoding a speech signal by reference to its articulatory features may be what characterises, at least in part, the special capacities of the left hemisphere (Liberman, 1974). On the other hand Marshall (1973) considers the crucial point to be that

> ... the phonological correlations in terms of which speech is perceived and the temporal segments of the acoustic wave which examplify phonological sequences is not one to one and thus some decoding mechanism, unique to the left hemisphere, is required to derive the one from the other. (p. 452)

The question of exactly what linguistic features the left hemisphere is specialised to detect has been pursued vigorously. This issue is beyond the scope of this book and the interested reader is referred to papers by Studdert-Kennedy and Shankweiler (1970), Shankweiler (1971), Blumstein (1974), Cutting (1974), Darwin (1974), Liberman (1974), Berlin and MacNeill (1976) and Berlin (1977). It may be noted, however, that a right ear advantage has been reported for stimuli other than speech. This implies that the left hemisphere's specialisation is not restricted to speech sounds.

### Dichotic studies with non-verbal stimuli

Papçun, Krashen, Terbeek, Remington and Harshman (1974) presented short Morse code sequences to naive and experienced Morse operators and found a significant right ear advantage in both groups of subjects. This suggests that the meaning of the stimuli was irrelevant to the asymmetry between left and right ears. However, longer Morse sequences yielded an advantage for the left ear, but only for the experienced operators. It was suggested that they were able to switch to using a more 'holistic' processing mode which is the preferred mode of operation of the right hemisphere. As with other authors who have had recourse to the analytic versus holistic distinction (Bever and Chiarello, 1974; Bever, Hurtig and Handel, 1976; Ross and Turkewitz, 1976; Gates and Bradshaw, 1977a) this is a post-hoc explanation, not an experimental test of the notion that different cognitive strategies are characteristic of left and right hemispheres.

Musical stimuli have been employed in a number of dichotic listening experiments. Kimura (1964) first reported an ear difference in the perception of dichotically presented melodies. The same subjects who showed an advantage for the right ear with pairs of digits showed a significant superiority for the left ear when the stimuli consisted of snatches of melody which had to be identified by means of a multiple-choice response method. This finding supports that of Milner (1962) who found that right brain damaged subjects were more impaired than patients with left sided damage on certain items of the Seashore test of musical abilities. Subsequently Shankweiler (1966) found right temporal lobectomised patients to be inferior to left temporal patients on a dichotic melodies test. Together with Kimura's finding this strongly implicates the right temporal lobe in certain aspects of music perception (see also Shapiro, Grossman and Gardner, 1981).

In an attempt to identify the musical dimensions which determine the left ear effect, Gordon (1970) presented competing melodies matched for rhythm and pitch to experienced musical subjects. In a second condition, single chords were heard at each ear. No asymmetry was found on the melody task but a significant left ear advantage emerged for the chords. Gordon suggested that the failure to find a left ear superiority for melodies, in contrast to Kimura's (1964) results, might have been due to differences in the rhythm and/or pitch of the stimuli employed by himself and by Kimura. He subsequently found (Gordon, 1978a) that these two features show different laterality patterns, pitch yielding no ear difference and rhythm an advantage for the right ear. Robinson and Solomon (1974), Natale (1977) and Gates and Bradshaw (1977a) also obtained a right ear advantage in recognition of rhythm but dichotic pitch perception has yielded contradictory results (for review see Craig, 1979a).

It is possible that the different results for dichotic recognition of melodies obtained by Kimura (1964) and Gordon (1970) were due not to stimulus differences in the two experiments but to differences in the musical experience of the subjects employed. Bever and Chiarello (1974) and Johnson (1977) found a left ear superiority for melodies among naive listeners but a right ear advantage for musically experienced listeners. Comparable findings were obtained in an electrophysiological study by Hirshkowitz, Earle and Paley (1978) but a significant group-by-ear interaction has not always been found (Gates and Bradshaw, 1977a; Zattore, 1979). The potentially confounding effects of musical experience with musical aptitude were dissociated in an experiment by Gaede, Parsons and Bertera (1978). Subjects low in musical aptitude showed larger between ear differences on tests of chord analysis than did subjects of greater aptitude but within the same aptitude level there was no effect of experience.

Gates and Bradshaw (1977b) have reviewed the literature concerning music and the cerebral hemispheres and caution against regarding one particular hemisphere as dominant for musical functions. Each half of the brain may make its own contribution towards different aspects of musical expression or appreciation. Other reviews on the role of the hemispheres in music are those by Wyke (1977) and Damasio and Damasio (1977).

Non-verbal tasks giving rise to left ear advantages have included discrimination or recognition of pitch (Curry, 1968; Schulhoff and Goodglass, 1969; Halperin, Nachson and Carmon, 1973; Oscar-Berman, Goodglass and Donnenfeld, 1974; Kallman and Corballis, 1975), environmental noises (Curry, 1967; Knox and Kimura, 1970; Carmon and Nachson, 1973b) and such stimuli as clicks (Murphy and Venables, 1970), sonor signals (Webster and Chaney, 1966) and square-wave patterns (Sidtis, 1980). As with other laterality paradigms the cognitive processing undertaken by the subject rather than the nature of the stimulus per se determines which, if either, ear will be superior on a given task (Tsunoda, 1971; Nachson, 1973; Van Lancker and Fromkin, 1973; Bartholomeus, 1974; Gates and Bradshaw, 1977a). For example, Spellacy and Blumstein (1970) presented dichotic pairs of consonant-vowel-consonant (cvc) syllables among which were randomly interspersed either real English words for half the subjects or musical or environmental sounds produced by a human voice for the other half of the subjects. The syllables heard at the two ears differed only in the initial consonant or only in the middle vowel. The subjects hearing occasional words showed a right ear advantage for the vowel-varied stimuli whereas those subjects who heard non-verbal sounds exhibited a left ear superiority for the same stimuli.

As a general rule stimuli which are heard within, or from part of, a linguistic context give rise to an advantage for the right ear whereas stimuli heard within a non-linguistic context are more likely to show a superiority favouring the left ear. Bartholomeus (1974) presented dichotic pairs of melodies sung by different people repeating different letter sequences. Using the same stimulus tapes but different subjects for each task, she found that recognition of melodies gave a significant left ear advantage, and letter sequences yielded a significant right ear advantage while voice recognition showed no difference between the two ears. Thus opposite ear superiorities may be found when subjects are constrained to process different aspects of the same stimuli. This implies that in the absence of specific constraints subjects are free to attend to different aspects of the stimulus.

## Reliability and validity of dichotic listening asymmetry

Much of the motivation in dichotic listening research lies in identifying the hemisphere responsible for speech or other functions. However, as with the tachistoscopic paradigm, left-right ear differences in dichotic listening scores are far more labile than one would expect if ear asymmetry is an index of some fixed structural attribute (Teng, 1981). In one study as many as 30 per cent of subjects exhibited a change in the side of the superior ear when re-tested within a period of one month (Pizzamiglio, Pascalis and Vignati, 1974). Even within a single testing session the magnitude or direction of asymmetry may change, perhaps due to changing strategies utilised by the subject (Perl and Haggard, 1975; Kallman and Corballis, 1975; Sidtis and Bryden, 1978). Over a number of sessions the proportion of subjects showing a right-ear preference for verbal stimuli tends to increase due to the greater probability of change among subjects showing an initial left ear advantage (Blumstein, Goodglass and Tartter, 1975; Shankweiler and Studdert-Kennedy, 1975).

Even with adequate reliability the validity of any test instrument is not guaranteed. One still needs to know what is being measured. While a mean right ear advantage has been correlated with left hemisphere dominance for speech as determined by the Wada sodium amytal test (Kimura, 1961a; Tsunoda, 1975) the validity of dichotic ear asymmetry as an index of cerebral speech representation in individual subjects is questionable. This is because in any sample of right handers it is usually found that a considerable proportion of subjects do not show the expected right ear superiority for verbal material. Bryden (1967) found this proportion to be approximately 15 per cent. Even allowing for a difference in the extent of lateralisation of executive aspects of speech, revealed by the Wada test, and receptive aspects, tapped by the dichotic listening technique, this figure of 15 per cent is too high to accord with the evidence from brain damaged populations. A closer correspondence between behavioural and neuropsychological estimates of speech lateralisation is obtained by only considering ear differences that reach a specified level of statistical significance (Wexler, Halwes and Heninger, 1981) but as Satz (1977) has argued, the probability of mis-classifying a right hander with a dichotic left ear advantage as right brained for speech is of the order of 90 per cent (see Chapter 5)!

In general even if it can be shown that one particular dichotic testing procedure can lead to an accurate prediction of the side of speech representation it does not logically follow that other procedures are equally valid or even tap the same processes. Nonetheless, the stimuli and tasks used in dichotic listening research have been almost as varied as the number of investigations undertaken with little or no attempt at proper validation. An exception to this criticism is the work of Geffen who together with her colleagues has devised a dichotic monitoring test. Subjects hear words in left and right ears and are required to make a manual response on detecting a specified target word in either ear. A greater number of detections in one ear than the other is said to reflect speech dominance of the hemisphere opposite the more accurate ear. The technique has proved reliable (Geffen and Caudrey, 1981) and has been validated against the assessment of language laterality by means of unilateral ECT (Geffen, Traub and Stierman, 1978) and, in four cases, against the Wada sodium amytal test (Wale and Geffen, 1981).

Despite the fact that dichotic listening techniques have often been adopted without proper validation, findings which show a difference in the direction and/or magnitude of ear asymmetry between groups of right and left (or non-right) handed subjects have been taken as indicating a difference in direction or magnitude of cerebral lateralisation. Discussion of this point is deferred until Chapter 13.


## TACTILE PERCEPTION

Although hemispheric specialisation of function has been studied mainly through the visual and auditory modalities, there has been some work carried out with regard to the sense of touch. The fact that the sensory and motor functions of each hand are represented predominantly in the contralateral cerebral cortex means that information available to one hand alone is processed largely in the opposite hemisphere. Thus the abilities of left and right hemispheres on tactile tasks can be assessed by comparing the performance of the right and left hands. Although a tendency towards greater tactile sensitivity of the left compared with the right hand has been reported for right handed adults (Semmes, Weinstein, Ghent and Teuber, 1960, but see also Rhodes and Schwartz, 1981) a sensitivity difference between the hands is probably not important in tasks employing supra-threshold stimulation.

Benton and his colleagues used an electromechanical device to stimulate the back of the hand. Three points lying in a straight line were stimulated in quick succession and the subject's task was to indicate from among four alternatives the orientation of the line in which the stimuli had been presented. More accurate perception of orientation was found for the left than for the right hand, at least among right handers (Benton, Levin and Varney, 1973; Varney and Benton, 1975; Benton, Varney and Hamsher, 1978). Earlier Carmon and Benton (1969) and Fontenot and Benton (1971) had found that patients with lesions in the left hemisphere were impaired on this task only on the right hand whereas patients with right sided lesions were impaired on both hands. This is the reverse pattern of results to that found for purely somatasensory defects by Semmes et al. (1960). These findings, together with those for normal subjects, therefore suggest that the right hemisphere plays an important role in mediating tactile perception of direction. This might explain why right brain damaged patients perform poorly in learning a tactual maze (Corkin, 1965).

Faglioni, Scotti and Spinnler (1971) found that, with vision excluded, patients with right hemisphere lesions were significantly impaired in comparison with left brain damaged patients in reproducing the position of crosses marked on a model. Similarly, right hemisphere patients were inferior at reproducing by touch alone the angle at which two movable rods were set relative to each other (De Renzi, Faglioni and Scotti 1971). In both these studies the results for tactual responding closely mirrored those for responding under visual guidance, which suggests that the right hemisphere plays a supra-modal role in appreciation of spatial relations generally (De Renzi and Scotti, 1969; De Renzi, Faglioni and Scotti, 1970).

On a tactual version of the Formboard Test, in which wooden forms have to be fitted into spaces of the same outline shape, patients with right posterior damage were found to be much slower than those with damage to the left hemisphere (De Renzi, Faglioni and Scotti, 1968). With regard to neurologically intact subjects, Witelson (1974) devised a 'dichotomous' tactile task which revealed a left hand superiority in perceiving meaning-less, three dimensional forms. Similar findings have been reported by others (Kleineman and Cloninger, 1973; Gardner, English, Flannery, Hartnett, McCormick and Wilhemy 1977; Dodds, 1978; Klein and Rosenfield, 1980). Thus it appears that the right hemisphere bears particular responsibility in the tactile perception of shape.

A superiority in either the perception of shape or in the perception of direction, if in fact these functions are dissociable, can account for the left hand advantage in Braille reading found for both experienced blind subjects (Hermelin & O'Connor, 1971) and blindfolded normal subjects

(Smith, Chu and Edmonston, 1977; Harriman and Castell, 1979) taught to read Braille. These findings are particularly interesting in view of the fact that reading is a verbal process and might therefore be expected to yield a superiority for the right hand. Oscar-Berman, Rehbein, Porfert and Goodglass (1978) obtained a right hand superiority in recognition of letters traced on the palm of the hand but a left hand advantage in discriminating the orientation of lines. Conceivably, the degree of difficulty in discrimination determines whether the spatial or verbal aspects of Braille dominate performance on the task and this determines the direction of asymmetry.

## ELECTROPHYSIOLOGICAL STUDIES

It is impossible from purely behavioural experiments conducted with neurologically intact subjects to specify with any accuracy the locus in the brain of those neural events which intervene between presentation of a stimulus and the occurrence of some response. Modern techniques of electroencephalography (EEG), however, hold out the promise of localising the electrical activity of the brain in so far as this can be detected at the scalp. Although identification of left or right hemisphere activity represents a fairly gross level of localisation it is sometimes possible to be more precise as to the area of brain that is active. In any case, electrophysiological indices of functional brain asymmetry are useful in complementing the purely behavioural data.

In order to record the brain's activity at the scalp at least two electrodes are required. The signal that is picked up is the difference in potential between the two electrode positions. This signal is amplified many times and used to drive a pen recorder which traces out a wave form plotting voltage against time. Ideally speaking, one of the electrodes should be at a site where no activity occurs at all and then the amplified signal would represent the total activity at the area of interest; in practice this state of affairs is difficult to achieve since no site is entirely free from underlying neural activity. In investigations of cerebral asymmetry, therefore, electrode leads from various positions over left and right hemispheres are often linked to a common reference site at which underlying activity is unlikely to be affected by variables manipulated in the experiment. Provided that the reference position is suitably chosen, this means that differences in potential between the common reference and each of the corresponding left and right hemisphere sites will be equal unless there is some asymmetry in activity between left and right hemispheres. If so, it shows up as a difference in the wave forms obtained from one or more of the left and right hemisphere leads. It is thus crucial to choose a common reference position that avoids any initial bias in potential difference between each of the hemisphere leads and the common reference.

The electrical wave forms recorded from the scalp can be broadly classified into two types. One type consists of event-related potentials, the other is ongoing EEG activity (Hillyard and Woods, 1979). Event-related potentials, as the term implies, refers to the changes brought about when a subject is presented with a particular stimulus. When the event precipitating the change is a visual or auditory stimulus under experimental control the resulting activity is referred to as the visual or auditory evoked potential. This broad classification of EEG activity into ongoing and event-related is purely arbitrary since, as Hillyard and Woods (1979) point out, even ongoing EEG activity might be considered event-related if only the event concerned could be specified. By definition, event-related potentials are time-locked to some stimulus or other specificable event. As these changes in electrical potential may be very small in relation to fluctuations in ongoing EEG activity the usual procedure is to use a computer to sum the potentials during the half-second or so following each presentation of the evoking stimulus so as to produce an average value. The principle here is that if particular changes in activity bear a constant relationship to the reference event they will show up against fluctuations of the ongoing EEG which, being 'random', should cancel out to zero when averaged over successive trials by the computer.

Despite the appearance given by the EEG of probing the machinery of the brain, electroencephalography is a relatively gross technique which tells us only that a population of neural units is active rather than quiescent. It is difficult to localise with anything but a fair degree of accuracy the spatial locus of this activity. When it is remembered that the brain is a three dimensional structure it will be appreciated that very precise localisation of the activity picked up by surface electrodes is rarely possible. It is also worth pointing out that EEG recording is technically difficult and fraught with potential artefacts due to muscle movement (Grabow and Elliott, 1974), eye movement (Anderson, 1977) and possible left-right differences in skull or brain mass underlying the electrodes (Rubens, 1977).

Since a decrease in the alpha component of the EEG implies an increase in the underlying cortical activity, relative suppression of the alpha rhythm over one or other side of the brain has been taken to indicate selective hemisphere involvement during different cognitive tasks. For example, differential suppression of alpha over the two hemispheres was observed by McKee, Humphrey and McAdam (1973) during the performance of verbal and musical tasks. Similarly, Nava, Butler and Glass (1975) found relatively greater suppression over the right hemisphere when subjects performed a face recognition task and greater suppression over the left hemisphere when subjects attempted problems of mental arithmetic. Such task-related changes in distribution of alpha activity have been reported by others (Morgan, McDonald and MacDonald, 1971; Galin and Ornstein, 1972; Morgan, MacDonald and Hilgard, 1974; Robbins, Dale and McAdam, 1974; Galin and Ellis, 1975; Galin, Johnstone and Herron, 1978; but see Mayes and Beaumont, 1977).

A frequent methodological problem in electrophysiological experiments has been that experimenters have either failed to exercise any control whatever over their subjects' cognitive strategies or else have relied simply on instructions to subjects to engage in a particular mental activity. Buchsbaum and Fedio (1970), for example, reported that the visual evoked response to lateralised presentation of words and nonsense patterns differed for the two types of stimuli, most markedly over the left hemisphere. However, they exercised no control over their subjects' processing. Beaumont and Rugg (1978), on the other hand, required subjects to perform a go/no-go discrimination task in which subjects responded either to the sound of visually presented letters or to their shape. A left-right asymmetry in the latency of the first positive and second negative peaks was recorded which was not attributable to lateralised stimulus presentation per se. Ledlow, Swanson and Kinsbourne (1978) carried out a similar study in that subjects were required to perform a same-different match to visually presented letters either on a physical or a nominal basis. The results of this experiment were consistent with the view that EEG asymmetry can reflect hemispheric processing differences as distinct from effects due to mere stimulus presentation (Willis, Wheatley and Mitchell, 1979; Shucard, Cummins, Thomas and Shucard, 1981).

Even changes in a subject's preparatory 'set' may be sufficient to reveal EEG differences between left and right hemispheres. Butler and Glass (1974) required subjects to perform various mental operations when presented with numerical information. Prior to each trial the contingent negative variation (CNV) was recorded. This is a shift in negative potential which occurs when subjects are presented with a stimulus signalling some imminent event to which they must respond. Butler and Glass found the CNV to be of greater amplitude over the left hemisphere in all of twelve right handed subjects and greater over the right hemisphere in a single left hander. In a subsequent experiment, carried out only among right handers, the warning stimulus was followed by a verbal stimulus in one condition and by pictures of faces in other conditions. The CNV following the warning stimulus was found to be of greater magnitude over the left hemisphere in the verbal condition and greater over the right hemisphere in the picture condition. Butler and Glass therefore concluded that the CNV asymmetry relates to differential hemispheric activation rather than to handedness. Some aspects of EEG activity, however, have been shown to vary with manual preference (Eason, Groves, White and Oden, 1967; Provins and Cunliffe, 1972).

The finding of Butler and Glass (1974) that the magnitude of the CNV was greater over the left hemisphere prior to a verbal task and greater over the right hemisphere prior to a facial discrimination task might lead one to expect that the direction of CNV asymmetry can be predicted on the basis of selective hemispheric involvement in particular tasks. Donchin, Kutas and McCarthy, (1977) asked subjects to perform either a 'functional' match or a 'structural' match between items presented tachistoscopically. Research with split-brain patients (Levy and Trevarthen, 1976), supported by results from normal subjects (Landis, Assal and Perret, 1979), suggests that the left hemisphere matches stimuli on a conceptual basis (for example, a picture of a knife and fork 'goes with' a picture of food) while the right hemisphere matches on a structural basis (that is two pictures which look physically similar will be matched). In the experiment which Donchin et al. carried out subjects were forewarned which match they would be required to make. In one set of trials the type of match required was varied from trial to trial while in a second set of trials the same match was required for a fixed number of stimulus presentations. It was found that CNV asymmetry was greater in the former condition but the direction of asymmetry did not differ for 'functional' and 'structural' matches. This could mean that the same hemisphere carried out both types of match. Certainly Levy and Trevarthen in their study with split-brain patients argued that the 'wrong' hemisphere was sometimes activated in preparation for a particular task. Alternatively, it could be that the CNV does not reflect all relevant aspects of preparatory cortical activation. This brings out the point that it is not always an easy matter to determine the functional significance of any given electrocortical event.

While not denying the clinical value of the EEG as a non-invasive technique it is probably fair to conclude that electro-physiological research has so far not contributed anything new to our knowledge of cerebral asymmetry but rather has corroborated findings from other areas of investigation. However, there are situations in which the EEG may disclose lateralised phenomena which, because of their very nature, may not have been suspected on other grounds. Morgan, MacDonald and Hilgard (1974), for instance, have related hypnosis to mediation by the right hemisphere and Cohen, Rosen and Goldstein (1976) claimed to show that sexual orgasm in humans is associated with increased amplitude of the wave form over the right but not the left hemisphere.

Recent reviews of electrophysiological analyses of hemispheric asymmetry have been compiled by Anderson (1977), Donchin, Kutas and McCarthy (1977), Thatcher (1977), Levy (1978), Marsh (1978), Hillyard and Woods (1979) and Rugg (1982). The paper by Donchin et al. is a particularly good methodological critique of research in this area.

## LATERAL EYE MOVEMENTS

When someone is asked a question, it is likely that before answering he will direct his gaze away from the questioner's eyes (Argyle and Cook, 1976) particularly if the question requires some thought. Day (1964) noted that people are fairly consistent as to the direction in which they shift their gaze to break off eye contact. Bakan (1971) proposed that the direction of lateral eye gaze reflects activation of the hemisphere opposite the direction of eye movement, movement to the left, for example, indicating activation of the right hemisphere. Since electrical stimulation of the exposed cortex of one hemisphere can induce a deviation in eye gaze towards the opposite side

(Penfield and Roberts, 1959) it is conceivable that naturally occurring lateral eye movements reflect, at least in some circumstances, asymmetrical activity in the two hemispheres. It follows from this proposition that manipulating differential hemispheric activity should induce changes in direction of lateral eye gaze.

Kinsbourne (1972) and Kocel, Galin, Ornstein and Merrin (1972) attempted to differentially engage the left and right hemispheres by asking different types of questions. The direction of a subject's first eye movement following a question was recorded and found to be related to the type of question being asked. Verbal questions tended to elicit eye movements to the right whereas questions of a spatial nature tended to elicit movements to the left. Similar results have been reported by many but not all authors (see Ehrlichman and Weinberger, 1978) and support, but do not prove, a hemispheric activation model of lateral eye movements.

The issue as to whether lateral eye movements are an enduring characteristic of an individual, as thought to be the case by Day (1964), or whether they reflect transient shifts in hemispheric activation, as conceived by Kinsbourne, was taken up by Gur, Gur and Harris (1975). These workers noted that in the studies by Kinsbourne (1972) and by Kocel et al. (1972) the experimenter was seated behind the subject, eye movements being recorded by means of a video camera, while in the situation described by Day (1964) subjects sat facing the experimenter. Gur et al. therefore carried out a study to examine the possibility that this factor was crucial in determining whether lateral eye movements vary as a function of the type of question asked. The results of this experiment showed that among right handers, gaze direction was not related to question type when the interviewer faced the subject but was so related when the interviewer sat behind the subject. Left handers showed a consistent direction of movement in the experimenter-in-front condition, but no systematic relationship between direction of eye movement and question-type in the experimenter-behind condition. It was therefore concluded that gaze direction is a function both of an individual's consistent tendency to rely on a particular half of the brain and of differential hemispheric arousal in response to specific experimental situations.

In order to explain the different findings obtained according to whether the interviewer sat behind or in front of the subjects, Gur (1975) suggested that when the interviewer sits opposite the subject the latter's anxiety level is increased which leads him to reply in a characteristic mode of thought. Hiscock (1977) therefore attempted to manipulate anxiety level experimentally and thereby observe the interaction of this variable with the presence or absence of an experimenter. It was found that anxiety level had no effect on the consistency or direction of eye movement or on the relative extent of left-right deviations of gaze in response to verbal and spatial questions. Nor was there any significant difference in direction of gaze for the two types of questions unless specific items were picked out which apparently reflected most clearly the verbal-spatial distinction.

If it is true that verbal questions primarily engage the left hemisphere while spatial questions tap the functions of the right hemisphere, it might be predicted that responses to such questions will be optimal when the appropriate hemisphere is activated rather than if the opposite hemisphere is aroused. Gur et al. (1975) analysed the data collected in their experiment with a view to seeing whether more accurate responses were given to verbal questions which were followed by eye movements to the right than to verbal questions eliciting leftward eye movements. Similarly, for spatial questions it was predicted that more correct responses would occur in association with eye movement to the left rather than the right. By and large, these expectations were confirmed.

The direction of subjects' lateral eye movements has been correlated with EEG alpha activity, consistent left movers showing a relatively greater degree of alpha activity (Bakan and Svorad, 1969). Relative alpha activity has also been correlated with susceptibility to hypnosis (Bakan, 1969). It was therefore hypothesised by Morgan, McDonald and MacDonald (1971) that subjects showing a predominance of left lateral eye movements in response to verbal and spatial questions would show a greater susceptibility to hypnosis than those with predominantly right lateral eye movements. This predication was confirmed although Gur and Gur (1977) believe that this relation holds only for right handed, left eye dominant males.

It is claimed that left lateral eye movements reflect relatively greater right than left hemisphere activation. It is also believed that activity in the alpha band of the EEG signifies a low level of cerebral activation. It can be predicted, then, that if left lateral eye movers are highly susceptible to hypnosis in comparison with right eye movers, highly susceptible subjects should show the opposite direction of alpha asymmetry to less susceptible subjects. However, Morgan, MacDonald and Hilgard (1974) were unable to relate hypnotic susceptibility to lateral asymmetry of alpha despite confirming that highly hypnotisable subjects show relatively more alpha activity than those less susceptible to hypnosis. On the other hand it is worth mentioning that the hypothesis of decreased left hemisphere activity during the hypnotic state has been supported by other workers using different response measures. Frumkin, Ripley and Cox (1978) found a reduced mean right ear advantage on a dichotic listening task during hypnosis compared with scores obtained before and after the hypnotic session. With a partial split-brain patient as their subject McKeever, Larrabee, Sullivan and Johnson (1981) observed that the usual difficulty of naming objects presented to the left hand was considerably reduced under hypnosis. This was seen tentatively in terms of reduced interference by the left hemisphere in right hemisphere speech mechanisms.

Gur, Gur and Marshalek (1975) discuss findings which suggest that people who typically move their eyes in a leftward direction prefer to sit on the right side of a classroom (looking towards the front). Although these findings were considered in the context of differential hemispheric activation, they might more parsimoniously be thought to reflect no more than the fact that people wish to be able to look at a blackboard in the middle of the room. Sitting on the right side of the room means that left eye movements bring the centre of the room into view whereas right eye movements do the opposite. People might simply be responding, albeit unconsciously, in a manner consistent with their usual direction of eye movements.

Because of the potential source of artefact of such factors as the spatial layout of a research laboratory Saring and Von Carmon (1980) had their subjects lie supine in a darkened room. It was argued that this should have maximised any lateral eye movement asymmetry. Although analysis of initial eye movements revealed no statistically significant effect of question type on direction of eye movement, the average frequency of all lateral eye movements in response to verbal questions was greater for movements towards the right.

Gross, Franko and Lewin (1980) asked whether *involuntary* eye gaze would influence the cognitive processing employed by subjects. In other words these workers examined the usual question in reverse by asking not whether direction of eye gaze would *reflect* differential hemispheric, but whether forced direction of eye gaze would *induce* asymmetrical activation and hence determine choice of processing mode. Subjects were asked to look either to the left or to the right while choosing the odd one out from among three Hebrew words. The odd word could be chosen according to semantic or non-semantic criteria. For example, of the three words 'watch', 'clock' and 'block' the odd word in terms of meaning is 'block' but 'watch' is odd in terms of rhyme. Gross et al. found that when right handers were asked to look to the left while hearing such sets of words, decisions were made on non-semantic grounds significantly more often than when subjects looked to the right. If semantic processing is considered characteristic of the left hemisphere these findings can be taken as supporting the idea that involuntary direction of gaze may influence asymmetrical hemispheric activation.

Although lateral eye movements appear at first sight to offer a simple and straightforward way of assessing which hemisphere is active at a given instant, the evidence relating eye movements to hemisphere function is at present rather insubstantial. The fact that opposite directions of movement may in some circumstances be associated with different questions, considered a priori to encourage different types of cognitive activity, is not sufficient evidence to warrant the conclusion that opposite hemispheres are being engaged. Different questions may vary not only in terms of verbal or spatial content but also in terms of degree of interest, emotion or imagery aroused in a subject. Directional asymmetry in eye gaze may be related as much to these or other variables as to differential hemispheric activation. What is desirable is to have independent evidence linking eye movements directly to functional cerebral asymmetry. Electrophysiological measures suggest themselves in this context but the problem of artefact, that is, of actual or potential eye movements producing an asymmetry in the EEG record (Anderson, 1977), would have to be circumvented. In the absence of such direct evidence, a study by Lefevre, Stark, Lambert and Genese (1977) is of some interest. These researchers presented a group of subjects with a verbal dichotic listening task and noted that subjects gave more rightward than leftward eye movements as well as showing the usual right ear advantage on the dichotic task. A second group of subjects presented with a non-verbal dichotic task yielded a left ear advantage and more eye movements towards the left than towards the right. While this does not prove that the first group of subjects were using their left hemisphere and the second group their right hemisphere the results are at least consistent with such a supposition. A sex difference was also noted: males made significantly more left movements than right movements overall while females showed little difference. By contrast, Beveridge and Hicks (1976) claimed that females tended to be left movers and males tended to be right movers.

The literature on lateral eye movements has been reviewed by Ehrlich-man and Weinberger (1978) and the reader should consult their paper for a thorough critique. These authors draw attention to a number of methodological problems as well as to difficulties of interpretation and theory. With regard to methodology two main problems stand out. These concern, first, the validity of items chosen so as to represent one or other cognitive mode and, second, the scoring of lateral eye movements. Ehrlich-man and Weinberger criticise the emphasis that has been placed on the first eye movements occurring after a question has been asked and the conse-quent ignoring of movements which occur prior to or during presentation of a particular item. Furthermore, eye movements include vertical as well as lateral deviations of gaze and the general tendency to pay little or no attention to vertical eye movements may have important implications for interpretation of the data on horizontal movements. Finally, as Ehrlich-man and Weinberger point out, differences in the distance between subject and experimenter may well account for certain of the inconsistencies in the results reported by different investigators. The topic of interpersonal distance has a copious literature in its own right (see Argyle and Cook, 1976) and it is to be expected that any effects due to hemispheric asymmetry will interact with social and personality factors in different ways under different circumstances.

## DUAL TASK EXPERIMENTS

The final method of investigating hemispheric functional asymmetry in neurologically intact subjects involves analysing the effects of requiring two tasks to be carried out simultaneously. The notion that the cognitive processing necessary for one task can be assessed by measuring the spare capacity available for allocation to a second task is one that has proved useful in several areas of psychology (Kalsbeek and Sykes, 1967). This idea is based on the assumption that the overall capacity of the organism is limited, an assumption that has, as it happens, been questioned (Neisser, 1973; Allport, 1980).

The dual task technique was first exploited within laterality research by Kinsbourne and Cook (1971) who asked their subjects to balance a dowel rod with the index finger of one hand whilst simultaneously repeating letters of the alphabet. Concurrent verbalisation was found to reduce balancing times for the right hand but not the left hand. This was explained by arguing that the neural mechanisms underlying both motor control of the right hand and the production of speech are located within the same cerebral hemisphere and thus the two tasks compete for limited neural space or processing capacity. In contrast, such intra-hemispheric competition does not occur when balancing is carried out by the left hand since this is controlled from the right hemisphere.

The simple task devised by Kinsbourne and Cook (1971) was subsequently taken up and modified by other investigations. The original findings have been replicated (Hicks, 1975; Johnson and Kozma, 1977) and an interpretation in terms of intra-hemispheric competition supported by evidence of impairment on various tasks carried out by the left hand when cognitive operations supposedly mediated primarily by the right hemisphere are performed (Kinsbourne, 1973; McFarland and Ashton, 1975; 1978a; Smith, Chu and Edmonston, 1977; Dalby, 1980).

The intra-hemispheric competition model predicts that since two tasks compete with each other both should be affected by interference (Bowers, Heilman, Satz and Altman, 1978). This has rarely been reported. Bowers et al., for example, found that although their cognitive tasks affected manual performance the reverse was not the case. A manual task was, however, found by Botkin, Schmaltz and Lamb (1978) to affect the number of digits that could be repeated backwards. Fewer digits were repeated when the right rather than the left hand was used to carry out a tracking task.

Lomas and Kimura (1976) examined the effect of concurrent vocalisation on different manual tasks. They found that speaking interfered selectively with the right hand only under certain conditions. These conditions were, firstly, when subjects had to depress each of four Morse keys rapidly in turn with each of the four fingers of one hand and, secondly, when a whole arm movement had to be made to depress each key in turn. Since a performance décrement for both hands, rather than a lateralised impairment, was found when a single button was pressed repetitively with one finger Lomas and Kimura suggested that 'it is the rapid positioning of a limb, or parts of a limb, ... which is related to the lateralised decrement produced by concurrent speaking.'

Cremer and Ashton (1981) required subjects to alternately tap with a rod two small metal targets. A concurrent verbal task interfered with the speed and consistency of tapping by the right hand while a visuo-spatial task interfered more with left hand performance. Reducing the size and increasing the distance apart of the targets did not remove these lateralised effects even though greater accuracy was required in hitting the targets. These findings were held not to support the view expressed by Lomas and Kimura.

The effect of increasing task difficulty may be confined to one hand or affect both hands. Hicks, Provenzano and Ribstein (1975) found that as subjects rehearsed increasingly difficult lists of words a deficit on a typing-like task showed up first on the right hand and then spread to include the left hand. Comparable results have been reported by other authors (McFarland and Ashton, 1978b; Bowers, Heilman, Satz and Altman 1978). However, Hicks, Bradshaw, Kinsbourne and Feigin (1978) found that concurrent verbalisation increased response times for both hands on a typing task but more so for the right hand. The magnitude of this asymmetrical effect increased with the difficulty of the typing task. The effect of task difficulty in terms of whether a purely lateralised effect or a bilateral effect (symmetrical or asymmetrical with respect to the two hands) is observed thus appears to depend upon the nature of the two competing tasks.

It may be helpful at this point to distinguish between two possible sources of interference between concurrent tasks. One source is competition for the same neural mechanisms of motor (including verbal) output. The other is competition between two tasks for attention or cognitive processing capacity (see also Lomas, 1980). Hellige and Longstreth (1981) asked subjects to tap with either the left or right index finger. Concurrent reading reduced the rate of tapping more for the right than for the left finger. This effect was greater for subjects reading aloud than for silent reading and larger again for subjects who were led to expect a test on what they had read. It was therefore concluded that lateralisation effects are mediated by both motor and cognitive aspects of the tasks. Similar results were reported by Bowers, Heilman, Satz and Altman (1978) who observed a bilateral but asymmetrical impairment on finger tapping when subjects merely had to listen to a story knowing that they would subsequently be asked to recall its contents.

Kinsbourne (1975a) argued that 'both hemispheres draw upon and often compete for a finite amount of attention invested in the organism as a whole'. Consequently attention may be distributed asymmetrically between the two hemispheres. The effect of this, according to Kinsbourne, is that there is a bias in responding to the side of space contralateral to the hemisphere which has the greater share of attention. Part of Kinsbourne's evidence for this view is that on tachistoscopic tasks for which no left-right hemifield asymmetry is normally observed concurrent verbalisation can induce a left-right difference in favour of the right visual hemifield.

On a task requiring the subject to detect and respond to a small gap in one of the sides of a square Kinsbourne (1973) found concurrent verbalisa-tion to lead to a right hemifield superiority and humming to lead to a bias in detecting gaps in the left visual hemifield. Other authors, however, have failed to replicate these findings (Gardner and Branski 1976; Boles, 1979).

Nonetheless there is sufficient evidence to suggest that concurrent performance of certain cognitive tasks can in some circumstances alter left-right perceptual asymmetries (Kinsbourne, 1970; 1973; Hellige, 1978; Hellige, Cox and Litvac, 1979; Allard and Bryden, 1979; Rizzolatti, Bertolini and Buchtel, 1979; Beaumont and Colley, 1980). The problem is that the additional task has sometimes been found to facilitate performance in the visual field opposite the supposedly activated hemisphere and in other cases has been found to impair performance. In cases of facilitation, this is 'attributable to the beneficial effect on performance of increase in arousal when this is moderate in degree' according to Kinsbourne and Hicks (1978). When the effect of an additional task is to disrupt rather than facilitate performance then the two tasks are said to compete for the same 'functional space' within the hemispheres. As Cohen (1979) has pointed out, Kinsbourne's theory has 'too much explanatory power and too little predictive power'. It is impossible to predict whether a particular task will help or hinder performance on a second task.

Hellige and Cox (1976) found a LVF superiority in tachistoscopic recognition of random forms. When subjects had to concurrently hold in memory two or four nouns on each trial a superiority of the RVF emerged. This would be predicted by Kinsbourne's theory of selective activation of the left hemisphere by a verbal task. However, a concurrent memory load of six nouns markedly reduced recognition in the right visual field resulting in a LVF superiority which would not be predicted.

The conditions under which a concurrent memory load either facilitates or impairs performance on a primary cognitive task was further investigated in a series of experiments by Hellige, Cox and Litvac (1979). To account for their complex set of results they proposed that a concurrent task may have either general effects, that is, influence performance by both hemispheres, and/or specific effects which are restricted to one particular hemisphere. In one of their experiments, for example, subjects had to compare tachistoscopically presented random forms with a set of forms held in memory. This yielded a LVF superiority in reaction time. When either 2 or 6 words had also to be held in memory then reaction times for both hemispheres improved but more so for the left hemisphere than for the right hemisphere, leading to a RVF superiority. It was argued that these findings reflected general activation of both hemispheres combined with specific activation of the left hemisphere. In another experiment subjects had to match visually presented upper and lower case letters according to their names (nominal identity match). There was a RVF advantage in reaction time for this task carried out alone. However, when subjects had to concurrently remember 2, 4 or 6 words the reaction time for both hemispheres improved, but that for the right hemisphere more than that for the left hemisphere, which resulted in a LVF superiority. The explanation offered was that the facilitating effects of general arousal in the left hemisphere were offset by the cognitive load of remembering the set of words.

The problem with Hellige's formulation is that it can be used to explain any particular pattern of results but cannot readily predict them since the nature and difficulty of both primary and concurrent tasks needs to be taken into account. For example, the effect of holding in memory a visual pattern made up by randomly filling in half the cells of a matrix (with either 3, 4 or 5 cells per side) did not produce 'general activation' in the same way as words were found to do. Instead, the performance of both hemispheres declined equally.

Whatever the difficulties of the model proposed by Hellige et al. it has the merit of illustrating how theoretical progress might be made through sets of carefully related experiments. Progress will inevitably be slow but the dual-task technique offers the possibility of determining something of the nature of the dynamic interaction between the two hemispheres. Due to the preoccupation of investigators with the issue of hemispheric asymmetry this most important problem has hitherto been immune from serious experimental attack.

All the methods in current use for investigating hemispheric asymmetry of function in normal subjects have now been described. Of the six methods, tachistoscopic half-field presentation and dichotic listening have generated the largest amount of research. This is partly because of the pre-eminent roles played by the senses of vision and hearing in Man and partly due to historical reasons. Tachistoscopic and dichotic techniques have been employed in laterality studies for about two decades while the other methods described above were devised more recently. Electroencephalography, it is true, had its origins in the nineteenth century but only in recent years has there been an acceleration of interest in lateralised electrophysiological phenomena.

Since visual field asymmetry and dichotic ear differences have both been claimed as indices of the same phenomenon – cerebral lateralisation of language – these two measures derived from the same set of subjects ought to correlate with each other. A significant correlation was indeed reported by Hines and Satz (1974) but other studies (Bryden, 1965; Zurif and Bryden, 1969) have yielded low and insignificant correlations. Fennel, Bowers and Satz (1977), for example, took pains to devise a tachistoscopic analogue of their dichotic task and still failed to find a significant correlation between the two sets of scores, although there was an increasing agreement over four testing sessions between the side of the ear advantage and the superior visual half-field.

## THE MEASUREMENT OF LATERALITY

There has been some discussion in the literature as to the most appropriate way of expressing the degree of lateral asymmetry which a subject shows in an experiment. A simple expression of the difference between left and right ears or visual fields for each subject can be misleading since the same difference in scores may be obtained by subjects who differ in their overall level of accuracy. Consequently, the same absolute difference score can represent different relative values. An example should make this clear.

Imagine that Smith obtains a score of 20 correct responses at the left ear and 30 correct at the right ear. His absolute difference score is therefore 10. If Jones obtains 30 correct responses at the left ear and 40 correct at the right ear then his absolute difference score is also 10. But Smith's right ear score is 50 per cent higher than his left ear score, while Jones's right ear is only one-third better than his own left ear. In order to reflect a relative difference between the ear or half fields a laterality coefficient may be computed according to the formula

$$LC = \frac{R - L}{R + L}$$

where R represents the number of correct responses for the right ear/visual field and L stands for the left ear/visual field. By this means an absolute lateral difference is converted into a ratio measure. In the example above, Smith would have a laterality coefficient of 0.20 while Jones would have a coefficient of only 0.14. It does not matter whether the score for the left side is larger or smaller than the score on the right since the sign of the coefficient, negative or positive, indicates the direction of any lateral advantage.

It was argued by Marshall, Holmes and Caplan (1975) that the degree of brain asymmetry underlying performance on a particular task is *theoretically* independent of the overall level of accuracy which a subject attains on that task. These authors discuss data presented by Harshman and Krashen (1972) which show that the absolute difference between scores for left and right ears does *in fact* correlate significantly with the total correct score for the two ears.

Although Marshall et al. consider that in certain circumstances one would expect laterality scores to correlate with overall accuracy scores, they favour the use of an index of laterality that is independent of accuracy in the sense that the values which the index might take within the total range of values possible is not 'constrained' by any given level of accuracy. The meaning of this should become clear in considering the following indices which Marshall et al. rejected:

1. Absolute difference scores; $(R_c - L_c)$
2. Per cent correct (POC); $R_c/(R_c + L_c)$
3. Per cent error (POE); $L_e/(R_e + L_e)$

where $R_c$ and $L_c$ refer to the per cent correct score at the right and left ear or visual hemifield respectively and $R_e$ and $L_e$ refer to the per cent error score at each ear or hemifield.

The absolute difference score can range from $+100$ to $-100$ and POC and POE can each range from 1 to 0. However, only if total accuracy is 50 per cent, i.e. $(R_c + L_c) = 100$, can $(R_c - L_c)$ take all values from $+100$ to $-100$. If, for example, $(R_c + L_c) = 120$ or $70$ then $(R_c - L_c)$ can never equal $+100$ or $-100$. Similarly, POC is constrained above 50 per cent accuracy and POE is constrained below 50 per cent accuracy. On the other hand, the value of $(R_c - L_c)/(R_c + L_c)$ is not constrained at or above accuracy levels of 50 per cent and the value of $(R_c - L_c)/(R_e + L_e)$ is not constrained at or below accuracy levels of 50 per cent. Marshall et al. therefore recommend that one or other of these coefficients (which they refer to as 'f') be used in laterality experiments – depending on the level of accuracy achieved. Repp (1977) has argued similarly but Richardson (1976) considers choice of the 'f' index to be as arbitrary as any other.

Despite the fact that it is the property of not being *constrained* by overall accuracy that most concerned Marshall et al. in choosing a laterality index subsequent authors have been interested in the statistical *correlation* between various indices of laterality and overall accuracy. The phi coefficient proposed by Kuhn (1973), for example, has been shown by Levy (1977c) to correlate with overall accuracy. (The value of phi is given by the formula

$$\Phi = \frac{R - L}{\{(R+L)[2T-(R+L)]\}^{\frac{1}{2}}}$$

where R and L = right and left ear scores respectively and T = total number of trials). Similarly Hellige, Zatkin and Wong (1981) found the absolute difference score, POC, POE, f and phi coefficient to correlate very highly with each other and with total accuracy on a dichotic listening task. It is, however, not surprising that this is the case since the same data are being used to derive both the laterality index and the total accuracy score. Thus the statistical significance of the correlations should not be calculated with respect to an expected value of zero (Stone, 1980). Yet this is what has been reported in the literature (Marshall et al., 1975; Birkett, 1977).

The search for a laterality index that is not biased with respect to accuracy recently led Bryden and Sprott (1981) to propose the adoption of a new index, lamda, based on the log odds ratios $(P_R/(1-P_R); P_L/(1-P_L)$ where $P_R$ and $P_L$ are the respective probabilities of a correct response at the left and right sides. As yet, however, there are no empirical data with which to evaluate the usefulness of this index.

The nature of the laterality index one chooses is intimately bound up with the sort of theoretical question one wishes to ask (Eling, 1981). It has been argued, for example, that laterality should be measured only on a nominal scale (Colbourn, 1978) such that only the direction and not the magnitude of any laterality effect is taken into account. Discussion of this point will be deferred until Chapter 14.

## SUMMARY

This chapter reviewed techniques devised for investigating cerebral functional asymmetry in neurologically intact subjects. Results using these techniques are in broad agreement with those of the split-brain studies and with findings obtained from patients with unilateral cerebral lesions. However, it is not always clear what is being measured in a given situation. Correlations between measures of tachistoscopic hemifield and dichotic listening asymmetry, for example, are low.

Laterality effects in normals are notoriously sensitive to experimental manipulations that ought not to be influential if what is being measured is some fixed attribute of cerebral organisation. The lability of such effects can be accommodated by information processing or dynamic models of asymmetry but not by purely structural models.

There has been some discussion in the literature as to how left-right differences in performance should be measured. Although different laterality indices correlate positively and significantly it has been argued that the index of choice is one that is not constrained at any level of accuracy.

# 5

# Language and laterality

The first observation that the left half of the brain is intimately concerned with the functions of speech is usually attributed to Broca (1861) although an unpublished manuscript noting a correspondence between defects of language and lesions of the left hemisphere was written by Dax in 1836. Broca's (1861) contribution (for a review of the very early literature see Benton (1964)) was to identify the third frontal convolution of the left hemisphere as the 'centre' for 'articulate speech'. Broca discovered during post mortem examination of the brain of a patient who had been dysphasic that this area, now known as Broca's area, was severely damaged. Subsequently Wernicke (1874) drew attention to defects in the comprehension of language following damage to the posterior third of the superior temporal gyrus on the left side, now referred to as Wernicke's area.

A band of fibres, the longitudinal arcuate fasciculus, runs between Broca's and Wernicke's areas. It has been argued that whereas damage to Wernicke's area produces defects of comprehension, and damage to Broca's area defects in the production of speech, damage to the arcuate fasciculus results in the syndrome of conduction aphasia. The patient is able to understand, but not repeat, what is said to him. This is attributed to the disconnection of the region responsible for understanding speech from the region responsible for organising the motor programmes for speech output (Geschwind, 1970; 1974).

Other regions of the brain, both cortical and subcortical, are implicated in language behaviour. In particular, the role of the thalamus has increasingly come under attention (see Brown, 1975; Ojemann and Mateer, 1979a, and papers in *Brain and Language*, Vol. 2. Number 1, 1975) in recent years. Yet as Brown (1979) points out we still do not have a coherent model of the neural bases of language. Despite considerable individual differences in the localisation of language in the brain (Brown, 1979; Ojemann, 1979; Gur and Reivich, 1980) the pre-eminent role of the left hemisphere in the majority of adults is not in doubt. This chapter reviews the evidence bearing on this aspect of the cerebral organisation of language with particular reference to handedness. Readers interested in other aspects of the neurology of language are referred to papers by Geschwind (1964); Lenneberg (1967); Brown, (1979a,b) and Marin, Schwartz and Saffran (1979).

Evidence concerning language lateralisation comes from a variety of sources. Unilaterally brain-damaged patients provide the classic source of data, beginning with Broca's observations in the nineteenth century. More recently, other techniques have been devised which allow estimates to be made as to the hemisphere responsible for speech. Some of these techniques are for use specifically with patients having, or suspected of having, some kind of damage to the brain. Others are used with neurologically normal subjects. The rationale behind them, and their limitations, were discussed in Chapter 4. All of these techniques can be used to explore the relation between handedness and hemispheric specialisation for language.

Originally it was believed that the preferred hand was a reliable indicator of the speech-dominant hemisphere. If an individual was right-handed, then speech was thought to be controlled by the left hemisphere; if he was left-handed, then the right hemisphere was considered responsible. We now know that this is not necessarily the case, at least for left-handers. Yet despite the development of the newer techniques for estimating speech dominance, the precise relation which left-handedness bears to the laterali-sation of language is still not clear. The issue is as much conceptual as empirical, having to do with the definition and measurement of handedness as much as with the demonstration of a hemisphere 'dominance' effect. This chapter reviews the evidence on the issue that has accrued from the different methodologies available. Findings obtained with brain-damaged patients are discussed first, followed by the results of studies carried out with normal subjects.

It should be appreciated that hemispheric specialisation for language is not absolute. Executive aspects of language appear to be more clearly lateralised than receptive aspects, but even this is not the whole story. The chapter therefore closes with a review of findings concerning the linguistic ability of the right hemisphere.


## UNILATERAL CEREBRAL LESIONS

Following Broca's (1861, 1865) analyses of aphasia in right-handed patients who had sustained lesions of the left frontal lobe, Jackson (1868) drew attention to the case of a left handed patient who had become aphasic following a presumed right-sided lesion. Gradually there arose the belief that speech is always represented in the hemisphere opposite the preferred hand. Although there were from time to time reports of 'crossed aphasia', in which the lesion is on the same side as the preferred hand (Bramwell, 1899), these were initially regarded as no more than occasional exceptions of the 'contralateral rule'. Even the finding that disturbances of speech were as likely to result from left as from right sided brain damage in patients who showed no strong preference for one or other hand was not considered to compromise this general rule (Chesher, 1936). However, evidence (Humphrey, 1951) suggesting that left handers were more variable than right handers in their hand preference for different tasks led Humphrey and Zangwill (1952) to analyse the incidence of language disorders in a small group of left handers having damage confined to one or other side of the brain. In contrast to the view that consistent left handers would become aphasic only with right sided lesions, Humphrey and Zangwill predicted that aphasia would occur from damage to either hemisphere. This expectation was confirmed, although there was a tendency for damage to the left hemisphere to result in more severe speech disturbances. In subsequent studies it was shown that there are more sinistral aphasics with left than right hemisphere lesions. (Goodglass and Quadfasel, 1954; Ettlinger, Jackson and Zangwill, 1956; Hécaen and Piercy, 1956; Brown and Simenson, 1957; Russell and Espir, 1961; Hécaen and Ajuriaguerra, 1964). This difference is in the approximate ratio of 2:1 (Piercy, 1964; Roberts, 1969).

Gloning, Gloning, Haub and Quatember (1969) compared matched groups of right and non-right handed patients on a number of verbal tests. No difference was found between the two groups with a lesion situated in the left hemisphere, but non-right handers, as a group, were significantly impaired on every test in comparison with right handers when the lesion was on the right side. There were some striking differences among the non-right handers with lesions of the right hemisphere, depending on whether the left or the right hand was used for writing. A higher frequency of transient expressive dysphasia was found among those who wrote with the right compared with the left hand. Among left handed writers defects of verbal comprehension were more frequent than among those who wrote with the right hand. These differences may reflect the influence of parental or pedagogic pressure to write with the right hand as compared with a natural predisposition to use the left hand.

Hécaen and Sauguet (1971) compared left and right lesion groups according to strength of manual preference, patients being divided into weak, medium and strong left handers. With right sided lesions, language defects were found only in the group of weak left handers and were virtually absent among the strongly left handed. Since weak left handers were observed to have sinistral relatives more frequently than strong left handers, Hécaen and Sauguet inferred that 'the intensity of left handedness is not directly related to bilaterality of language representation or to familial left handedness. Indeed, the opposite is the case'. More recently, however, Hécaen, De Agostini and Monzon-Montes (1981) have confirmed that as far as most language functions are concerned, cerebral bilaterality *is* associated with familial sinistrality rather than strength of hand preference. Nonetheless, some language functions, such as naming of objects, appear to depend upon the left hemisphere in both familial and non-familial sinistrals.

The fact that dysphasia in non-right handers occurs after right as well as left sided lesions implies either that language is represented in both hemispheres of non-right handers and/or that such people are more likely than dextrals to show right hemisphere lateralisation of speech. If speech is indeed represented bilaterally in non-right handers then the frequency of aphasia consequent upon unilateral cerebral lesion should be higher in this group than among right handers. Such has been reported to be the case by some authors (Subirana, 1969; Luria, 1970) but not by others (Russell and Espir, 1961; Newcombe and Ratcliff, 1973). As the negative findings were obtained in studies of the effects of circumscribed missile wounds of the brain it may be that the positive findings are related to diffuse cerebral damage encroaching more frequently on the language areas of the brain.

A better prognosis for recovery among sinistrals than among dextrals (Subirana, 1958; Zangwill, 1960; Luria, 1970; Hécaen and Sauguet, 1971) can also be explained by the view that language is bilaterally, though not necessarily equally, represented in the two hemispheres of left handers.


## THE WADA TEST

If speech is represented bilaterally in left handers, then it should be possible for this to be demonstrated in individual left handers.

Injection of sodium amylobarbitone (a barbiturate anaesthetic agent) into the common carotid artery results in the drug being circulated through the cerebral vascular system. It has the effect of temporarily depressing the functions of the cortex, initially of the hemisphere on the side of the injection and subsequently of the opposite hemisphere. If a patient is requested to hold up one arm and to count aloud during injection of the drug, the arm on the side contralateral to the side of injection collapses as the drug takes effect. Similarly, if the drug is transported to the speech dominant hemisphere the patient stops counting. He may be mute for some minutes and have difficulty in naming objects and carrying out commands for a little while thereafter. By injecting the drug into the carotid artery on different days it is possible to observe the effect of disruption of left and right hemispheres alternately. This is the basis of the test for hemispheric speech representation first described by Wada (1949).

A serious limitation of the Wada technique is that the risks inherent in puncturing the carotid artery preclude its use without obvious medical justification and testing is therefore restricted to those patients who are candidates for brain surgery. A second limitation is that the duration of the effect with standard doses is confined to only a few minutes before the drug diffuses around the circle of Willis and into the opposite hemisphere. This severely constrains the type of testing that can be carried out. Sometimes there is a complete loss of consciousness, particularly, some have claimed, after injection on the left side (Serafetinides, Hoare and Driver, 1961; but see Rosadini and Rossi, 1967). This is in accord with data suggesting that impairment of consciousness occurs more commonly following acute cerebrovascular accident to the left hemisohere compared to the right (Albert, Silverberg, Reches, and Berman, 1976).

Rasmussen and Milner (1975) summarise data collected in their clinic over a number of years' use of the Amytal test (see Wada and Rasmussen, 1960; Branch, Milner and Rasmussen, 1964). Considering cases where there was no evidence of early injury to the brain, which may result in a shift of hemispheric dominance for speech (see Chapter 7), the test was carried out on 140 right handers. Of these, 134 (96 per cent) patients had speech controlled from the left half of the brain and 6 (4 per cent) had right hemisphere speech. Among 122 non-right handers, left and mixed handers being considered together, 86 (70 per cent) patients had speech controlled from the left hemisphere, 18 (15 per cent) had right hemisphere speech and in 18 (15 per cent) speech was bilaterally represented.

The figure of 4 per cent of right handers with right sided speech is almost certainly an over-estimate of the extent of right sided speech among right handers in general. The incidence of aphasia from right sided lesions in right handers among unselected series of patients is closer to 1 per cent than to 4 per cent (Levy, 1974a). Furthermore, Rossi and Rosadini (1967), who also used the Amytal technique, reported that only one patient out of 74 right handers, all of whom received injections on both sides, did not have exclusively left hemisphere speech dominance. The patients studied by Rasmussen et al. were all pre-selected for Amytal testing because there was some suspicion, based on the results of psychological tests or on the extent of left handedness in the patient's family, that the patient might have shown some departure from the typical dextral pattern of left hemisphere dominance. Such pre-selection is sufficient to explain the relatively high proportion of dextrals with right hemisphere representation of speech in Rasmussen and Milner's data.

The data for non-right handers collected by Rasmussen and his colleagues show approximately two thirds of left handers to have left sided speech while the remaining third are divided between those with speech on the right side and those with speech represented bilaterally. Among the latter there was sometimes a qualitative difference in the effects of amytal injection into the left and right carotid arteries. Anaesthetisation of one hemisphere produced defects in naming objects but no discernible difficulty in tasks of serial ordering such as counting, saying the days of the week or reciting the alphabet. Depression of the opposite hemisphere had the converse effect.

Rasmussen et al. combined their data for left and mixed handers, since their previous analyses had shown little or no difference between the two groups of patients. As more data accumulate it may turn out that there are in fact subtle differences and that these relate to the presence or degree of sinistrality in the patient's family or to the position of the hand during writing (see below).

## ELECTRICAL STIMULATION OF THE EXPOSED CORTEX

The work of Foerster in mapping the human brain by electrical stimulation of the exposed cortex was developed and extended by his pupil Penfield and associates, culminating in the monograph by Penfield and Roberts (1959).

Electrical stimulation has either positive or negative effects as far as speech is concerned. A positive effect means that application of the electrode elicits some vocalisation whereas a negative effect either disrupts ongoing speech or produces an inability to vocalise or to use words properly. It should be noted, however, that a positive effect has never been found to produce a single word, far less a complete utterance, the usual effect being a sustained or interrupted cry.

Penfield and Roberts (1959) give the following data for those of their patients who received electrical stimulation in Broca's area and/or in inferior parietal and/or posterior temporal regions of the left hemisphere during surgery carried out for the relief of focal epilepsy. Of 65 right handed patients who had post-operative aphasia, 54 had shown some effect of stimulation on their speech and a further 10 out of 15 right handed patients showed aphasic arrest during cortical stimulation but had no post operative aphasia. Stimulation affected speech in 3 left handers, of whom 2 had post-operative aphasia, while another 7 left handers showed no effect of stimulation and no aphasia after surgery. All of these 7 patients had had cerebral birth injuries which might have induced a shift of dominance to the right hemisphere. With stimulation of corresponding regions of cortex in the right hemisphere only one out of 14 right handers showed any effect of stimulation and none showed post-operative aphasia. Six left handers were operated on the right hemisphere of whom one was affected by stimulation and also showed aphasia post-operatively.

Penfield and Roberts concluded from their data that 'the left hemisphere is usually dominant for speech regardless of the handedness of the individual with the exception of those who have cerebral injuries early in life'. This conclusion is thus somewhat at variance with that usually encountered in the literature. However, the relatively small number of left handers without early birth injury who were stimulated does not permit of seriously challenging the accepted view. Moreover, as Penfield and Roberts admit, the effects of cortical stimulation are not consistent since 'electrical interference in a given area is only effective about 50 per cent of the time'. It cannot be assumed, therefore, that stimulation would not have had some effect on another occasion in those patients who were stimulated but showed no interference in their speech.

## UNILATERAL ELECTRO-CONVULSIVE THERAPY

Psychiatric treatment for depression may involve administration of electric shock to the (anaesthetised) patient's head. Although the rationale for this treatment is somewhat arcane, dramatic results have been claimed. The shock is usually administered either to both sides of the head or to the side considered on a priori grounds to be non-dominant with regard to language functions.

A number of studies have indicated that, in right handers, shock to the left side of the head impairs verbal learning, memory and word finding (Zamora and Kaebling, 1965; Gottlieb and Wilson, 1965; Halliday, Davidson, Brown and Kreeger, 1968; Fleminger, Horne and Nott, 1970; Pratt and Warrington, 1972) while shock to the right side is more likely to impair non-verbal functions (Cohen, Noblin and Silverman, 1968; Berent, Cohen and Silverman, 1975; D'Elia, Lorentzson, Raotma and Widepalm, 1976; Berent, 1977).

The first attempt to assess cerebral dominance for language in a group of non-right handed patients appears to have been that of Pratt, Warrington and Halliday (1971) who tested 12 right handers and 12 so-called left handers. The criterion for inclusion in the 'left handed' group was a greater preference of skill for the left hand in any one of four specified activities. The test which best discriminated between the effects of left and right sided electro-convulsive therapy (ECT) after two treatments to each side was naming of objects after hearing a verbal description of them given by the examiner seven minutes after shock administration. Eleven out of the 12 right-handers achieved higher scores after shock to the right side of the head, one patient showing no left-right difference. Among the 'left' handers 8 patients obtained higher scores after right sided ECT, 2 were better after left sided shock and 2 patients showed no difference. Re-testing of the latter patients some twenty minutes after ECT indicated right hemisphere dominance for each of them. In a later study carried out with left-handers classified in the same way as before Warrington and Pratt (1973) calculated language to be lateralised to the left hemisphere in 26 patients, to the right hemisphere in 9 patients and uncertain or bilateral in 2 patients.

Other investigators have looked at the effects of unilateral ECT in relation to handedness. However, even considering all studies together, the numbers of non-right handers have been too few, and the criteria for classification of hand preference too simplistic, to allow any precise conclusions as to the relative proportions of left and right hemisphere speech dominant individuals to be drawn. The most that can be said is that the majority of individuals said to be non-right handed still show left hemisphere language representation (Annett, Hudson, and Turner, 1974; Fleminger and Bunce, 1975; Clyma, 1975).

As a non-invasive technique ECT is a potentially useful method for estimating language laterality in individuals presumed to be without gross organic brain damage. However, differences between left and right sided treatments in individual patients may be small (Annett et al. 1974) and, of course, the procedure can only be used with those psychiatric patients for whom ECT has been chosen as the treatment of choice. Moreover, the possibility that clinical depression may be particularly associated with impaired function of the right hemisphere (see Chapter 12) means that there may be an inherent bias towards poor performance of the right hemisphere in depressed patients. The use of appropriate control procedures is therefore doubly important in estimating speech lateralisation in such patients.

## LANGUAGE LATERALITY IN NEUROLOGICALLY INTACT SUBJECTS

### Electrophysiological studies

During the past decade or so a number of investigators have used electrophysiological techniques to study hemispheric specialisation of function, usually in right handers. Schafer (1967) reported finding a significant asymmetry in the electrical activity of the brain immediately prior to speech production. There were characteristic variations in waveform over the left temporal region as a function of different spoken letters, but no parallel changes were recorded over the corresponding region of the right hemisphere. The specificity of such 'cortical command potentials' (Ertl and Schafer, 1967) led Schafer to suggest that they might reflect the process of selecting particular speech sounds. Subsequently McAdam and Whitaker (1971) demonstrated that prior to the production of various test words summed negative wave potentials were of greater magnitude over the left than the right hemisphere. However, this work was criticised by Grabow and Elliott (1974) on the grounds that tongue movements may have contaminated the EEG recordings. Grabow and Elliott therefore carried out an experiment similar to that of McAdam and Whitaker and found that movements of the tongue to left or right induced asymmetric scalp potentials which were not observed when subjects merely had to think of a word.

Cohn (1971) presented evidence that summated auditory evoked potentials show a greater amplitude of initial output over the right hemisphere with click stimuli but higher amplitude over the left hemisphere when subjects listen to monosyllabic words. Similar findings with nonsense syllables as stimuli were reported by Morrell and Salamy (1971) and Morrell and Huntingdon (1972). However, differential waveforms over the two hemispheres were recorded by Wood, Goff and Day (1971) only when their syllables required a linguistic analysis. This result was interpreted in terms of 'meaningfulness' by Matsumiya, Tagliasco, Lombroso and Goodglass (1972) who showed in a separate study that inter-hemispheric asymmetry was greatest when subjects had to understand each stimulus word, and minimal when they could ignore the semantic content.

Teyler, Roemer, Harrisson and Thompson (1973) recorded evoked potentials when subjects simply thought of the meaning of words which could be used either as nouns or verbs. Significant differences were found within each hemisphere for the two forms of the same word but an overall greater magnitude of response was recorded over the left hemisphere.

The above results confirm for right handers the asymmetrical involvement of left and right hemispheres in matters of speech production and perception. The role of left handedness in this connection has not often been investigated although in more recent studies, relating to other aspects of verbal function, the question of handedness has been raised. These studies were discussed in Chapter 4.

### Dichotic listening and visual half-field studies

Several early experiments utilising the dichotic and tachistoscopic procedures with verbal tasks have shown that, as a group, left handers exhibit reversed or attenuated ear (Satz, Achenbach, Pattishall and Fennell, 1965; Satz, Achenbach and Fennell, 1967; Curry, 1967; Knox and Boone, 1970) or visual field (Bryden, 1965; Orbach, 1967; McKeever and Gill, 1972; McKeever, Van Deventer and Suberi, 1973; Hines and Satz, 1974) asymmetry as compared with right handers. Similar effects appear with the dual task (Hicks, Provenzano and Ribstein, 1975; Lomas and Kimura, 1976) and lateral eye movement paradigms (Hicks and Kinsbourne, 1978). These group effects are due to the fact that within-subject laterality differences are often smaller among sinistrals than among dextrals (Curry, 1967; Curry and Rutherford, 1967; Klisz and Parsons, 1975) and sinistrals show more reversals of the 'typical' dextral pattern (Curry, 1967; Curry and Rutherford, 1967; Bryden, 1975).

In the light of certain of the clinical evidence it might be expected that one factor influencing the direction or degree of perceptual asymmetry obtained in laboratory experiments among left handers would be their degree of sinistrality. The results are conflicting. Some authors have reported finding a greater proportion of subjects showing a left ear advantage among strongly, compared with weakly, sinistral subjects (Satz, Achenbach and Fennell, 1967; Lishman and McMeekan, 1977; Geffen and Traub, 1979) although Dee (1971) reported the reverse. In terms of group mean scores McKeever and Van Deventer (1977b) found no effect of degree of either left or right handedness but Searleman (1980) found strong left handers to show significantly smaller phi-scores, indicating reduced asymmetry, than weak left handers. Conversely, strong right handers showed significantly higher phi-scores than weak right handers.

One reason for the conflicting results may have to do with other, uncontrolled, variables in some of these studies. It is, for example, becoming clear that a family history of left handedness is in some way related to the cerebral organisation of functions.

### THE ROLE OF FAMILIAL HANDEDNESS

### Clinical studies

Hécaen and Sauguet (1971) classified brain damaged left handers according to whether or not they had one close left-handed relative. Disturbances of spoken language were found with lesions of the right hemisphere only among those with at least one sinistral relative. The severity of impairment did not differ from that seen after left-sided lesions. Among patients with damage to the left hemisphere those with left handers in their immediate family were superior on a number of written, verbal tasks. Zangwill (1960) had earlier expressed the opinion that

> just as left handers are as a rule more variable in their hand preferences, so too are they less completely lateralised at the cerebral level.

Hécaen and Sauguet, however, argued that their findings suggest that 'cerebral bilaterality is present only in the familial type of left handers'. Consistent with this view, Hécaen, De Agostini and Monzon-Montes (1981) recently reported that disorders of language follow lesions of either hemisphere more often among left handers with at least one sinistral relative.

Unlike Hécaen and Sauguet (1971), Newcombe and Ratcliffe (1973) showed familial left handers to be impaired on tests of language more than non-familial sinistrals. They also found 5 out of 12 non-familial sinistrals to be dysphasic after right sided lesions in comparison with the absence of aphasia in 6 familial sinistrals. Thus the findings of Newcombe and Ratcliffe do not confirm those of Hécaen and Sauguet. Since the patients in Newcombe and Ratcliff's study sustained penetrating missile wounds of the brain whereas those of Hécaen and Sauguet suffered from naturally occurring lesions it is possible that aetiological differences account for the discrepant findings.

### Dichotic listening studies

Lishman and McMeekan (1977) analysed dichotic listening data from neurologically intact subjects in terms of familial sinistrality and found, particularly for females, that among strongly left handed individuals with left handed relatives the mean laterality ratio was significantly reduced in comparison with left handers without familial sinistrality. It was concluded that bilateral speech representation applied only to strong left handers with left handed relatives. Since bilateral speech need not imply that speech functions are distributed equally between the hemispheres this does not necessarily conflict with the finding that the *proportion* of subjects showing a left ear advantage was highest among the strongly left handed.

Dee (1971) observed that left handed subjects with sinistral relatives showed, as a group, a left ear advantage while subjects without left handedness in their pedigree showed a right ear advantage on a verbal dichotic task. Comparable findings were reported by Zurif and Bryden (1969). However, Geffen and Traub (1979) found an increased incidence of right ear advantage on their dichotic monitoring task in left handed males with at least one sinistral relative in comparison with non-familial sinistrals. Briggs and Nebes (1976) found no effect of familial handedness but nor did they find any effect of handedness per se.

## Tachistoscopic studies

Results using the tachistoscopic paradigm and a variety of verbal tasks are as conflicting as those obtained with the dichotic technique. Zurif and Bryden (1969) report that a group of familial sinistrals showed no significant visual field differences but non-familial sinistrals showed a strong and consistent significant right hemifield superiority on each of a number of verbal tasks. McKeever, Van Deventer and Suberi (1973) reported left handers with a positive family history to show a RVF superiority where non-familial left handers showed no significant difference between fields. Higgenbottam (1973) obtained comparable results in as much as the greatest RVF superiority was shown by his familial compared with non-familial sinistrals. By contrast, McKeever, Gill and Van Deventer (1975) found minimal half-field asymmetry in a group of 5 familial left handers as did McKeever and Jackson (1979). No asymmetry for non-familial sinistrals and a RVF superiority for familial left-handers were reported by Schmuller and Goodman (1979).

One of the reasons underlying the inconsistent results with regard to the effect of familial sinistrality might be that this factor is likely to be confounded with family size (Bradshaw, 1980). The chance of recording a positive family history of sinistrality was found by Bishop (1980b) to increase with the number of siblings respondents said they had. Another factor is that neither definitions of a positive family history of left handedness, nor methods of enquiry (e.g. self-report by family members versus questionnaire responses by subject) have been consistent across studies.

## Familial sinistrality in dextrals

Among right handers, Hines and Satz (1971) obtained a significant tachistoscopic RVF superiority, at least at certain rates of stimulus presentation, among right handers having left handed relatives; a non-significant LVF advantage was obtained for right handers without a positive family history of sinistrality. However, Hines and Satz (1974) did not replicate the effect of familial sinistrality. McKeever and his co-workers obtained a significant RVF superiority only in dextrals without left handedness in the family, those with sinistral relatives showing the same pattern as left handers, that is non-significant field differences (McKeever, Gill and Van Deventer, 1975; McKeever, Van Deventer and Suberi, 1973). Hannay and Malone (1976) found right handed females without familial sinistrality to show, for certain retention intervals, a significant RVF superiority while right handers with left handed relatives showed a slight, non-significant RVF superiority.

There is a measure of agreement from the above studies that in right handers the presence of left handedness in the family is associated with a shift in tachistoscopic asymmetry on verbal tasks away from a strong RVF superiority. Comparable findings have been reported for a dichotic listening task with children (Harter-Craft, 1981) and for adult dual-task performance (Hicks, 1975, but see Wolff and Cohen, 1977).

If this is seen as suggesting reduced left hemisphere language representation in those with sinistral relatives it would be consonant with the findings from brain damaged right handers indicating greater bilaterality of verbal functions (Hécaen et al., 1981) and increased recovery from aphasia (Luria, 1966; 1970; Hécaen and Sauguet, 1971) in association with familial sinistrality.

### Familial handedness and non-verbal performance

Hécaen et al. (1981) reported with regard to spatial performance that deficits occurred only after right sided lesions in familial left handers. Thus in contrast to bilateralisation of language functions in this group spatial performance is more distinctly lateralised. However, tachistoscopic half-field studies with normal subjects suggest that for sinistrals the presence of familial left handedness reduces perceptual asymmetry for non-verbal as for verbal tasks (Gilbert, 1977; Albert and Obler, 1978) or shifts the asymmetry in the direction opposite to that for dextrals (Schmuller and Goodman, 1980). Similar effects for both dextrals and sinistrals were noted on a lateralised tactile spatial task by Varney and Benton (1975).

Keller and Bever (1980) asked right handed subjects to categorise musical intervals presented monaurally and obtained a significant REA in non-musicians for this task and a non-significant left ear advantage among musicians. Having left hand relatives affected this ear difference among musicians but not among non-musicians. Musicians without sinistral relatives showed a greater and statistically significant left ear advantage compared to those with a positive family history of left handedness who showed a non-significant REA.

On balance then, familial sinistrality seems to be associated with bilateral representation of non-verbal as well as verbal functions.


## INHERITANCE OF LANGUAGE LATERALISATION

The evidence reviewed above, albeit inconsistent, concerning the influence of familial sinistrality on both lateral asymmetry in perception and the effects of brain damage, suggests the possible role of genetic factors in determining speech lateralisation. In view of the relationship between handedness and the cerebral organisation of speech recent genetic theories of handedness (reviewed in Chapter 2) do, in fact, make explicit statements regarding the genetic control of speech lateralisation. Observations of pre-natal neuro-anatomical asymmetry in language-related areas of the brain constitute indirect evidence in favour of a genetic blue-print but it is not yet certain that the anatomical asymmetry actually relates to functional asymmetry after birth. Data on the hereditability of hand preference do not address directly the issue of speech laterality. What is required, therefore, is a study aimed specifically at determining the inheritance or otherwise of speech lateralisation.

Bryden (1975) used dichotic listening as an index of speech lateralisation in 49 families. In terms of a laterality score there was a positive and significant correlation between the scores for spouses and between scores for mothers and their children. There was a negative correlation between the scores for siblings (p < 0.10). Bryden tentatively argued that the latter correlation indicated that in terms of dichotic listening asymmetry one is 'not dealing with something under genetic control'. However, it is difficult to conclude anything at all from a correlation, particularly one which is barely significant. In any event, if a negative correlation for siblings implies non-genetic control of dichotic listening asymmetry what are we to make of the positive (and significant) correlation for spouses? What conceivable genetic mechanism could account for this?

Genetic models are often tested using twins on the basis that monozygotic (MZ) twins have identical genetic make-up whereas dizygotic (DZ) twins do not share genetic material to the same extent. Phenomena under genetic control, therefore, are expected to show greater similarity in MZ than in DZ twins. Springer and Searleman (1978) gave a dichotic listening test to 35 DZ and 53 MZ pairs of right handed twins. No evidence was obtained to suggest a higher concordance in MZ than in DZ twins for any of several measures of laterality, including direction of ear asymmetry (regardless of magnitude) and strength of handedness. The data therefore suggest that variation in direction and degree of dichotic listening scores is non-genetic in origin. But, if so, the fact that twins with a left hander in their immediate family showed a significantly smaller ear asymmetry (and lower overall score) than twins without a family history of sinistrality is not readily explained, at least within traditional Mendelian models of inheritance.

## DOMINANCE PROPORTIONS

The question arises as to the relative proportions of left and right handers in the general population with left sided, right sided and bilateral speech representation. The techniques used with normal subjects give estimates that do not closely correspond to those derived from the clinical literature. For example, of a group of normal right handers as many as 15 per cent usually fail to show a right ear advantage on a verbal dichotic test (Bryden, 1978). If a left ear advantage, per se, were considered indicative of right-sided speech then up to this proportion of right handers would be classified on such a test as having right hemisphere speech. This proportion is very much higher than estimates based on the effect of unilateral cerebral lesions or the Amytal test (but see Wexler, Halwes and Heninger, 1981).

Given the level of reliability of dichotic listening scores (see Chapter 4) the lack of correspondence between the normal and the clinical literature should not be surprising. Satz (1977) has exposed the folly on probabilistic grounds of inferring right sided speech dominance in an individual with a left-sided ear advantage. He estimates (using parameters derived from the brain damaged population) that though the probability of left sided speech given a right ear advantage is 97 per cent, the probability of right sided speech given a left ear advantage is only 10 per cent. Hence, in view of the dubious validity of interpreting dichotic and tachistoscopic asymmetry in terms of left and right hemisphere speech dominance, recourse must be made to estimates based on samples of brain damaged patients, but here, too, problems arise. These are ably discussed by Levy (1974a).

Hemisphere dominance for speech can be inferred from either the presence or absence of aphasia following a unilateral cerebral lesion. Thus a left sided lesion in association with aphasia implies (at least some degree of) left hemisphere speech control. Similarly, the absence of aphasia following an appropriately localised right sided lesion also implies left sided speech. Dominance proportions, then, can be inferred in at least three ways:

1. All individuals with unilateral lesions are assessed for the presence of aphasia, or
2. Only individuals with lesions in the so-called language areas are considered, or
3. The relative frequency of left and right sided lesions among the total aphasic population is calculated.

A logical problem arises with all three methods. It is that if speech can be bilaterally represented, then damage to the speech areas of one side could conceivably occur without any ensuing aphasia. Dominance would then be ascribed incorrectly only to the non-lesioned hemisphere. However, the amytal studies appear to rule out the force of this logical objection since in cases of bilateral speech some difficulty has been noted with injection on either side. Indeed, that is the very reason for assuming bilateral speech in the first place. The observation required to give empirical 'flesh' to the logical 'bones' of the possibility just outlined would be that of observing no speech disturbance with injection on either side. Provided the dose of the anaesthetic is sufficiently large this does not seem to have happened.

Having dismissed the logical difficulty there is an empirical problem as far as methods 1 and 2 above are concerned. If there is a bias in the proportions of left and right lesions entering the series of patients in the literature then this will influence the dominance proportions inferred for the normal population. There is evidence that there is such a bias (Satz, 1979). This is due probably to the fact that people suffering from aphasia are more likely to come to the attention of a neurologist than individuals suffering from certain kinds of right hemisphere damage which may pass unnoticed by the individual himself, such that he does not seek medical attention or is not referred for neurological investigation.

Levy (1974a) suggests as a solution to the problem of bias that one should estimate dominance proportions from the relative frequency of left and right sided lesions among non-aphasic brain-damaged patients. Using this method and the data of Goodglass and Quadfasel (1954) she estimates language laterality in left handers to be predominantly or exclusively left-sided in 60 per cent of cases and right sided in approximately 40 per cent. These proportions are not, in fact, significantly different, statistically speaking, from the proportions based on the same series and calculated according to methods 2 and 3 above, which each give estimates of 53 per cent left hemisphere speech among left handers.

Roberts (1969) calculated from a number of different series that approximately 63 per cent of left handed aphasics had left-sided lesions. Considering the data available Levy considers 56 per cent as a reasonable compromise estimate of the frequency of left sided speech in sinistrals. (Her method of analysis does not, it will be noted, allow her to identify the proportion of sinistrals with bilateral speech.) With regard to right handers, Levy quotes a personal communication from Bogen to the effect that only two cases of aphasia in association with left hemiplegia occurred in 600 consecutive cases of stroke-induced aphasia, giving an estimate of 0.33 per cent right sided speech representation in right handers. She considers this to be a more realistic figure than other estimates which might be based on the literature since there is probably a tendency to over-report cases of crossed-aphasia (i.e. presumptive lesion and preferred hand on the same side) in dextrals because of their rarity.

Levy's computations ignore the data from the sodium amytal test which suggest that at least some proportion of left handers have speech represented bilaterally. Satz (1979) has proposed a method for determining the acceptability or otherwise of models of speech representation in the left and right handed. He considers three possibilities. These are that: 1. speech is entirely unilateral; 2. it is entirely bilateral; or 3. it may be unilateral in some cases and bilateral in others. He tabulates data from 12 published studies which show the frequency of aphasia after left and right sided lesions. The overall reported incidence of aphasia in the left handed varied from a low of 0.3 (Newcombe and Ratcliff, 1973) to a high of 0.9 (Chesher, 1936), the mean value being 0.6. Among right handers the lowest frequency reported was 0.33 (Penfield and Roberts, 1959), the highest was 0.38 (Hécaen and Ajuriaguerra, 1964) and the mean was 0.35. Thus the mean frequency of aphasia for left handers was almost twice as great among left handers as right handers.

Given the assumption that aphasia always occurs after damage to the speech-dominant hemisphere (which, of course, will not be the case where damage is restricted to regions outside the language areas of the brain) it is possible to estimate the expected upper limit (EUL) for the frequency of aphasia generated by a particular model of speech dominance in the population. If the EUL is exceeded by the empirically determined proportions then the model can be rejected. Satz considered the unilateral model first and took as examples of left and right sided speech in right handers the values of 96 and 4 per cent respectively. (He quotes this figure as 'empirically estimated', citing as a reference a paper presented at a conference by one R. Milner. As these figures are identical to those given for the Amytal test by Rasmussen and B. Milner (1975) it is presumably these data to which he refers). Assuming that of 100 brain damaged individuals 50 have left sided and 50 have right sided lesions, the expected upper limit for the frequency of aphasia will be 48 plus 2, that is, 50 per cent. In fact, on a strictly unilateral model the exact figures used for estimating the EUL are immaterial. Assuming L and R lesions occur with equal frequency, and that a lesion in the speech hemisphere will always produce aphasia, the EUL will always be 50 per cent.

Since the observed frequency of aphasia in right handers is well below this figure, the model of unilateral speech dominance for right handers is, according to Satz, at least viable. The same cannot be said for left handers since the mean observed frequency is greater than 50 per cent. The strictly unilateral model can therefore be rejected for left handers. There remain the possibilities that all left handers have bilateral speech representation (which would give an EUL of 100 per cent) or that some left handers have bilateral speech and others have unilateral speech.

No one has suggested that all left handers have bilateral representation of speech but the distribution of speech representation in sinistrals can be estimated on the grounds of the Amytal test as 70 per cent left sided, 15 per cent right sided and 15 per cent bilateral. On these figures the EUL of the frequency of aphasia is 35 per cent (those with left speech representation) + 7.5 per cent (those with right sided speech) + 15 per cent (those with bilateral speech) = 57.5 per cent. According to Satz (1980) this incidence is exceeded in the majority of papers which report relative data for left handers. He therefore suggested that a model based on a higher proportion of sinistrals with bilateral speech than 15 per cent would provide the only model not rejected by the observed incidence.

The question of exactly which left handers have speech located on one side of the brain or the other and which of them have speech represented bilaterally cannot yet be answered unequivocally. If the conclusion of Hécaen and his colleagues (Hécaen and Sauguet, 1971; Hécaen, De Agostini and Monzon-Montes, 1981) that familial left handers have bilateral speech, is accepted there is still the problem of distinguishing between those sinistrals with speech on the left side and those with speech on the right. The amytal data do not appear helpful in this context. The results reported have combined data from mixed and left handers, since according to Rasmussen and Milner (1975) there is no difference between them.

According to Annett (1975) the incidence of left or right sided speech representation among sinistrals should be considered a function of the proportion of sinistrals in the population at large. This in turn depends upon the criterion used to establish left handedness. Her theory (discussed in Chapter 2) postulates that the general population may be conceived as made up of two overlapping distributions (see Fig. 1, Chapter 2). These two distributions (of differences in skill between left and right hands) consist of those individuals in whom the right shift factor biasing the left hemisphere towards subserving speech, and coincidentally favouring greater skill of the right hand, is present (RS +) and those in whom it is absent (RS −). Given a fixed overlap between these two sub-populations the criterion of sinistrality can vary. Now, individuals with right-sided speech constitute one half of those in whom the right shift factor is absent, since in the absence of this factor speech dominance is distributed between left and right hemispheres according to chance expectation. However, the proportion of RS − individuals among left handers must be proportional to the left handers in the total population. This is because increasing the proportion of left handers in the overall population by shifting the criterion to the right takes in a larger segment of the RS − distribution, but this segment constitutes a changing proportion of the number of individuals to the left of this criterion (i.e. the left handers). Thus the extent of right brainedness among left handers is seen to be a function of the position of the criterion used to classify sinistrality.

## HAND POSTURE

The suggestion was discussed in Chapter 2 that the manner in which a pen is held in the hand during writing may be used as an index of speech lateralisation. This idea was put forward by Levy and Reid (1976; 1978) who employed tachistoscopic verbal and spatial tasks and found in 70 out of 73 subjects that the direction of presumed cerebral lateralisation was predicted by handedness and hand posture. Specifically, for the verbal task, subjects had to report 3-letter nonsense syllables presented in either the left or right hemifield and, for the spatial task, they had to locate the position of a dot in a rectangle exposed within one or other visual field. It was claimed that among subjects with a normal (non-inverted) writing posture, language and spatial functions were lateralised to the contralateral and ipsilateral hemispheres respectively. The reverse pattern held for subjects who wrote with an inverted posture.

Support for Levy and Reid has not been unequivocal. Volpe, Sidtis and Gazzaniga (1981) observed four left handed patients who had been given the Wada Sodium Amytal test. Three of these patients had left hemisphere speech and wrote with their left hand in the upright position; the fourth patient wrote with a hooked posture and had right hemisphere speech. These results are entirely contrary to those predicted by Levy and Reid's hypothesis. Using normal subjects Beaumont and McCarthy (1981) found a dichotic right ear advantage among right handers but no asymmetry among left handers. There was no difference in ear scores among the latter subjects between those who wrote with the hooked and upright postures.

Lawson (1978) used a face recognition task which yielded a LVF advantage for right handers and a RVF superiority for left handers. Within each handedness group the difference between female inverters and non-inverters was in the direction opposite to that predicted by Levy and Reid but, for males, the difference was in the predicted direction. In terms of vocal reaction time Bradshaw and Taylor (1979) found that inverted sinistrals showed a weaker RVF superiority than non-inverters. According to Levy and Reid the latter should have shown a LVF superiority.

Levy and Reid (1978) also argued, on the grounds of their tachistoscopic data, that inverted sinistrals are less lateralised for language than are non-inverted sinistrals. That is, the specialisation of the left hemisphere for language among inverters was assumed to be less complete than the right sided specialisation of non-inverters. Todor (1980) predicted that this would hold true for other tasks mediated by the language hemisphere and he therefore required his subjects to carry out a sequential motor task. He hypothesised that left handers who employed the upright posture (indicative of contralateral cerebral lateralisation of language) should perform better with the left hand than inverted sinistral writers. He found this to be the case. Warshal and Spirduso (1981) argued that if inverters are less lateralised for language than non-inverters then differences between the left and right hands during performance of concurrent verbal-manual tasks (see Chapter 4) should be smaller for inverters. This prediction, too, was supported but the pattern of results shown by right handers (non-inverters) was closer to that of left inverters than to that shown by left non-inverters, which would not have been expected on Levy and Reid's hypothesis.

Since inversion of the writing hand has been hypothesised to correlate with reduced lateralisation of brain functions it would be expected that there is a relation between inversion and familial sinistrality which also appears to relate, albeit inconsistently, to reduced perceptual asymmetries. McKeever (1979) but not Searleman (1980) reports that inversion is significantly more frequent among left handers with sinistral relatives than among non-familial left handers. However, in the study by McKeever, hand posture per se did not discriminate between left handers on a laterally presented word recognition task although the presence or absence of a family history of left handedness did so. Familial left handers showed a greater RVF superiority, the group-by-visual field interaction being significant (at least with unilateral trials). On a colour naming task, by comparison, inverted writers showed a significant RVF superiority where non-inverters showed no significant difference (interaction significant), but as inverters were more often of the familial type of left hander this result might equally as well be due to familial sinistrality.

Levy and Reid's model has been evaluated by Weber and Bradshaw (1981). These authors find empirical support to be lacking for the basic postulates of the theory which Levy and Reid derived from their original observations. A particular difficulty would seem to be that the proportion of left handed inverters found in the general population does not correspond to the proportion of sinistrals estimated on other grounds to have left hemisphere speech (Searleman, Tweedy and Springer, 1979) although it must be admitted that the range of estimates of both proportions has varied quite widely and there is at least some overlap. Another difficulty is that Levy and Reid make no suggestion as to how sinistrals with bilateral speech representation might be identified. However, in a recent reply to Weber and Bradshaw (1981) it was suggested by Levy (1982) that the earlier findings she reported with Reid may relate not so much to language (or speech) representation in general as to reading and writing – that is, visual aspects of language, in particular. There is little or no reason to suppose that these are represented bilaterally in any or all left handers.

It is not obvious how the neural control of speech differs as between well-lateralised individuals and those with bilateral speech representation. Although it is clear that the left hemisphere is dominant not only for speech but also for verbal memory and reasoning generally (Meyer and Yates, 1955; Meyer, 1959; Milner, 1962, 1971; Lansdell, 1968, 1969, 1973; Benton, 1968; Newcombe, 1969, 1974; Buffery, 1974) the nature of hemispheric control of speech output is still somewhat obscure. The vocal musculature is innervated bilaterally. That is to say, anatomic connections are present between the tongue, larynx, lips (and other structures) and the two sides of the brain (Espir and Rose, 1976). It has therefore been suggested that the lateralisation of language to one hemisphere is to prevent inter-hemispheric competition in the control of speech. Indeed, at one time it was believed that requiring a natural left hander to use his right hand for writing would induce speech defects, such as stuttering, as a result of co-opting the left hemisphere into language functions in competition with the right hemisphere (Travis, 1931). Although this idea of hemispheric competition is not very precise, a number of studies have been carried out to investigate speech lateralisation among stutterers.

## STUTTERING

Jones (1966) carried out the Wada test with four patients who had stuttered from childhood, and found all four to have bilateral speech representation. After surgery (for lesions in the approximate speech areas) the stammer cleared and Wada testing revealed no difficulty with speech after injection on the side ipsilateral to the surgical removal. With the anaesthetic introduced on the opposite side, the usual speech impairment was observed. This might be taken to imply that the surgery had prevented one side of the brain from attempting to assume control of the mechanisms for speech output. However, other reports using the Wada technique have provided little evidence to suggest that bilateral speech production is a significant feature of all cases of stuttering (Andrews, Quinn and Sorby, 1972; Luessonhop, Boggs, Labowit and Walle, 1973; Dorman and Porter, 1975).

Studies using the tachistoscopic (Moore, 1976), dichotic listening (Curry and Gregory, 1969; Brady and Berson, 1975; Rosenfield and Goodglass, 1980) and electroencephalographic (Moore and Lang, 1977) techniques have sometimes, but not always (Slorach and Noehr, 1973; Pinsky and McAdam, 1980), suggested that a greater proportion of stutterers than controls have some language processes lateralised in the right hemisphere. Sussman and McNeilage (1975a) argued that whereas receptive aspects of language are lateralised to the left hemisphere, production of language is not so clearly lateralised in stutterers as it is in normals.

Wood, Stump, McKeehan, Sheldo and Proctor (1980) compared patterns of regional cerebral blood flow in two cases of stuttering. While off medication, both subjects showed high levels of blood flow in anterior regions of the brain at the right side; with stuttering controlled by haloperidol, the flow of blood was greater in the left hemisphere. However, the situation is not simply that the right hemisphere is more implicated than usual in the production of speech. The two patients of Wood et al. showed no stuttering, and the usual blood flow asymmetry favouring the left side, when they had to read aloud a passage of prose. Thus it seems to be not so much the articulation of words per se that engages the right hemisphere in these patients but the spontaneous putting together of meaningful speech.

## SPEECH AND MOTOR FUNCTIONS OF THE LEFT HEMISPHERE

The association between handedness and the cerebral lateralisation of language offers much scope for speculation. It has been suggested, from an evolutionary point of view, that language may have arisen out of primitive man's use of manual gestures to communicate with his fellows (Hewes, 1973). Since handling of tools and weapons requires precise manipulation of the fingers and thumb, a dextral bias in hand preference for wielding these implements might have predisposed our earliest ancestors towards the use of the right hand for gestural communication. The emergence of speech might then have developed from neural systems for motor control already lateralised to the left half of the brain.

The term apraxia (or dyspraxia) refers to difficulty in carrying out purposeful movements or sequences of movements, as in the manipulation of common objects. The disorder arises as a result of brain damage, particularly of the left parietal lobe, and for the label apraxia to be applicable, the patient's difficulties must not be due to problems in comprehending the examiner's instructions. Central to any definition of apraxia is the idea that any paralysis or weakness of limbs is insufficient to account for the movement disorder. Certain broad categories of apraxia are recognised (Hécaen and Albert, 1978). Ideomotor apraxia refers to an inability to correctly perform simple gestures such as a salute, making the sign of the cross or pretending to stir a cup of coffee. Ideational apraxia is seen when a patient cannot carry out a complex sequence of movements, even though he is capable of carrying out each movement individually. For example, he may strike a match perfectly well but in attempting to light a candle he may try to light the wick with the match unlit or strike the match against the candle. Constructional apraxia is an impairment in the construction of two- or three-dimensional figures as in drawing or using matches or building blocks. Dressing apraxia refers to difficulty in putting on clothes; the patient may manipulate them haphazardly, unable to relate them spatially to his own body, or he may be unable to put them on in the correct sequence.

The literature on apraxia is highly confusing and contradictory. This is in part because different investigators have held different ideas as to the independence or otherwise of the different sub-types of dyspraxia or as to the essential nature of the various defects. Some have seen apraxia as primarily a defect of execution, others have considered the problem to be one of planning or of conceptual organisation. It is not the intention here to discuss the controversies or to review the literature. The interested reader is referred to reviews by Geschwind (1967), Ajuriaguerra and Tissot (1969), Hécaen and Albert (1978) and, for constructional apraxia, Benton (1969) and Warrington (1969). More recent papers concerning the distinction between ideational and ideomotor apraxia are those by Lemkuhl and Poeck (1981) and De Renzi, Faglioni and Sorgato (1982). For present purposes it is sufficient to note that in right handers ideational apraxia occurs as a result of lesions located posteriorly in the left hemisphere. The apraxia is always bilateral, that is, occurs for both left and right hands.

The clinical literature, then, suggests that in the execution of certain types of movement or sequences of movement the left hemisphere is implicated to a greater extent than the right hemisphere. This view is supported by the results of experimental studies carried out with non-apraxic brain damaged patients. Although a unilateral lesion of the left or right hemisphere may impair the accuracy or speed of unilateral movements of both the contralateral and ipsilateral arm (Wyke, 1968; Heap and Wyke, 1972; Haaland and Delaney, 1981) left sided lesions appear to produce greater deficits (Wyke, 1971a; Kimura, 1977a; Kolb and Milner, 1981a) or bilateral impairment on tasks in which only a contralateral deficit is observed after a right sided lesion (Wyke, 1967; 1971b).

The importance of the left hemisphere for sequential movements was highlighted in a study by Kimura and Archibald (1974). Left brain damaged patients were impaired relative to those with right side damage, not only on traditional tests of apraxia but also in imitating a sequence of meaningless manual movements. Left lesioned patients were not impaired in copying a single hand posture nor in flexing a single finger at the middle joint without simultaneous flexion of the other fingers. This last task, in fact, had earlier been found to be performed better by the left hand of neurologically intact right and left handers (Kimura and Vanderwolf, 1970).

Although sequencing errors may occur as a result of either left or right sided lesions (Kim, Royer, Bonstelle and Boller, 1980) it is generally held that left hemisphere damage more frequently leads to impairment. A number of experiments carried out with normal subjects support the idea that the left hemisphere is dominant for certain aspects of movement control (Wolff, Hurwitz and Moss, 1977; Summers and Sharp, 1979; Taylor and Heilman, 1980).

What is the relevance of all this to understanding the nature of left hemisphere speech specialisation? Speech is, of course, in its very nature a highly organised and complex sequence of sounds, but there is more to speech and language than sequential motor activity (Poeck and Huber, 1977). The correspondence within the same hemisphere of the mechanisms underlying speech and movements of the hands would gain added significance if there were evidence to connect speech and manual activity more directly. Such evidence is beginning to emerge.

Kimura (1973a) observed the movements which right handed people made with their hands during five minutes of conversation and during five minute periods of silent verbal and non-verbal problem solving. Hand movements were classified into two main classes for analysis: 1. self-touching movements and 2. free movements. During speaking, but not in the silent conditions, free movements were made more by the right than by the left hand (see also Kimura and Humphrey, 1981). In a second study left handers were observed in the speaking condition and while humming. In the latter condition, free hand movements were symmetrical while in the speaking condition those sinistral subjects with a right ear advantage (on a dichotic listening test) showed the same excess of right hand activity as the majority of right handers (Kimura, 1973b). Subjects who showed a dichotic left ear advantage tended to produce more free movements of the left hand. Kimura argued that the lack of movement asymmetry in the humming condition showed that more than mere vocal activity was involved in producing the difference between the hands. Since there was an association between the direction of ear difference on the dichotic listening task, and asymmetry of hand movement, she concluded that activation of the speech system in one hemisphere is associated with concomitant activation of certain other motor systems in that same hemisphere. This, of course, assumes that the direction of dichotic ear asymmetry is a valid index of the hemisphere in which speech is represented.

Kinsbourne and Cook (1971) asked subjects to balance a dowel rod on the index finger of one hand. The right hand was better at balancing unless the subject had to concurrently repeat the alphabet, in which case both hands performed equally well. Lomas and Kimura (1976) repeated this experiment using different manual tasks and found that speaking interfered selectively with the right hand only under certain conditions. These conditions were, firstly, when subjects had to depress each of four Morse keys rapidly in turn with the four fingers of one hand and, secondly, when a whole arm movement had to be made to depress each key in turn. Since a bilateral rather than a lateralised impairment was found when a single button was pressed repetitively with one finger, Lomas and Kimura suggested:

> ... it is the rapid positioning of a limb, or parts of a limb, ... which is related to the lateralised decrements produced by concurrent speaking. If that is true, an important contribution of the left hemisphere to speaking may also be in the control of rapid placement.. (of) the articulatory musculature (p. 31).

This view is supported by the results of an experiment by Sussman (1971). Subjects were presented with a computer controlled tone in one ear and had to match this tone through movements of the tongue which were transduced so as to produce a sound which was relayed to the other ear. Significantly better performance was found when the target tone was presented to the left ear, and the tongue controlled the sound heard in the right ear, than in the reverse condition. This laterality effect was also obtained when the cursor for tracking the target was attached to the jaw (Sussman and MacNeilage, 1975b) but not if the jaw was used to track lateralised visual input (MacNeilage, Sussman and Stolz, 1975) nor if it was the right hand that was used to track auditory input (Sussman 1971). These findings suggest a close relation between self-generated auditory input and output of the speech musculature.

The act of speaking requires very precise articulation of the vocal apparatus. At any one moment the positioning of the folds of the vocal tract is determined not only by what has just been said but by what is about to be said (Springer, 1979). Speech thus requires very fine sequential organisation not only at a psychological level but at a physiological level also. The results of Sussman and his colleagues suggest that the left hemisphere is particularly adept at monitoring and controlling the precise movement of the speech articulators.

The motor mechanisms of the left hemisphere as they relate to speech are further indicated by studies of non-verbal oral-facial movements (such as protrusion of the tongue or lips) in aphasic patients. It has been shown that in performing sequences of three such movements aphasics are worse than non-aphasic left brain damaged patients who in turn are worse than patients with right brain damage. The latter are unimpaired (Mateer and Kimura, 1977; Mateer, 1978).

Electrical stimulation of the exposed cortex in a small group of patients has been reported to disrupt production of single oral-facial movements at exactly the same electrode sites that produce speech arrest during naming of objects. Stimulation also disrupted sequential oral-facial movements, but not repetitive movements of the same gesture, at other sites within the classical language zone but not at control sites outside this area. Furthermore, stimulation at sites where oral-facial movements were disrupted also impaired the patient's ability to identify stop consonants (e.g. p, b, d) embedded in a nonsense syllable. There was no site at which phoneme identification was impaired where there was not also a disruption of oral-facial movement, although at one site (in one patient) movement was disrupted without any impairment of phoneme identification (Ojemann and Mateer, 1979b).

The evidence reviewed above suggests that the left hemisphere bears a special responsibility for certain aspects of movement of both the hands and the vocal apparatus. There appears to be an overlap between the neural systems engaged during vocal and manual activity. At an anecdotal level this is not surprising. Charles Darwin (1872) pointed out that

> children learning to write often twist about their tongues as their fingers move, in a ridiculous fashion.

Could it be, as Kimura (1977b) suggests, that

> brain regions considered to be important for symbolic-language processes might better be conceived as important for the production of motor sequences which happen to lend themselves readily to communication?

One line of evidence connecting manual activity and symbolic language processes concerns cases of 'signing aphasia' in the deaf. This refers to the deficit shown by those individuals who have learned to use their hands to communicate in sign language and subsequently sustain brain damage which impairs this ability. Kimura (1977b) tabulates seven such cases reported in the literature and it is notable that in every case the lesion responsible could with reasonable confidence be localised to the left cerebral hemisphere. Although it is claimed that these cases indicate an impairment in executing symbolic gestures it may in fact be the case that this aspect of their difficulty is secondary to a deficit in dealing with sequences of movements in general, the apparent linguistic defect deriving from this (Kimura, Battison and Lubert, 1976). However, there is evidence from hearing subjects that aphasia-producing lesions of the left hemisphere are associated with defects not just of execution but of comprehension of manual symbolic gestures (Gainotti and Lemmo, 1976; Seron, Van der Kaa, Remitz and Van der Linden, 1979). This implies that it is the appreciation of symbolic significance that is impaired.

## LANGUAGE AND THE RIGHT HEMISPHERE

Despite the undoubted pre-eminence of the left hemisphere for language in the majority of right handers, it is possible that the right half of the brain can participate in certain language functions as indicated by the split-brain studies reviewed in Chapter 3.

Hughlings Jackson (1874) was of the opinion that though the left hemisphere was responsible for what he termed propositional speech the right hemisphere could undertake 'the automatic revival' of words, particularly under conditions of great emotion. (This may be why the speech of aphasics is often peppered with expletives and stereotyped phrases.) In line with Jackson's view it has sometimes been observed that the ability to sing may be preserved despite severe expressive aphasia (Yamadori, Osumi, Masuhara and Okubo, 1977), injection of amylobarbitone into the left carotid artery in a left speech-dominant patient (Gordon and Bogen, 1974) and even in the total absence of the left half of the brain following hemispherectomy (Gott, 1973). This suggests at least some role for the right hemisphere in certain language functions.

With regard to written aspects of language, a distinction has again been drawn between conscious and automatic writing. Left sided lesions may impair the former, leaving intact such well practised actions as writing one's name and address (Luria, Simernitskaya and Tybulevich, 1970; Simernitskaya, 1974).

Studies carried out with Japanese subjects suggests that the right half of the brain is preferentially involved in reading a certain type of material. Japanese orthography is unusual in that three types of symbols – Katakana, Hirakana and Kanji – are used. The first two are phonetic symbols standing for syllables, whereas Kanji are non-phonetic symbols, or ideograms, representing complete ideas. Tachistoscopic studies have shown a right visual hemifield superiority in recognition of Katakana and a left hemifield advantage for Kanji (Hatta, 1976; 1977; Saganuma, Itoh, Mori and Kobayashi, 1977; Endo, Shimizu and Hori, 1978). A dissociation between these two scripts has been observed in Japanese aphasics (Sasanuma, 1975) and in a patient with alexia and agraphia. The latter showed severe reading and writing difficulties in Kana but Kanji presented less of a problem (Yamadori, 1975). The conclusion from all these studies, that processing of Kanji is less dependent on the integrity of the left hemisphere than processing of Kana, is supported by electrophysiological findings (Hink, Kaga and Suzuki, 1980) and by results obtained with three patients who sustained partial commissurotomy in the course of removal of pineal gland tumours (Sugishita, Iwata, Toyokura, Yoshioka and Yamada, 1978). However, the suggestion that experience with these two different systems of writing leads to the Japanese having a different cerebral organisation to that of Westerners (Tsunoda, 1975; Hatta and Dimond, 1980) lacks convincing experimental support.

Even in the West we use ideographic symbols in some circumstances. When numbers are written as figures a given digit stands for a word. When subjects are asked to judge which of two simultaneously displayed numbers is numerically the larger, irrelevant variations in the physical size of the digits influence response times when the numbers are printed in figures (e.g. 9) but not when they are printed alphabetically (nine). This finding (Besner and Coltheart, 1979) suggests that there are two independent processing systems. A right visual field superiority on the task when digits are displayed was found by Katz (1980) but a LVF superiority by Besner, Grimsell and Davis (1979).

## Language following hemispherectomy

A unique opportunity to study lateral specialisation of function in the human brain is provided by those patients who have been treated by left or right hemispherectomy for malignant tumour or for convulsions associated with infantile hemiplegia. In this procedure the patient loses the entire cortex of one half of the brain, though in cases of recurrent tumour the hemisphere may be removed in two or more stages. Although there are differences in the method of removal favoured by different surgeons (Goodall, 1957), it is usual to spare as much of the thalamus, basal ganglia and hippocampal cortex as possible and for this reason Austin and Grant (1955), among others, have pointed out that the operation might be better described as hemi-decortication, reserving the term hemispherectomy for those cases in which subcortical structures are also destroyed.

Because of the association between aphasia and the left side of the brain, neurosurgeons have been understandably reluctant to remove the entire left hemisphere of adult right handers. Consequently, there are very few cases of this type in the literature. The first reported case (Zollinger, 1935) was of a woman who could answer 'alright' in response to questions immediately after the operation but otherwise could say little more than 'yes', 'no', 'goodbye', 'please' and a few other words. However, beginning from the first or second post-operative day there were gradual signs of improvement until her death three weeks after surgery. The second case, that of Crockett and Estridge (1951), suvived for a period of four months after left hemispherectomy. Immediate post-operative speech was limited to 'yes' and 'no' but after two weeks the patient was able to say 'no, I don't want any' and 'put me back to bed'. During the following two weeks his speech improved and he began to use a few more simple words, but subsequently 'a block in his speech appeared' and he only said 'aw-caw' and 'yes' and 'no'. Such limited expressive speech following left hemispherectomy was also observed in the third recorded case (French, Johnson, Brown and Van Bergen, 1955).

A more recent case of dominant hemispherectomy in a right handed adult is that reported by Smith and Burkland (1966) and Smith (1966). The patient was a 47-year-old man at the time of operation. Nine months previously he had had a neoplasm removed from the left hemisphere which had resulted in some post-operative speech disturbances. Prior to hemispherectomy, however, he was speaking more or less normally. Smith and Burkland state that 'the patient spontaneously articulated words and short phrases fairly well immediately after the operation' but he could not repeat single words until the tenth post-operative week. This patient subsequently exhibited considerable improvement in his speech capabilities (Smith, 1966, 1974; Zangwill, 1967).

As well as the four adult cases of left hemispherectomy there are two reports of left hemispherectomy carried out during early adolescence. Hillier (1954) describes the case of a young boy who underwent total left hemispherectomy at the age of 14 years, following an operation less than one year earlier for removal of a tumour from the left parietal lobe. The first operation left him with a 'mixed' aphasia but this cleared before a reappearance of his symptoms, including 'very severe aphasia', led to the hemispherectomy. The patient was said to be unable to speak after this operation but by the sixteenth day he was able to use words like 'mother', 'father' and 'nurse'. It is reported that on discharge from hospital on the thirty-sixth post-operative day 'he appeared to have perfectly normal powers of comprehension of the spoken word' and was able to refer to the medical and nursing staff by name. At this time his vocabulary was said to show 'daily improvement'.

A similar case was reported of a left hemispherectomy performed on a 10 year old girl two years after a malignancy had been removed from the left lateral ventricle. This patient also showed speech limited initially to simple words and phrases during the early post-operative period but her auditory comprehension appears to have been good (Gott, 1973). Interestingly, singing of familiar songs was still possible.

Although cerebral organisation in adolescence may be characterised by greater functional plasticity than is the case among adults, the picture after left hemispherectomy is essentially one of gross linguistic improverishment in which comprehension is better preserved than expressive speech.

## The effect of right cerebral lesions on language functions

Crossed aphasia in dextrals, that is aphasia resulting from a right sided lesion, is extremely rare (Brown and Hécaen, 1976) with confirmation at autopsy of a strictly unilateral lesion reported for only four cases at the time of writing (Brust, Plank, Burke, Guobadia and Healton, 1982). Even now cases of crossed aphasia in right handers are likely to be reported in the literature on account of their rarity (e.g. Brown and Wilson, 1973; Zangwill, 1979; Wechsler, 1976; April and Han, 1980) although there is little to suggest that there is any qualitative difference between the aphasias produced by left and right sided lesions (Carr, Jacobson and Boller, 1981).

Although frank aphasic symptoms are only rarely seen after lesions of the right hemisphere, or even after total right hemispherectomy (Damasio, Almeida and Damasio, 1975; Smith, 1974) in patients with left hemisphere specialisation for language, careful studies are now pointing to certain linguistic impairments in association with right-sided brain damage. Eisenson (1962) was one of the first to note such effects and reported that right lesioned patients used more descriptive terms and more qualifiers than left brain damaged patients. Marcie, Hécaen, Dubois and Angelergues (1965) noted impairment in some patients on tests of sentence production, vocabulary selection and syntactic transformation after right hemisphere lesions. More recently, Caramazza, Gordon, Zurif and De Luca (1976) found right brain damaged patients to be impaired relative to controls on tests requiring the answer to such questions as 'who is shorter?', given that 'A is taller than B'. Gardner, Silverman, Wapner and Zurif (1978) found right hemisphere lesions to impair appreciation of antonymic contrasts (opposites) and Winner and Gardner (1977) report that right-sided damage interfered with the ability to understand metaphors.

In general, then, recent evidence points to conceptual or linguistic difficulties of a fairly high level. Having said this, it does not follow that there are not difficulties at a more 'elementary' level of language. In particular it seems that the ability to interpret information conveyed by prosodic aspects (i.e. intonation) of speech is impaired by right hemisphere damage (Schlanger, Schlanger and Gerstmann, 1976) as may be the ability to employ appropriate intonation oneself (Ross and Mesulam, 1979). These observations are consistent with findings with normal subjects suggesting a role for the right hemisphere in appreciation of the prosodic factors in speech (Haggard and Parkinson, 1971; Zurif, 1974; Blumstein and Cooper, 1974; Craig, 1979a; Dwyer and Rinn, 1981).

### Recovery from aphasia

Whatever its normal role in language functions, there is evidence that in some cases, though not all perhaps, the right hemisphere participates in the recovery of language following aphasia. This is shown most clearly by those instances in which a lesion of the right hemisphere has re-instated an aphasia produced earlier by left sided brain damage (Nielsen, 1946) or when amytal injection into the right carotid artery has disrupted the speech of patients previously rendered aphasic by left hemisphere damage (Kinsbourne, 1971). Supporting evidence for the view that the right hemisphere may participate in the recovery from aphasia comes from findings obtained with the regional cerebral blood flow technique (Meyers, Sakai, Yamaguchi, Yamamoto and Shaw, 1980) and from studies of dichotic listening performance which have shown a disproportionate increase in the left ear score with recovery (Pettit and Noll, 1979), at least among certain categories of aphasic patient (Castro-Caldas and Botelho, 1980).

### Studies with normal subjects

The extent of language function in the right hemisphere under normal circumstances is not easy to infer from the clinical studies. Experiments have therefore been carried out in an effort to examine this question in subjects with intact brains.

#### The Moscovitch model
Moscovitch (1972; 1973) presented subjects with a set of unrelated letters which were heard through both ears. Following this, a single probe letter

was presented to one or other visual half-field and the subject's task was to respond, using the left hand, as fast as possible to indicate whether the visually presented letter was or was not a member of the auditorily presented memory set. The left hand was used in order, it was assumed, to engage the right hemisphere in response output. The argument was that if the right hemisphere can store a representation of the auditory set and compare it with the visual probe letter then there should be an advantage in reaction time to letters presented in the LVF compared to the RVF. In the event, an RVF advantage was generally observed implying, according to Moscovitch, that the letters presented to the LVF had to be transferred to the left hemisphere for processing. However, a LVF advantage (for 'same' responses) was observed when the auditorily presented memory set consisted of only a single letter.

It is possible that the subject retained a single letter in the form of a visual representation of the acoustic stimulus and that he subsequently matched this representation against the visually presented probe stimulus. The observed left field advantage may therefore reflect not a 'linguistic' process but a process of matching based on purely visual features of the letters to be compared. Hence Moscovitch (1976) extended his earlier experiment by presenting a single letter binaurally followed by a visually presented letter. His aim was to determine whether single letters are dealt with by the right hemisphere in a visual or a phonological code.

Moscovitch argued that if the subject uses a linguistic strategy, that is compares the names of two letters, then it should be more difficult for him to discriminate between acoustically (e.g. I-Y) than visually (e.g. V-U) similar letter pairs. This should be reflected in longer reaction times (RT) to acoustically similar letters. Conversely, use of a matching strategy based on physical features should result in extended RTs to visually similar letters. His results pointed to the latter strategy. However, when the two letters to be matched were presented simultaneously, rather than successively, the pattern of results differed for the two visual fields in such a way as to suggest that the right hemisphere was still using a visually-based strategy while the left hemisphere was using an acoustically based strategy. Forcing the subject to use an acoustic strategy, by making the task one in which he had to indicate whether the terminal phoneme of one letter's name was the same as that of another (e.g. B and G have the same terminal sound, B and M do not), still led to a RVF advantage. Moscovitch claimed that together with his other findings this showed that the right hemisphere had little or no 'linguistic abilities'.

On the basis of his own findings Moscovitch argued that in the normal brain the right hemisphere has little or no language. However, he was struck by the apparent paucity of linguistic skills in the right hemisphere of normal subjects compared to that reported for commisurotomised patients. He also noted that the effects of a left hemisphere lesion may be more devastating in terms of its effect on language than even total left hemispherectomy (Smith, 1974). Moscovitch therefore suggested that the reason why the right hemisphere shows so little language ability under normal circumstances is that it suffers from inhibitory control by the left hemisphere. In the absence of the left hemisphere, and after section of the corpus callosum, this inhibition is reduced or eliminated. An appropriately localised unilateral lesion of the left hemisphere represents a situation intermediate between the normal state of affairs and disconnection of one hemisphere from the other by commissurotomy or hemispherectomy. The language behaviour observed is therefore at a correspondingly intermediate level.

Moscovitch found support for his proposition in the fact that split-brain patients sometimes begin to write with the left hand what is clearly a correct response to a stimulus seen in the left visual field, but that then the left hemisphere takes over control and the response is finished incorrectly since the left half of the brain has not seen the stimulus (Levy, Nebes and Sperry, 1971).

It is possible to argue against Moscovitch on various grounds (see Selnes, 1974; Ulrich, 1978) and his theory is certainly difficult to test experimentally. Yet the notion of inter-hemispheric inhibition is one that has been adopted by other authors to explain how language becomes lateralised to the left hemisphere (Gazzaniga, 1974). It is known that myelinisation of the fibres of the corpus callosum is not complete until about 10 years of age (Yakovlev and Lecours, 1967). This is roughly halfway between 5 years of age (Krashen, 1973) and puberty (Lenneberg, 1967) which have been taken to mark the end points of the process of lateralisation. Furthermore, behavioural evidence suggests that the callosum becomes increasingly functional in an inhibitory (Dennis, 1976) as well as a facilitatory (Galin, Johnstone, Nakell and Herron, 1979) manner throughout childhood.

The argument proposed by Moscovitch is not as far-fetched as it perhaps appears. It has been known for some time that the majority of fibres from the left and right eyes converge upon single cells in the visual cortex of the cat and monkey (Hubel and Wiesel, 1967). These binocular cells can be made to fire by impulses arriving by way of fibres from either eye but there is normally a bias in favour of one eye or the other, the so-called ocular dominance of the cell. If the animal is raised with a surgically induced squint, or with one eye sutured from birth, then there is a dramatic reduction in the number of binocular cells. Even after the eye is opened most cells can only be driven by input presented to one eye. What happens, therefore, to the connection between the cell and the ineffective eye? Some workers believe that a process of inhibition takes place which prevents the cell from firing in response to impulses travelling along the pathway from the ineffective eye. Since evidence suggests that humans also possess binocular cortical neurones and that their connections with the left and right eyes are susceptible to modification, as with the cat and monkey (Aslin and Banks, 1978), it is not difficult to see the possible extrapolation from the cellular to the hemispheric level (see also Kinsbourne, 1975b).

*Lexical decisions in left and right visual fields*

Although Moscovitch argued that the right hemisphere showed little or no language ability, other workers have since been less dismissive. One class of experiment which has been used to investigate this issue involves what is known as a lexical decision task. Letter strings are presented to the subject who has to decide as quickly as possible whether the letter string constitutes a word or not. Leiber (1976) carried out such a study and found that reaction times were equal in the left and right visual fields when the stimuli were not words, but a RVF superiority obtained when the letter-string made up a word. She argued, therefore, that since a good proportion of the non-words actually looked like words and could be pronounced without difficulty, the RVF superiority obtained with real words was related to their semantic aspect. (In a subsequent experiment Axelrod, Haryadi and Leiber (1977) used letter strings which were always pronounceable but the approximation of the letter clusters to real English varied such that some letter combinations occurred very commonly in English (called high frequency approximations) while others occurred much less frequently (low frequency approximations). The findings were that RTs to high frequency approximations were responded to faster in the RVF. This suggests that predictable letter sequences may be treated as words even in the absence of semantic information.)

An interesting finding was reported by Bradshaw, Hicks and Rose (1979). These workers presented letter strings to the left or right visual field for different durations. At the shortest duration (20 msecs) subjects were unable to identify any words, but lexical decisions were more accurate in the left than the right visual field. As the exposure duration was increased to the level at which words could be positively identified a RVF superiority emerged. The LVF advantage at the shortest duration implies that the right hemisphere may recognise genuine words when it sees them, even though it cannot necessarily identify them, which suggests that in this respect the right half of the brain is not inferior to the left. This would be consistent with certain other findings.

Marcel and Patterson (1978) presented one word, followed by a pattern mask (see Chapter 4) such that the word could not be identified, and the mask was followed by a letter string about which subjects had to make a lexical decision. Where the first (masked) word (e.g. bread) was semantically related to the following word (e.g. butter) reaction times in the lexical decision task were shortened. Since the masked word could not be identified this semantic priming effect calls into question the usual explanations of masking (Turvey, 1973) and suggests instead that a mask does not necessarily interrupt visual processing but prevents the appearance of a 'word' in consciousness. The important feature of Marcel and Patterson's findings for present purposes is that semantic priming affected RTs in the left and right hemifields to an equal degree. This again suggests that the right hemisphere is not without some word recognition ability.

## Deep dyslexia

The above findings are consonant with recent research conducted with patients suffering from a certain type of rare reading disability arising as a result of damage to the brain. The syndrome of 'deep dyslexia' (see Coltheart, Patterson and Marshall, 1980) is characterised by certain kinds of errors which the patient makes when unable to read aloud particular words. In a high proportion of cases errors bear an obvious semantic relationship to the target words. For example, asked to read aloud 'gnome' one patient said 'pixie' (Marshall and Newcombe, 1971). In addition, the patient is unable to read pronounceable non-words (Coltheart, 1980a). These and related observations have been interpreted to mean that 'deep dyslexics' are unable to employ a phonological code, that is, they cannot decode a word in terms of its sound structure, and have to rely on a purely visual code in retrieving words from their internal dictionary or lexicon (Warrington and Shallice, 1979; Coltheart, 1980a). It has been argued that this visual route to word recognition is an attribute of the right hemisphere (Saffran, Bogyo, Schwartz and Marin, 1980). However, the implicit assumption that this hemisphere is *incapable* of phonological analysis is challenged by a recent finding with normal subjects (though supported by split-brain investigations – see Chapter 3).

For both left and right visual fields it takes longer to reject as 'illegal', in a lexical decision task, pseudo-homophones (letter strings that do not constitute words but sound like real words, e.g. 'bloo', 'rayne') than letter strings that look like real words but do not sound like real words. Homophonic real words (e.g. rain, reign), however, take longer to be responded to than non-homophonic (real) words in the RVF but not in the LVF (Barry, 1981). Perhaps, then, the right hemisphere undertakes phonological analysis only when a search through the lexicon reveals no entry for a particular letter string and a phonological 'check' is made. (This, of course, would require that the results of such a check are themselves compared with the *visual* representation, for otherwise pseudo-homophones would never be correctly rejected as non-words).

One of the factors in the argument for right hemisphere reading by deep dyslexics is that they are usually able to read words that are highly imageable, that is readily give rise to a visual image, but they are often unable to read words that are highly abstract (Richardson, 1975). A number of studies with normal subjects have shown that ear (McFarland, McFarland, Bain and Ashton, 1978; Kelly and Orton, 1979 – but see Lambert and Beaumont, 1982) or visual hemifield asymmetry is negligible or reduced for concrete highly imageable words as compared with abstract, non-imageable, words (Ellis and Shepherd, 1974; Hines, 1976, 1977; Hatta, 1977; Day, 1977, 1979; Marcel and Patterson, 1978;. Elman, Takahashi and Tohsaku, 1981) although negative results have also been reported (Orenstein and Meighan, 1976; Hines, 1978; Bradshaw and

Gates, 1978; Tzeng, Hung, Cotton and Wang, 1979; Schmuller and Goodman, 1979; Young and Bion, 1980b; Shanon, 1979b). The suggestion has therefore been made that either the right hemisphere can process concrete words and/or that such words survive, better than abstract words, transmission across the corpus callosum from the right to the left hemisphere (Lambert and Beaumont, 1981).

The idea that the right hemisphere may be involved in processing highly imageable words is supported by data from temporal lobectomised patients which show a deficit in recall of concrete, highly imageable, words, leaving abstract words unaffected, when the lesion is on the right. Contrariwise, left sided lobectomy affects abstract words but has little effect on recall of concrete words (Jones-Gotman and Milner, 1978).

While the above findings suggest that highly concrete words are equally well recognised in left and right hemispheres, it is conceivable that the relevant factor is not concreteness or imageability, per se, but some other factor which co-varies with imageability. One such factor is the age at which words are learned. Words with an early age of acquisition tend to be highly concrete (imageable) while words acquired at a later age tend to be abstract. Ellis and Shepherd (1974) first drew attention to this but a number of experiments by Young and his colleagues have failed to show any influence of age of acquisition of words on dichotic listening (Young and Ellis, 1980) or tachistoscopic hemifield asymmetry (Ellis and Young, 1977; Young and Bion, 1980b) even when it is the age at which words are first read rather than heard that is under investigation (Young, Bion and Ellis, 1982). However, in all these experiments only high imagery words varying in age of acquisition were used, which does not allow for the possibility that there is an interaction between visual field, imagery and age of acquisition. In experiments by Beaton, Sykes, King and Jones (1982) low imagery words were used as well and the results suggested that there was indeed an interaction between these variables. In terms of reaction time, the right hemisphere emerged as sensitive to manipulations in age-of-acquisition but not imagery. These findings raise the possibility that the results with deep dyslexics, which appear to show a facilitating effect of highly imageable words on reading performance, might just as plausibly be attributed to the fact that such words tend to be learned at an early age in life.

Overall, it seems fair to conclude that the right hemisphere does possess some word processing capacity, probably more so for the written than for the spoken word, for concrete or early-learned rather than abstract or later-learned words and for receptive rather than executive aspects of language. (For reviews of this topic see Zangwill, 1967; Searleman, 1977; Hécaen, 1978). The implications of this right hemisphere linguistic capacity in the re-training of dysphasic (Glass, Gazzaniga and Premack, 1973; Sparks, Helm and Albert, 1976) and brain-injured dyslexic patients (Carmon, Gordon, Bental and Harness, 1977) have already received some attention.

## SUMMARY

This chapter has reviewed evidence concerning language lateralisation in left and right handers. Although it is undoubtedly the case that the left hemisphere has primary responsibility for language in the vast majority of right handers the proportion of left handers for which this is true is not entirely clear. The tachistoscopic and dichotic listening techniques lack reliability and, by and large, have not been adequately validated against other criteria. (An exception to this is the dichotic monitoring technique developed by Geffen which was discussed in Chapter 4 but this has not yet been applied to large scale studies of left handers.) EEG, ECT and regional cerebral blood flow techniques are similarly unhelpful in this respect. Lesion studies have their own problems of interpretation, while other methods which might conceivably be used to established cerebral dominance, such as dichaptic stimulation and the dual-task technique, have not been validated nor widely adopted in experiments with large numbers of left handers. There remains the sodium amytal test which is accurate but suitable only for use with candidates for brain surgery. With the proviso that such patients may have atypical cerebral organisation for speech the amytal test provides the best estimate we have of speech lateralisation in left and mixed-handers. The figure of 70 per cent left sided speech in non-dextrals accords reasonably well with Levy's estimate of 63 per cent derived from the lesion data. It should be noted, however, that the figure of 70 per cent applies only to one particular (and rather loose) criterion of non-right handedness.

The problem of determining by non-invasive means which left handers have right sided speech still resists solution. The suggestion that hand posture during writing provides a ready indication has not lived up to its early promise. Equally the idea that familial sinistrality is found in association with bilateral speech has given rise to conflicting findings.

It is possible that speech and language developed from pre-existing neural structures lateralised to the left cerebral hemisphere. Left hemisphere control of sequential motor activity which lent itself readily to a symbolic gestural system may have been the evolutionary precursor to present day lateralisation of language.

It is becoming increasingly evident that the left hemisphere's control of language even among the fully right handed is not absolute. The nature of the right hemisphere's contribution has yet to be determined precisely, but it may turn out that this half of the brain has a more elevated role to play than is customarily believed.

# 6

# Biological and comparative aspects of asymmetry

The previous chapters of this book reviewed evidence of asymmetry of function between the left and right sides of the brain. There are, however, physical differences of one kind or another between the two cerebral hemispheres. The relation between these physical differences on the one hand and the functional differences on the other is at present unclear, but presumably the functional specialisation of the hemispheres is genetically laid down in the structure of the brain. What then are the origins of this biological pre-programming, and what are the evolutionary pressures that have led to cerebral asymmetry in humans? At present the answers to these questions remain unknown.

One approach to the problem of the origin of asymmetry is to consider whether this feature of brain organisation distinguishes man from all other vertebrates. Asymmetry is most marked for language, which is usually thought to be a uniquely human characteristic. It does not follow, however, that there are not precursors of asymmetry, or even asymmetry itself, in the brains of other species. If structural or functional precursors of human cerebral asymmetry can be demonstrated in species other than man, this might furnish some clues as to the evolution of asymmetry in humans. What follows is a review of physiological and anatomical brain asymmetries in man and of investigations into physical and functional brain asymmetry in animals.

## HEMISPHERIC BLOOD FLOW

There are a number of reports of hemispheric differences in the vascular system of the brain.

Di Chiro (1962) examined carotid arteriographs noting which of the three main veins draining the hemispheres had the largest diameter. Generally, the vein of Labbé was larger on the left and the vein of Trollard on the right. Di Chiro's patients were classified according to handedness and speech lateralisation as determined by the Amytal technique. Although he does not discuss the matter, it is possible to examine Di Chiro's data to see whether the general pattern of results is better predicted by a knowledge of the patient's handedness or by their side of speech representation. A larger vein of Labbé on the left is found slightly more often in right handers alone than in all patients with left hemisphere speech regardless of handedness.

Carmon and Gombos (1970) measured systolic pressure in the ophthalmic artery as an indirect measure of pressure in the internal carotid artery which supplies blood to the hemisphere. Among right handers, pressure was higher in the right ophthalmic artery; among left handers it was higher on the left. Since measurement of brachial artery pressures (taken from the arms) showed no correlation with handedness it was argued that hand use, per se, did not produce the differences found for opthalmic artery pressure.

## REGIONAL CEREBRAL BLOOD FLOW

During mental activity there are concomitant changes in neuronal activity, in particular an augmentation of the oxidative metabolism of neural tissue. This increased metabolism is accompanied by increased local generation of carbon dioxide which indirectly affects the contractile elements of the blood vessels so that vasodilation (an increase in the diameter of the blood vessel) takes place. Consequently, there is an increase in the flow of blood through the vessel. Thus the local blood flow is adapted to the increase in functional demands.

### Techniques for measuring regional cerebral blood flow

Changes in local or regional cerebral blood flow can be followed by the introduction of a radioactive isotope into the arterial blood system. The isotope may get into the blood stream either through direct injection into an artery of a solution containing a radioactive tracer or through inhalation of a radioactively labelled gas such as Xenon-133. The isotope is inherently unstable and is transformed into a more stable form with the emission of radiation. This radioactive activity can be measured by scintillation detectors placed on or around the skull and the course of the decaying isotope through the blood vessels thereby monitored.

Initially, Xenon-133 was administered by carotid injection (Lassen and Ingvar, 1972). More recently the inhalation technique was developed by Obrist and his colleagues (Obrist, Thompson, Wang and Wilkinson, 1975). The subject inhales the radioactive gas mixed with air. The Xenon diffuses rapidly in blood and enters brain tissue. The advantages of the inhalation technique are firstly, that it is non-invasive (which eliminates the risks involved in puncturing an artery) and, secondly, that it provides simultaneous measurement of regional cerebral blood flow in both hemispheres. By contrast, injection of Xenon-133 through the carotid artery allows measurement from only one cerebral hemipshere at a time.

Positron emission computed tomography (PECT) is a technique which combines the principles of the Xenon-133 clearance method with computerised tomography. Analogues of metabolically significant molecules are tagged with positron emitting radioisotopes and injected intravenously. The distribution of the tagged molecule in the brain can be determined because the emitted positrons annihilate when they collide with circulating electrons and the two gamma photons, emitted in each annihilation, travel in opposite directions. The coincidental emissions are recorded by detectors at the scalp and a three-dimensional representation of the distribution of activity in the brain is reconstructed.

The assumption governing neuro-behavioural research using PECT is that increased metabolism indicates activation of brain regions and, conversely, that increased activity of brain regions will be reflected in increased neural activity and hence metabolism and blood flow. (The possible rôle of inhibitory influences has been generally ignored.) Measures of cerebral metabolic rate thus provide a means of inferring changes in neuronal activity in localised regions of the brain (Phelps, Mazziotta and Huang, 1982).

A number of other techniques have been developed for constructing images of the brain. Single photon emission computerised tomography (SPECT) employs injection of single photon (gamma ray) emitting isotopes. The relevant isotopes have much longer half-lives than those used with PECT which do not require an on-site cyclotron for their production.

The use of powerful magnets for producing resonance of nuclei with changes in the polarity of the magnetic field is a recent development in neuro-imaging that enables brain imaging without the use of radioactivity. So far magnetic resonance imaging has only been of use in an anatomical context but the technique has potential for mapping regional metabolic activity (Gur, 1973).

## Hemispheric activation and cerebral blood flow

The normal pattern of regional brain metabolism and blood flow has been studied with the Xenon-133 clearance technique and with PECT. Although initial studies with the Xe-133 technique reported greater overall flows in the right hemisphere (Carmon, Harishanu, Lowinger and Lavy, 1972; see also Dabbs, 1980) subsequent work with both this (Obrist, Thompson, Wang and Wilkinson, 1975; Risberg, Halsey, Blauenstein et al., 1975; Gur and Reivich, 1980) and the PECT technique (Greenberg, Reivich, Alavi et al., 1981; Phelps, Kuhl and Mazziotta, 1981) indicate that the normal brain is characterised by overall symmetry in blood flow and metabolism during the resting (i.e. non-activated) state. Both techniques also show increased blood flow in anterior compared with posterior regions in normal but not schizophrenic (Ingvar and Franzen, 1974; Farkas, Wolf, Fowler et al., 1981; Buchsbaum, Ingvar, Kessler et al., 1982) brains.

Consistent with the assumption that localised increases in brain activation are accompanied by increased regional cerebral blood flow, Carmon, Lavy, Gordon and Portnoy (1975), using the Xenon-133 injection technique, reported a greater increase in left hemisphere flows during a verbal memory task compared with listening to music. The converse was found for those right hemispheres studied. Using the Xenon-133 inhalation technique Risberg, Halsey, Blauenstein et al. (1975) found greater left hemisphere increases in blood flow during a verbal analogies task compared with resting values and a visuo-perceptual task. Gur and Reivich (1980) replicated these results for the left hemisphere and the verbal task but for the visuo-perceptual task half of the subjects showed an increase in right hemisphere flow while the other half showed an increase in left hemisphere flow. It was argued that this supported the view that the task could be performed by strategies mediated either by the left or by the right hemisphere. Exchanging the visuo-perceptual task (Street's Incomplete Figures) for a spatial orientation task led to the expected lateralised effects being obtained (Gur, Gur, Obrist et al., 1982). These findings have recently been replicated in a PECT study (Gur, Gur, Rosen et al., 1983).

In the study by Gur et al. (1982) both resting cerebral blood flow and the effects due to presumed hemispheric activation (by the experimental tasks) were influenced by the subjects' sex and handedness. In the future it is likely that neuro-imaging techniques will help to clarify the rôle of these and other variables as they relate to cerebral asymmetry of function, since the techniques can provide simultaneous data on many brain regions. Research can therefore centre on the interaction between different regions of the brain. This will represent a move away from the localisation of putative brain centres to the determination of networks of brain activity involved in various behavioural phenomena.

Readers interested in regional cerebral blood flow studies are referred to volume 9 (1980) of *Brain and Language*..

## ANATOMICAL ASYMMETRY OF THE HUMAN BRAIN

In recent years there has been a growth of interest in anatomical differences between the left and right sides of the brain.

Drawing on earlier studies (for review see Von Bonin, 1962), Geschwind and Levitsky (1968) chose to examine the area known as the temporal plane, or planum temporale. They inserted a knife blade into the Sylvian fissure and dissected the brain in the plane of this fissure thus exposing the upper surface of the temporal lobe. The planum temporale is that area bounded anteriorly by Heschl's gyrus (the primary auditory receiving area), posteriorly by the posterior border of the Sylvian fossa and laterally by the Sylvian fissure (see Figure 7). Geschwind and Levitsky found that this area was larger on the left in 65 out of 100 brains, larger on the right in 11 cases and approximately equal in 24 brains. The finding of a larger left temporal plane in the majority of cases has since been confirmed by other investigators (Witelson and Pallie, 1975; Teszner, Tvaras, Gruner and Hécaen, 1972; Rubens, Mahowald and Hutton, 1976). It is, however,

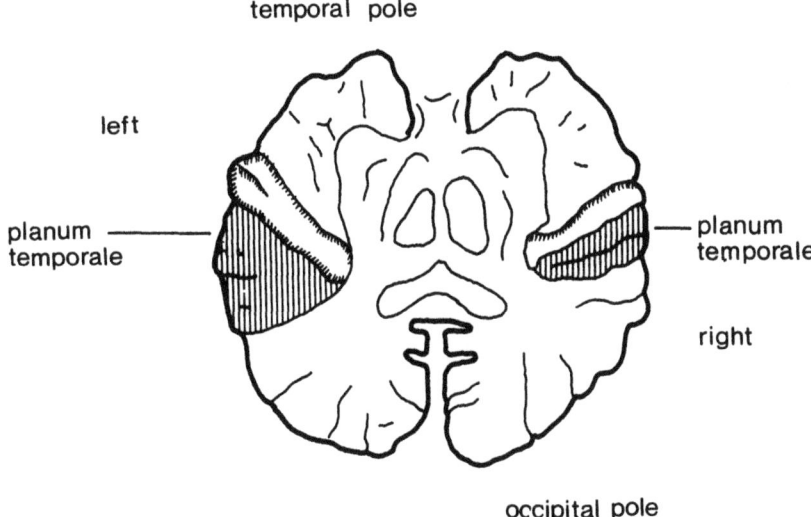

temporal pole

left

planum temporale

planum temporale

right

occipital pole

**Fig. 7** Diagram showing larger planum temporale on the left side. The upper surface of the temporal lobe has been exposed by cutting horizontally through the Sylvian (Lateral) fissure

difficult to be certain that the asymmetries reported are not artefactual (Witelson, 1977a).

The posterior extent of the planum temporale has usually been determined by inserting a knife blade into the Sylvian fissure and advancing it until the posterior wall is reached. However, Rubens, Mahowald and Hutton (1976) found that in 25 of 36 brains the Sylvian fissure on the right side angulated more sharply upward than on the left. Consequently, the 'posterior wall' of the Sylvian fissure may not represent the true termination of this fissure. Since it is not yet known how often the auditory association area is limited in its posterior extent by this angulation, gross measures of anatomical asymmetry may not represent a valid index of asymmetry of auditory association cortex (Rubens, 1977). There may also be a problem in determining the anterior boundary of the planum. The planum temporale is usually defined as lying posterior to the transverse gyrus of Heschl. Not infrequently two gyri are encountered and the problem lies in deciding whether the second gyrus should be considered as part of the planum or as a second gyrus of Heschl. The measured area of the planum obviously reflects the outcome of this decision which, in the absence of histological information, is essentially arbitrary.

An attack on these problems was made by Galaburda, Sanides and Geschwind (1978) who measured the distinctive cellular areas of the auditory regions of a small series of brains. In three out of the four specimens the cytoarchitectonic area they refer to as the temporo-parietal region was larger on the left side. This architectonic asymmetry was present in the three brains which showed an asymmetry of the planum temporale and was absent in the specimen showing no conspicuous left-right difference. Thus asymmetry of the temporal plane, at least in these three brains, was shown to be related to an asymmetry at the level of cellular architecture.

The studies mentioned above all involved dissection of the brain postmortem. For some purposes it is clearly desirable to have information concerning brain asymmetry to correlate with other data from living people. Le May and her colleagues have developed angiographic techniques in which a coloured dye is injected into the carotid artery and hence transported through the cerebral arterial system. The dye then shows up on X-ray photographs. The middle cerebral artery enters the Sylvian (or lateral) fissure from below and its branches leave the Sylvian fossa by looping under then over the parietal operculum (the area of the parietal lobe immediately above the fissure) to emerge on the surface of the hemisphere. Le May and colleagues measured the size of the arterial arch in the Sylvian fossa as an indirect measure of the height of the Sylvian point (the posterior tip of the Sylvian fissure) so as to infer the size of the surrounding gyri. They found (Le May and Culebras, 1972; Hochberg and Le May, 1974) a narrower arterial arch, indicating a lower Sylvian point, in the left hemisphere in 38 out of 44 patients. A lower Sylvian point on the left indicates a less steeply angled fissure on this side with respect to the horizontal plane. This implies a greater development of the left than of the right parietal operculum. Consistent with the arteriographic evidence, coronal sections of a number of other brains also indicated a lower Sylvian point on the left in the majority of cases.

The asymmetries described above relate to the cortical areas lying within and surrounding the lateral (Sylvian) fissure. As such, the greater tissue on the left side forms part of, or is close to, those areas of the brain known to be concerned with language functions. The asymmetry of the temporal plane has been reported to be present in foetal and neonatal brains (Witelson and Pallie, 1973; Wada, Clarke and Hamm, 1975; Chi, Dooling and Gillies, 1977). It is tempting to speculate, therefore, that the specialisation of the left hemisphere for language in some way depends upon the anatomic asymmetry. However, it is worth noting that as well as those anatomic differences between the hemispheres discussed above, other asymmetries have been observed for which functional correlates do not suggest them-selves so readily.

Le May (1976) using computerised axial tomography (CAT scan) found the anterior portion of the right hemisphere (as measured in the lowest section of the scan) to be wider than the corresponding region on the left in 70 per cent of 174 brains and wider on the right in only 8.6 per cent. Wada, Clarke and Hamm (1975) also reported (as did Witelson and Pallie, 1973) that the area of the frontal operculum is larger on the right, although their methods of measurement were such that the three-dimensional view may not necessarily accord with this. Falzi, Perrone and Vignolo (1982) measured the surface area of the anterior speech region (defined as the pars opercularis and pars triangularis of the third frontal convolution) in 12 brains. No surface asymmetry was apparent. However, when the tissue in the depths of the sulci were also considered a clear asymmetry emerged, the left side being larger than the homologous area on the right. More generally, a recent report of regional cerebral blood flow studies (Xe[133] inhalation technique) has suggested that there is more grey matter relative to white in the left compared with the right hemisphere of right handers (Gur, Packer, Hungerbuhler, Reivich, Obrist, Amarner and Sackeim, 1980). As well as differences in the extent of cortex at the two sides of the brain it has been claimed that there are left-right anatomic differences in the mass of the posterior nucleus of the thalamus (Eidelberg and Gala-burda, 1982). The significance, if any, of all these asymmetries is at present unknown.

## ANATOMICAL ASYMMETRY IN RELATION TO HANDEDNESS

Anatomical asymmetries of the brain have been correlated with handed-ness (see Witelson, 1980).

McRae, Branch and Milner (1968) reported that the occipital horns of the lateral ventricles were more likely to be longer on the left side than on the right in 87 right handed subjects. These observations are consistent with other findings showing that the left ventricle is usually larger than the right in right handers (Le May, 1977; Le May and Geschwind, 1978).

Among 13 left handers, the length of the occipital horns tended to be equal or, if they were not, there was an equal chance of the right or left horn being longer (McRae et al., 1968).

In an arteriographic study, Hochberg and Le May (1974) identified on x-ray films the Sylvian point as the most medial margin of the last branch of the middle cerebral artery as it left the Sylvian fissure. A second point, x, was marked one centimetre lateral to the Sylvian point and a perpendicular was drawn from the Sylvian point to the superior orbital line (just above the eyes). The angle made by this perpendicular as it met a line joining x to the Sylvian point was measured. In 71 right handers this angle was greater (by 10 degrees or more) on the right side than on the left side; 27 patients had comparatively equal angulation on the two sides and 8 had a greater angle on the left. Among 13 patients said to be left handed, the corresponding figures were: 6 patients with greater angulation on the right side, 20 patients with equal angulation on both sides and 2 with greater angulation on the left. Thus in this group of patients the trend was for left handers to show reduced asymmetry compared with right handers. Since the angle of the Sylvian arch may be taken as an indication of the mass of the tissue of the parietal operculum, the results suggest a greater equality of cortical tissue in this region among left than right handers. Other anatomical measurements of the brain (Le May, 1976; Le May and Geschwind, 1978) and the skull (Le May, 1977) point in the same direction, namely reduced asymmetry in left handers.

The demonstration of an association between cerebral anatomic asymmetry and handedness does not, of itself, indicate whether the asymmetry is causally related to handedness. Nonetheless, it constitutes indirect evidence that the anatomic asymmetry is related in some way to differences in function between the hemispheres. More direct evidence of an association comes from a study by Ratcliffe, Dila, Taylor and Milner (1980). These workers report that in patients with left hemisphere speech, as determined by the Amytal test, the branches of the middle cerebral artery form a narrower arch on the left than the right side as they pass under the parietal operculum. In patients with atypical speech representation (i.e. bilateral or right sided) the mean asymmetry in the angle of arch at the two sides was significantly reduced in comparison with patients who have left hemisphere speech. However, there was considerable individual variation and some patients showed a marked dissociation between morphological and functional asymmetry.

If temporal lobe asymmetry is considered to be related to the specialisation of the left hemisphere for language then certain problems arise. Firstly, the proportion of brains showing anatomic asymmetry is less than the relative frequency of left hemisphere dominance for language. Secondly, how can a quantitative difference between the two hemispheres explain the qualitative difference in their functions? One approach to this question might be to argue that the more important a particular cognitive function, then the greater is the commitment of cortical tissue to the function in question. For example, the area of cortex devoted to the sensory and motor functions of the hands exceeds that concerned with the entire trunk of the body (Penfield and Rasmussen, 1950). The importance of language to humans is thus reflected, on this view, in the greater extent of tissue at the side of the brain in which language is represented.

It is interesting to speculate on the possible evolutionary significance of anatomic asymmetry. Le May (1976) discusses observations she has made on endocranial casts of the fossilised skulls of early man and notes that there are asymmetries similar to those in modern man. Abler (1976) measured the length of the perimeter of skulls of 100 Melanesian Islanders and the casts of early fossil skulls. In both cases the right side was longer. Similar asymmetry was not seen in 26 pongid (chimpanzee, orang-utan and gorilla) skulls. This suggested to Abler that the cranial asymmetry he observed was specifically human. There are, however, indications that cerebral anatomic asymmetry is not confined to Man.

## ANATOMIC ASYMMETRY IN ANIMALS

Groves and Humphrey (1973) found in a small group of mountain gorilla skulls that the mean size of the frontal portion was significantly longer on the left. This asymmetry was not seen in the skulls of lowland gorillas, and Groves and Humphrey wondered whether some asymmetry in chewing could have led to the skull asymmetry. Lowland gorillas have less strongly developed masticatory apparatus, which would fit with the lack of skull asymmetry in these animals. But if skull asymmetry is in some way related to greater left sided chewing there is still a need to explain this phenomenon. Is some neural asymmetry responsible? Interestingly, Schaller (1963) observed that in a group of 8 mountain gorillas all began chest-beating displays with their right hand which is at least suggestive of asymmetry at a cerebral level.

Le May and Geschwind (1975) measured the height of the Sylvian point in X-rays of the brains of apes and monkeys and found this to be higher by 3 mm. or more on the right side, as in humans, in 10 out of 12 orang-utans, 4 out of 9 chimpanzees, 2 out of 7 gorillas and in 3 out of 30 new and old world monkeys. Yeni-Komshian and Benson (1976) reported the mean length of the Sylvian fissure to be longer on the left in chimpanzees, as in humans, but not in monkeys.

A problem in viewing the human anatomical asymmetry as the basis for lateralisation of language is raised by these suggestions of asymmetry in the region of the Sylvian fissure in the chimpanzee and the orang-utan, but not the monkey. The problem is that if human anatomic asymmetry in the temporal lobes is related to language functions then, since apes do not have language, they should not show anatomic asymmetry. One way out of this dilemma would be to suggest that the human asymmetry provides not so much for language, per se, but for some more fundamental cognitive skill on which language depends. The animal asymmetry would then suggest that chimpanzees and orang-utans, but not monkeys, have this same fundamental cognitive skill. If so, it is difficult to envisage what this might be, such that it differentiates between chimpanzees and monkeys.

It could be argued that chimpanzees have been taught the rudiments of language (Hill, 1978), and that orang-utans but not monkeys could be taught. A difficulty here is that not only does such an argument presuppose the outcome of any attempt to teach orang-utans and monkeys 'language' but it ignores the fact that the signs acquired by apes have relied upon visual and motor systems, whereas the anatomic asymmetry is in the region of the brain devoted to auditory functions. Furthermore, while preliminary reports suggest functional asymmetry in the monkey brain for visual discrimination responding on the basis of prior auditory signals (Dewson, 1977) no serious claims of asymmetry have been made on behalf of the chimpanzee. To be sure, it has been reported that two chimpanzees, including the famous Washoe, show more lateral eye movements towards the left than the right after breaking eye contact with their human caretaker (O'Neil, Stratton, Ingersoll and Fouts, 1978) but the idea that this relates to cerebral asymmetry is highly speculative. Finally, if anatomic asymmetry in animals does in fact turn out to be related to aspects of auditory communication it will be surprising that the asymmetry is found in a creature whose social habits are such that it relies very little on vocal communication with its peers (the orang-utan) while asymmetry is absent in a creature (the monkey) that depends quite extensively on vocalisation (Nottebohm, 1979).

It could be suggested in view of these problems that anatomic asymmetry in animals brains is related not to the precursors of language but to certain other functions which may be lateralised in the brains of certain primates. Is there, then, any evidence for functional asymmetry in infra-human species?

## HEMISPHERIC FUNCTIONAL ASYMMETRY IN ANIMALS

It has been claimed that the Japanese macque monkey (*Maccaca fuscata*) can better discriminate between calls produced by its own species if the sounds (together with noise) are presented to the animal's right ear rather than its left ear (Peterson, Beecher, Zoloth, Moody and Stebbins, 1978). However, this claim must be regarded as tentative since only 5 animals were used in the experiment. Generally, the search for functional hemispheric asymmetry in monkeys has proved fruitless (Hamilton, 1977; Warren, 1977). Even testing monkeys on a task that might be expected, by analogy with humans, to show asymmetry, that of recognition of other monkey faces, has not revealed any left-right difference (Overman and Doty, 1982). This is not to say that individual animals may not show superior performance on some tasks with one hemisphere rather than the other. In the cat, for example, this appears to be the case (Webster, 1977a; Whitfield, Diamond, Chineralls and Williamson, 1978). In any event, small hemispheric differences in individual animals are not the same as the species-specific asymmetry seen in Man where the majority of the population show the same functional lateralisation.

The extent to which individual behavioural asymmetry in animals relates to 'accidents' of paw preference, or spatial orientation tendencies developed throughout the animal's lifetime must be carefully assessed before any observed behavioural asymmetries are attributed to pre-existing biological asymmetries. Stamm, Rosen and Gadotti (1977) have shown that electrophysiological asymmetry in the monkey's frontal cortex can be modified by training procedures, and Denenberg, Garbanati, Sherman, Yutzey and Kaplan (1978) have found that handling of rats in infancy determines whether left and right sided cortical ablation has differential effects on the rats' behaviour.

The rat has proved a useful experimental animal for the study of behavioural asymmetries in relation to biochemical left-right differences in the brain (Zimmerberg, Glick and Jerussi, 1974; Glick, Jerussi and Zimmerberg, 1977; Glick, Weaver and Meibach, 1980; Robinson and Bloom, 1977; Robinson, 1979; Myslobodsky and Shavit, 1978). It appears that there is a tendency for a majority of laboratory rats to rotate towards the right side rather than the left as a result of amphetamine administration and to choose the right rather than the left lever in an operant conditioning situation (Glick and Ross, 1980). These behaviours can be predicted by the direction of asymmetry in the dopamine content of the nigrostriatal system of the brain. However, the right-sided bias may hold only for female rats; preliminary results suggest that male rats show little or no laterality effect (Robinson, Becker and Ramirex, 1980). It may be that there is a sex difference in dopamine asymmetry in the rat striatum.

The only clear example to have been reported of a species-specific behavioural asymmetry bearing any conspicuous similarity to cerebral dominance in humans relates to the neural control of song in the canary and certain other birds (Nottebohm, 1971; 1972; 1977; 1979). The canary's vocal organ, the syrinx, is made up of two halves, each innervated independently by the tracheosyringealis branch of the hypoglossal nerve. Sectioning the left side of the hypoglossus abolishes most of the components of the bird's song while section of the right hypoglossus has little effect. At a higher level of the nervous system, section of the part of the brain referred to as the hyperstratum ventrale (pars caudale) has comparable effects. Section of this structure on the left has far more dramatic effect on the bird's song than section on the right side. This cerebral dominance of the left side, like that in humans, is reversible. If the lesion is made early in the bird's life the right side takes over control.

The avian brain has produced further indications of species-specific asymmetry. Andrew, Mench and Rainey (1980) found that young domestic chicks with only the right eye open learned a visual discrimination more quickly than chicks with only the left eye open. As there is virtually complete crossing over of the optic pathways at the chiasm in the chick this suggests a relative superiority of the left hemisphere on this task. Rogers and Anson (1979) have suggested that this lateralisation of function in the chick brain may result from asymmetrical influences during embryonic development. Apparently, the right eye is exposed to light entering the egg while the left eye is shielded by either the yolk sac or the left side of the body. Rogers (1981) reports that chicks raised in the dark do not show species-specific lateralisation of certain behaviour, although individual chicks may exhibit lateralised effects. This, of course, only puts the question back one stage further. Why is the position of the chick embryo in the egg such that the left hemisphere has the advantage?

The issue of cerebral asymmetry in non-human species (see Walker, 1980; Denenberg, 1981 for recent reviews) is currently one of fervent debate. Against the comparatively few positive reports must be set a very large number of negative reports (Bureš, Burešová and Křivánek, 1981). However, this may reflect little but the fact that, generally speaking, investigators have not expected until recently to find any cortical or behavioural asymmetry in infra-human species and have therefore not performed appropriate experiments. The important point has been made that any such endeavour should utilise ecologically meaningful tasks which might, on a priori grounds, be expected to reveal asymmetries more readily than tasks designed for the practical convenience of the experimenter. In particular, the importance of olfaction in the life of so many animal species has been ignored in experiments to date (Warren, 1981).

It has long been recognised that there are sub-cortical (habenular) asymmetries in certain species of fish (Morgan, 1977) although the functional significance of such asymmetry is unknown. The climate now seems right for invesgitators to accept at least the possibility of asymmetry at a higher level in species other than Man. A comparative perspective on asymmetrical functioning in the brain should considerably advance our theoretical approach to cerebral asymmetry in humans.

## SUMMARY

This chapter began by looking at cerebral vascular and anatomic asymmetries in Man. As yet, it is only possible to speculate on the relation between such asymmetries and psychological functioning. One possibility, at present untested, is that take-over of speech functions after early damage to the left hemisphere is more complete in individuals in whom the normal left-right asymmetry of the planum temporale is reversed or attenuated. However, the presence of anatomic asymmetry in the brains of certain of the great apes raises questions with regard to the psychological significance of anatomic asymmetry in humans.

As well as anatomic asymmetry in animals there are indications of functional asymmetry in some species. This provides a comparative perspective from which to view lateral specialisation in the human brain.

The development of neuro-imaging techniques in recent years is likely to lead to considerable advances in our understanding of the relation between asymmetrical brain structure and function on the one hand and behaviour on the other.

# 7

# The ontogeny of cerebral specialisation

Is specialisation of function between the left and right sides of the brain something that develops in the course of an individual's life, or is it there from the start in the form of an innate predisposition? This is the question to which this chapter is addressed.

The earliest views on the problem were based on observations of the effect of lateralised cerebral lesions. It has been known for a long time that recovery from brain damage is more complete in young people than in adults. This has generally been taken to reflect greater neural plasticity in the young. This concept of plasticity is related to that of hemispheric equipotentiality – the idea that each hemisphere can, if necessary, subserve the functions of the other. Since neural plasticity of function is said to decline with age, each hemisphere is thought to be increasingly less capable of subserving the functions of its partner should the need arise. From this, the view has developed that early in life both hemispheres participate equally in the acquisition of new skills, but that gradually certain functions become concentrated in one hemisphere rather than the other. According to this account, language becomes progressively lateralised to the left hemisphere as the right hemisphere ceases to participate in verbal behaviour. This view of the progressive lateralisation of language was articulated most fully by Lenneberg (1967), and can be considered to have been the conventional view until very recently.

The fact that during childhood each half of the brain is capable of subserving the functions of the opposite half, if required to do so, does not logically entail that both hemispheres participate equally in all functions early in ontogeny. This leads to the question of whether some degree of hemispheric specialisation of function can be demonstrated at or around birth or very early in infancy. In recent years, therefore, investigators have been interested in two issues. One has to do with the age at which evidence of lateral asymmetry is first found, and the other, related to the progressive lateralisation hypothesis, is whether asymmetry of function, once established, increases with age. If so, it might be argued that as one hemisphere gradually evolves its own specialised functions it ceases to play any part in those functions for which it is not specialised.

## THE EFFECTS OF EARLY BRAIN DAMAGE

Lesions sustained very early in life do not permanently impair the faculty of speech, and damage to the left or right side of the brain before speech has developed may retard, but not prevent, the subsequent acquisition of language (Krynauw, 1950; Basser, 1962). Lesions sustained after the onset of speech, but before the teenage years, may lead to transient disturbances of speech, whether the damage affects the left or the right hemisphere (Wilson, 1970; Hécaen, 1976). Thus there has arisen the view that early in life both hemispheres participate in the process of language acquisition but that gradually all verbal functions become lateralised to the left hemisphere. Lenneberg (1967) argued from the differential effects of left and right sided lesions at different ages that this process is more or less completed by puberty, but Krashen's (1973) analysis of the relevant data led him to conclude that lateralisation of language to the left hemisphere is complete by the age of 5 years. On the other hand, Brown (1975) and Brown and Hécaen (1976) have gone so far as to suggest that lateralisation for receptive language is a process which continues throughout life. They propose that there is both increasing localisation of language functions within the left hemisphere and increasing lateralisation of language to the left hemisphere in passing from childhood through adolescence and early adulthood to senescence. In partial support of this theory Johnson, Cole, Bowers, Foiles, Nikaido, Patrick and Woliver (1979) found that on a dichotic listening task left ear scores, but not scores for the right ear, decreased as a function of age in subjects whose age ranged from 50–79 years.

If damage occurs to the left hemisphere early in life the right hemisphere will in some circumstances take over the control of speech. This illustrates what is known as cerebral plasticity of function.

Rasmussen and Milner (1975) have provided data on speech representation in adult patients who sustained left hemisphere lesions before the age of 6 years. On the basis of the sodium amytal test, the proportion of patients having right sided speech was found to be approximately 12 per cent. Where the lesion was sustained after the age of 6 years, only 4 per cent of patients had speech controlled from the right side of the brain. This suggests that speech transfers from the left to the right hemisphere much more readily in those cases where the injury occurs early. In other words, the immature nervous system appears to show greater functional plasticity than does the adult brain (but see St James-Roberts, 1981). This decline of hemispheric functional plasticity with age can be illustrated by the case of two jellies left standing in different containers. Gradually they begin to set and as they do each loses the potential to assume the shape of the other, a potential that was present when they were both in liquid form.

The factors that predispose towards a shift of speech to the right hemisphere after early injury to the left side of the brain include the extent and severity of damage and the locus of injury. According to Rasmussen and Milner, damage which spares the frontal and parietal speech areas does not lead to a shift of lateralisation, while damage to either of these areas will induce a right hemisphere take-over. Occasionally speech will be subserved by both halves of the brain, in which case different language functions may be segregated to each side.

Despite the ability of the right hemisphere to take over control of speech, careful experimental investigations have disclosed that when one hemisphere is required to undertake the functions of its partner, as in cases of hemispherectomy, such compensation may be less than perfect, even when the damage precipitating the take-over occurs very early in life. (Griffith and Davidson, 1966; Kohn and Dennis, 1974; Dennis and Whitaker, 1977; Day and Ulatowska, 1980). Similarly, early unilateral injury or disease less extreme than total hemispherectomy may lead to cognitive deficits related to the laterality of the lesion being detectable years later. Left-sided damage impairs mainly verbal functions and right-sided damage impairs mainly non-verbal functions (Annett, 1973; Kershner and King, 1974; Rudel, Teuber and Twitchell, 1974; Rankin, Aram and Horwitz, 1981). However, unilateral cerebral lesions may not affect verbal and non-verbal functions to the same degree.

McFie (1961) studied IQ scores of 28 hemispherectomy patients operated on for infantile hemiplegia and compared these with cases of children with later unilateral injury. He came to the conclusion that 'the majority of patients with infantile hemiplegia show a verbal intellectual deficit . . . irrespective of the hemisphere damaged'. McFie interpreted his results to mean that there is a limit to the capacity of the remaining hemisphere to sustain the functions that would normally be exercised by both hemispheres. He argued that this limitation 'may bear particularly heavily upon language as opposed to non-language functions'. On the other hand, Carlson, Netley and Hendrick (1969) observed among their hemispherectomy patients that there were more individuals with a higher verbal than performance IQ whereas the opposite was the case in a group of neurologically intact controls matched for age, sex and full scale IQ. The hemispherectomised patients were not significantly inferior in verbal IQ to the control patients. Indeed, there is at least one published instance of a hemispherectomy patient having a verbal IQ well above average. Smith and Sugar (1975) report the case of a patient who had his left hemisphere removed at the age of $5\frac{1}{2}$ years. At the age of 26 this patient had a verbal IQ of 126 and a performance IQ of 102. In short, it seems that if one hemisphere is required to sustain the functions normally accomplished by the entire brain then verbal skills are developed at the expense of non-verbal skills.

It is often assumed that evidence showing cognitive deficits related to laterality of lesion implies that there is very early complementary speciali-sation of function of the two sides of the brain. However, as Witelson (1977a) has pointed out, this conclusion does not necessarily follow from the evidence. The hemispheres may function symmetrically early in life but the presence of a lateralised lesion may impair the subsequent acquisition of cognitive skills, which develop in an asymmetrical fashion some years later. Thus the presence of a laterality effect at the time of testing does not establish that cognitive functions were represented asymmetrically at the time the lesion was sustained. Neither does loss of plasticity necessarily indicate progressive lateralisation of function, nor equipotentiality – the ability of one hemisphere to carry out functions normally undertaken by the other – imply that early in life *both* hemispheres participate in all functions equally.

It would clearly be desirable to carry out longitudinal studies of patients having early cortical damage, with a view to discovering the temporal relationship between early lateralised damage and the presence of later cognitive defects. At the present time such longitudinal data are not available and we have to be content with studies relating patterns of cognitive impairment to the age at which the cortical lesion was sustained.

Woods (1980) compared individuals who had sustained early brain lesions with their siblings as controls. He classified his patients according to whether or not they had shown evidence of cerebral damage occurring before their first birthday. Analysis of IQ scores showed that lesions of either the left or of the right hemisphere before the age of one year impaired both verbal and performance IQ, as did later lesions of the left hemisphere, whereas damage to the right hemisphere after the first birthday impaired only performance IQ. These findings might be inter-preted to mean that there is greater hemispheric equivalence of function during the first year of life than subsequently. Alternatively, it could be argued that due to the greater plasticity of the immature nervous system there is a greater degree of functional reorganisation following damage to the brain in the infant's first year. As a consequence there might subse-quently be a reduced propensity for left and right hemispheres to acquire skills asymmetrically.

The traditional view, based on the childhood aphasia literature, is that at birth the hemispheres subserve all functions equally. In contrast, Kins-bourne (1975c) has argued that those lesions most often sustained by children are precisely those that are least likely to be confined to a single hemisphere. Since it is in practice difficult to be certain that a lesion is strictly unilateral, it is possible that many cases of so-called right-sided lesions reported in the literature on childhood aphasia include many instances where the lesion was not in fact restricted to one side of the brain (see also Woods and Teuber, 1978). The aphasia may therefore have been due to encroachment of the pathology in the left hemisphere even if the neurological signs could be referred to the right hemisphere. With the advent of modern radiographic procedures it should in future be possible to be much more confident that a lesion is strictly unilateral. In the meantime, Kinsbourne's views are in line with a growing tendency in recent years to regard the two halves of the brain as being asymmetrical with respect to function at or shortly after birth. The fact that the anatomical asymmetry between left and right sides of the planum temporale in adults has been observed also in the brains of new born (Witelson and Pallie, 1973) and even pre-term infants (Wada, Clarke and Hamm, 1975; Chi, Dooling and Gilles, 1977) has been raised in support of this idea. However, as discussed in Chapter 6, the significance of left-right morphological asymmetry is obscure and its presence in foetal as well as adult brains does little or nothing to clarify the issue. We must look instead for behavioural evidence of cerebral asymmetry in infants.

## EVIDENCE OF FUNCTIONAL ASYMMETRY IN THE INTACT INFANT BRAIN

It would, of course, make no sense to seek hemispheric specialisation for behaviours that have not yet developed. One could not, for example, investigate cerebral speech representation in infants who cannot yet speak. On the other hand it is not unreasonable to consider whether there are 'language-related' processes functional in very young infants and, if so, whether these are lateralised to one hemisphere rather than the other.

A phenomenon relevant to our understanding of speech is that of categorical perception. Intuitively, it might be imagined that a sound which is always perceived in a particular way – as for example the 'd' sound in 'deep', 'deck' and 'duke' – is always represented in the same way in the acoustic waveform reaching the ear. However, this is not so. The temporal distribution of acoustic energy in different frequency wavebands is not identical in each instance of the sound 'd' in each of the words 'deep', 'deck' or 'duke'. The fact is that at any given instant the speech signal contains information about more than one phoneme or unit of speech. With regard to the words above, the sound following the 'd' is different in each case and this is reflected in the energy distribution corresponding to the 'd', yet we perceive the 'd' sound as the same in the three instances. Thus we show perceptual invariance in response to a physical variance. Sounds which vary physically along a particular acoustic dimension are perceived identically until a certain value along that dimension is reached beyond which we perceive another sound. This is the phenomenon referred to as categorical perception and the point along the dimension at which perception changes is called the category boundary (Springer, 1979). What needs to be emphasised is that we find it difficult to tell the difference between pairs of stimuli which, though physically differing along the relevant dimension, do not cross the category boundary. On the other hand, stimuli which are closer together on the relevant dimension, but are such that one member of the pair falls on one side of the category boundary while the second member falls on the other side, are readily discriminated.

The phenomenon of categorical perception of speech sounds is fundamental to our understanding of the speech code. As such one would expect it to be present in children prior to the emergence of speech proper. This line of reasoning led Eimas, Siqueland, Juszyck and Vigorito (1971) to conduct an experiment to investigate whether infants do in fact show evidence of categorical speech perception. Their technique involved what is known as the non-nutritive sucking paradigm. The paradigm involves presentation of a stimulus contingent upon a specified rate of sucking. If the rate increases over baseline levels, the stimulus is presented. In due course, once the stimulus-response contingency is established, sucking rate decreases as the stimulus ceases to act as a reinforcer maintaining high rates of sucking. If a novel stimulus is introduced, however, a high sucking rate is quickly re-established before again falling off. An increase in the sucking rate following presentation of a novel stimulus, relative to the rate when the response to a previous stimulus has declined, may be taken as indicating that the infant can discriminate the two stimuli. In the study by Eimas et al., then, it was expected that if infants perceive speech sounds categorically their sucking rates would indicate good discrimination across category boundaries and poor discrimination within each category. This is what was found.

The relation of phoneme perception to the left hemisphere is suggested by the results of a study by Entus (1977). The method she employed was to combine the non-nutritive sucking and dichotic listening paradigms. Identical stimuli were presented to an infant's left and right ears and after sucking rates first increased and then declined to some criterion level the stimulus in one ear was changed. Entus found that sucking rates recovered more dramatically if the changed stimulus was presented to the right rather than to the left ear. Interestingly, for musical stimuli greater recovery rates were observed for novel stimuli heard in the left ear. Recently, however, Vargha-Khadem and Corballis (1979), using exactly the same stimuli and equipment as Entus, failed to replicate the latter's findings. On the other hand Glanville, Best and Lawrenson (1977) who used heart rate rather than sucking rate as their dependent measure obtained results similar to those of Entus.

Support for the view that phoneme discrimination is mediated primarily by the left cerebral hemisphere even in infants comes from electrophysiological research. In an extended series of investigations Molfese and his colleagues have consistently obtained greater auditory evoked potentials over the left hemisphere for phonetic material, including words, but not for musical sounds or bursts of noise (Molfese, Freeman and Palermo, 1975; Molfese, 1977).

Although the findings with infants are consistent with those obtained with adults, it is possible that different mechanisms are responsible in the two cases. Davis and Wada (1977) recorded a greater amplitude of evoked potential over the left temporal region with clicks as stimuli and over the right occipital region for flash stimuli in 16 infants aged from 2–10 weeks. Since the stimuli were devoid of meaning it is difficult to suggest what kind of 'information processing' might have given rise to the observed EEG asymmetries. Davis and Wada suggest the possibility that events which can be referred to a prior history give a greater amplitude over the left hemisphere while those not so referrable give a greater right hemisphere response. It should be possible to test this suggestion by experimentally manipulating the subjects' experience with the experimental material. Even if the suggestion of Davis and Wada is not supported, these researchers make an important point. It is that laterality effects might come about as a result of unknown variables, such as interest value, level of arousal or familiarity, which should be excluded before one accepts that cerebral specialisation in infants is the same as in adults.

## LATERAL ASYMMETRIES IN PERCEPTUO-MOTOR RESPONSE

In addition to research concerned directly with the brain's response to specific kinds of stimuli, work has been carried out into early asymmetries in motor or perceptuo-motor behaviour. At least part of the motivation underlying these investigations is that such lateral differences in behaviour may be early precursors of cerebral lateralisation of function.

It has been shown in a series of studies reviewed by Turkewitz (1977) that infants as young as 24 hours old are more sensitive to auditory and tactile stimulation presented on the right than on the left. For example, if the corner of an infant's mouth is touched lightly he will turn towards the source of stimulation, but the response can be elicited more reliably by touching the right side of his mouth. Similarly, an infant will turn his eyes towards the direction of a sound but the threshold for this response is lower for a sound presented to his right compared with his left ear. These lateral asymmetries in response, however, may not be innate but rather the consequence of another right-sided bias whose origin is not yet understood.

Turkewitz observed that infants aged from birth to a little over 100 hours almost invariably spend their time lying on their right side. One consequence of this is to attenuate sound arriving at the right ear. Since prior exposure to continuous loud noise lowers sensitivity to subsequently presented stimuli it occurred to Turkewitz that reduced sensitivity of the left ear might account for the lateral bias in the eye-turning response to auditory stimulation presented at the left or right ear. He therefore carried out an experiment in which infants' heads were gently maintained in the mid-line position for 15 minutes before they were tested with auditory stimuli. The results showed that this procedure eliminated the bias towards the right shown by control infants who were allowed to remain in the right-turn position. Surprisingly, it was later found that maintenance of the head in the midline position also eliminated the bias towards right-sided turning in response to somesthetic stimulation of the region around the mouth. It might have been expected that lying on the right would have stimulated the right side of the face, including the perioral region, which would presumably reduce sensitivity of the right side in the same way that continuous auditory stimulation of the left ear was thought to reduce its sensitivity. Consequently, holding the infant's head in the mid-line position, according to this argument, should have augmented rather than eliminated the right-turning bias.

In fact, the situation is even more complex. When an infant lies with its head turned to one side, there is differential tonus of the neck muscles on the two sides. By appropriate experimental manipulations, Turkewitz was able to show that either differential muscle tonus or somesthetic stimulation of the cheek was sufficient to induce an asymmetrical turning response. However, although pre-test stimulation of the right cheek resulted in greater responsiveness to stimulation of the right than of the left perioral region, pre-stimulation of the left cheek had the effect of eliminating any lateral asymmetry in turning. This suggests that pre-test stimulation influences a system which is already biased towards the right, rather than itself being the cause of the asymmetry in turning response.

It is not clear why somesthetic stimulation appears to sensitise one side of the face, making it more responsive, while the opposite seems to be the case for auditory stimulation. What is becoming clear is the reciprocal nature of the relation between head posture and lateral sensitivity in response. If infants less than 12 hours of age are maintained in the midline position for 15 minutes or so, then they are as likely to turn left as right on being released. This is so even though infants of this age continue to show a bias in turning on to their right side if their head is held only momentarily in the central position. Infants older than 12 hours, however, exhibit the usual right-sided posture even after 15 minutes maintenance in a mid-line position. Thus lateral differences in sensory stimulation appear to interact at a very early stage in life with an innate turning asymmetry. The cause of the latter is unknown, but tests on children born by Caesarean section have disposed of the possibility that it is in some way related to passage down the birth canal.

Leiderman and Kinsbourne (1980) found a right sided bias at least in children born to two right handed parents. The absence of clear asymmetry in children where one of the parents was left handed suggests that the bias in right handers is not due to some lateral bias in uterine position – unless this is itself related to parental handedness.

Whatever the cause of the right-side bias in turning it underpins Kinsbourne's (1978) hypothesis that left hemisphere language develops out of orienting and attentional mechanisms. Left hemisphere control of speech, according to Kinsbourne, is a special case of the more general bias towards left-sided control of responding which he believes may have evolved as a consequence of asymmetrical ascending impulses from the brain stem. The right sided bias in turning produces pointing by the right hand during early attempts at naming objects. Hence left hemisphere activation accompanies the earliest signs of language. This may be why language is lateralised to the left hemisphere (Kinsbourne and Lempert, 1979).

## LATERAL PREFERENCE IN INFANTS AND YOUNG CHILDREN

As might be expected, manual preference in infants and children has received a certain amount of attention (Gesell and Ames, 1947; Hildreth, 1949) although the developmental sequence (see Palmer, 1964) is not well documented.

Young (1977) provides a review of work concerning early asymmetries in behaviours such as reaching or grasping, which have been found by Michel (1981) to correlate at age 16 and 22 weeks with the direction of head-turning bias in early infancy. A number of researchers have observed that an initial preference for reaching with the left hand up to the age of approximately 28 weeks gives way to a later preference for reaching with the right hand. Unfortunately, the early studies suffer from a number of methodological shortcomings and from a lack of statistical analyses which make it difficult to have confidence in their conclusions. Recent work suggests that though initially the left hand might reach more, objects are grasped for longer with the right hand (Bresson, Maury, Le-Bonnier and De Schonen, 1977). Interpretation of this finding is equivocal as the left hand might release its grip earlier than the right hand simply in order to perform some other task (Caplan and Kinsbourne, 1976). It seems, though, that the right hand is used more than the left to manipulate objects at least by the age of 15 months.

Ramsay (1979) observed in a cleverly set-up experimental situation that the right hand was favoured for hitting a xylophone even if the baton was initially placed in the child's left hand. Also, if the child was given two sticks, one in each hand, the stick in the right hand was commonly used to hit the stick held stationary in the left hand but rarely was the reverse the case. This is consistent with a preference for the right hand shown in tapping and similar tasks by children as young as 5 years of age and below (Ingram, 1975a; Finlayson, 1976; Wolff and Hurwitz, 1976).

With regard to other limb preferences, Peters and Petrie (1979) observed that when infants were lowered onto a hard surface the stepping reflex (alternate flexion and extension, i.e. bending and stretching of each leg) began with the right leg in 18 out of 24 infants. This asymmetry was maintained throughout each of four sessions beginning when the infants were an average of 17 days old and ending when the average age was 105 days. As the babies were lowered to the surface by their mothers, who were all right handed, it is possible that some subtle bias in the behaviour of the mother was responsible. However, this possibility was considered and eliminated in an experiment by Melekian (1981) who confirmed the original findings.

The relationship, if any, between such early lateral asymmetries in motor behaviour and preference for one limb over the other in adulthood has yet to be established. There is a need for large scale longitudinal investigations plotting the time course of asymmetries in preference and skill for different tasks. These two aspects of motor control – preference and skill – are probably related in that the greater the level of skill required for a particular task, the more consistent is the preference for a particular hand. Young (1982) has concluded from his review of the evidence that the degree of asymmetry in both preference and skill between the hands for difficult tasks is relatively constant with age. On easier tasks, however, the degree of preference shown for a particular hand may vary with age. This explains some of the discrepant findings in the literature, since some authors have found hand preference in infants and children to change with age (Cohen, 1966; Černácek and Podivinský, 1971; Seth, 1973; Ramsay, 1979) while others have not (Annett, 1970c; Ramsay, Campos and Fenson, 1979).

## TACTILE ASYMMETRY IN CHILDREN

Various tasks have been used in an attempt to examine tactile asymmetry in childhood. Witelson (1974, 1976a), using a tactile non-verbal matching task designed to be analogous to the dichotic listening technique, found a left hand superiority among boys aged 6–13 yrs. Flanery and Balling (1979) also found greater left hand accuracy on a similar dichaptic task, but one carried out without the aid of vision, and no age-by-hand interaction. That is, the difference between the hands was similar for each age group, the youngest of which was in the first grade at school in America. When the data were analysed in terms of laterality coefficients, however, a trend towards increasing asymmetry with greater age was evident, largely on account of a greater proportional change in the performance of the left compared with the right hand. Cioffi and Kandel (1979) found a left hand superiority in perceiving nonsense shapes and a right hand superiority in the perception of two-letter words in children from the age of 6 years onwards. Some inconsistent sex differences were found but there was no indication of a developmental increase in the size of the hand advantage for either class of stimulus.

A number of investigators have looked at Braille reading by blind and sighted children. Although this clearly involves the accurate perception of the spatial relationships among the raised dots it is also a linguistic task. The fact that the results do not consistently show a left hand superiority across different ages (Hermelin and O'Connor, 1971; Rudel, Denckla and

Spalten, 1974; Rudel, Denckla and Hirsch, 1977) may therefore be due to the relative balance between linguistic and non-linguistic cognitive processes changing with the level of task investigated, the grade of Braille used, or the subjects' familiarity with Braille. It is difficult, therefore, to draw any firm conclusions from this line of research escept that greater control must be exercised over the cognitive processing demands of the tasks employed.

## DICHOTIC LISTENING

The dichotic listening technique has been the most widely favoured method in attempting to investigate hemisphere specialisation in childhood. The results of much of this research have been ably reviewed by Witelson (1977a) and Young (1982) and only a brief summary of the major issues arising from this work will be made here.

### Verbal stimuli

The first point of note is that, consistent with certain of the findings on infant asymmetry, there appears to be some degree of cerebral specialisation for language as early as 3 or 4 years of age (Kimura, 1963b; Nagafuchi, 1970). Although a significant right ear advantage has not always been observed at such an early age, Witelson points out that of 31 studies reviewed by her a significant left ear advantage for processing verbal sounds was not found at any age. In studies published since Witelson's review a right ear superiority for linguistic stimuli has also been the common finding (Young, 1982). This consistency is quite remarkable given the lability of laterality effects and the vagaries of children as subjects. But although the fact of early functional asymmetry is not in dispute, the question of whether the degree of this asymmetry changes with age is very much a topic of debate (see Porter and Berlin, 1975). Some workers have found the magnitude of the right ear advantage to increase with age (e.g. Satz, Bakker, Teunissen, Goebel and Van Der Vlugt, 1975), others have reported a consistent difference between the two ears (Geffen, 1976; Hiscock and Kinsbourne, 1980a; Bryden and Allard, 1981; Eling, Marshal and Van Galen, 1981), while still others have noted a decline in the difference between the two ears with age (Geffen and Wale, 1979).

Bryden and Allard (1978) have argued that the free recall procedure, in which the subject gives his responses in any order he chooses, is not a sound methodological way of determining degree of cerebral specialisation, since there is a tendency for subjects, if not constrained otherwise, to report material from the right ear first, at least if stimuli are presented at fairly rapid rates. This being so, the right ear superiority shown in a verbal dichotic listening task may reflect nothing other than this tendency. Any changes in ear asymmetry with age might therefore be seen as developmental trends in reporting from the right ear first. While this itself is a phenomenon which requires explanation, interpretation of dichotic listening studies with children (and, indeed, with adults) is difficult if the free recall procedure is used unless methods of analysis are adopted which take acount of this potential artefact. The most obvious means of doing so is to consider separately those trials in which the subject's first report is from the right ear and those trials on which the initial ear of report is the left. Using this method of analysis Bryden and Allard (1978) noted a progressive increase with age in the proportion of subjects showing a superiority of the right ear in initial report of consonant-vowel syllables. This supports the findings of Satz et al (1975) who used a longitudinal rather than a cross-sectional design, and sophisticated statistical techniques (multivariate analysis) to partial out the effects of age, sex and ear of report. These authors concluded that, with other variables taken into account, there was an increasing asymmetry in favour of the right ear with increasing age from 5 to 11 years, although the difference between ears was significant only for the two oldest groups of subjects (aged 9 years and 11 years). However, the ability to deploy attentional strategies varies with age (Geffen, 1978; Sexton and Geffen, 1979; Geffen and Wale, 1979); it is therefore clearly necessary to control for such strategies. When this is done no developmental increase in the size of the right ear advantage is found (Bakker, Hoefkens and Van Der Vlugt, 1979).

One problem in interpreting the meaning of any change in ear asymmetry with age is that the overall accuracy of subjects on the experimental task will usually also change with age, so ceiling and floor effects will influence the degree of asymmetry that might be expected at different ages. A further difficulty is that different processes such as phonetic discrimination, syntactic ability and semantic memory may lateralise at different ages (Porter and Berlin, 1975) and it is not always clear what processes are being tapped in any particular experiment. It is also becoming evident that the presence, absence or degree of ear asymmetry in children is influenced by relatively subtle differences in methodological procedures, stimuli and task requirements (Geffen, 1978; Geffen and Wale, 1979; Bryson, 1980). This makes it difficult to compare results across different studies and age groups.

### Non-verbal stimuli

With regard to the question of hemispheric asymmetry of non-verbal processes in children the available data are relatively sparse. Knox and Kimura (1970) found a left ear advantage in dichotic recognition of environmental sounds in children aged 5, 6, 7 and 8 years although the ear difference was not significant at any age. Working with children aged 7 to 12 or so, Bryden and Allard (1981) found a left ear advantage (with controlled order of report) for non-verbal sounds, but this was only significant for the oldest group of children. However, Piazza (1977) obtained significant left ear superiorities, with no developmental increase, in recall of similar material at each of the ages of 3, 4 and 5 years among subjects who yielded significant right ear advantages for verbal material.

In her report, Piazza mentions that for these children reciting aloud affected tapping rates of the right hand more than the left hand. Conversely, humming impaired left hand more than right hand tapping. The lateralised effect of a concurrent verbalisation task supports the findings of Hiscock and Kinsbourne (1978; 1980b) who found decreased tapping by the right hand in children as young as 3 years of age. The degree of asymmetry did not increase in groups of children of increasing age. This lack of a developmental increase was supported by longitudinal data, which similarly showed that with a year interspersed between testing sessions there was no increase in asymmetry from the first to the second time of testing the same subjects.

## TACHISTOSCOPIC HEMIFIELD PRESENTATION

Tachistoscopic studies carried out with children suffer from the same shortcomings and methodological difficulties as those undertaken with adults. It might even be argued that there is a particular difficulty with children, namely the requirement to maintain central fixation. It is perhaps naive to suppose that young children will conscientiously maintain their gaze on a central fixation spot. This is relevant to any discussion of developmental changes in degree of asymmetry, since any observed change in the magnitude of visual field differences may relate to changes in fixation strategy rather than degree of cerebral asymmetry. The need for some form of fixation control in studies with children is thus even more pressing than in studies with adults. Although many experimenters have examined visual hemifield asymmetry in children, only a relatively small number have controlled their subjects' fixation.

### Verbal stimuli

The situation with regard to lateral tachistoscopic stimulation is that a considerable number of studies have found a mean right visual field superiority in word recognition at varying ages from about 6 years upwards (Forgays, 1953; Olson, 1973; Marcel, Katz and Smith, 1974; Marcel and Rajan, 1975; Turner and Miller, 1975; Carmon, Nachson and Starinsky, 1976; Butler and Miller, 1979; Davidoff, Beaton, Done and Booth, 1982). By and large, the magnitude of this asymmetry has shown no developmental increase across different ages (but see Forgays, 1953; Turner and Miller, 1975) although the proportion of subjects showing a right field superiority may change (Tomlinson, Keasey and Kelly, 1979).

The age at which a significant visual field difference first occurs varies in different experiments. Given the different tasks that have been used this is not surprising. However, interpretation of a particular set of findings in terms of hemispheric specialisation is highly problematic. Words and letters may be processed either visually or verbally and without constraining the experimental task in some way it is not possible to say which strategy is being adopted. This confounds attempts to investigate the emergence and development of cerebral asymmetry, since changes in laterality effects with age may reflect differences in processing which are not necessarily related to changes in hemispheric specialisation. An example should make this clear.

Barroso (1976) spoke aloud the name of an object and then flashed a picture to the left or right hemifield. The subject's task was manually to respond 'same' or 'different' depending on whether the picture was of the spoken object. Adults showed a right field superiority on this task. On the other hand, when the task was that of matching two simultaneously presented pictures, adults gave a left field advantage. The adult pattern of results was shown by 10 and 12 year old children but not by children of 6, 8 and 9 years of age. Does this mean that children do not have lateral specialisation of function for non-verbal processing until they are 10 years old? Obviously not, since before that age they may alternate between physical and nominal matching strategies. Matching according to the name of the objects depicted may occur on some trials, perhaps favouring the right visual field. On other trials matching might be carried out by converting the name of the spoken object into a visual image and matching that to the picture presented on the tachistoscope. This could give rise to an advantage for the left visual field. The net effect of these different strategies might then be to eliminate any significant field difference when scores are averaged over trials. On this view, the significant right hemifield superiority at age 10 might simply reflect the fact that one particular strategy has come to predominate. A similar argument might be applied to the absence among younger children of a left field advantage in matching the simultaneously presented pictures.

## Non-verbal stimuli

Tachistoscopic non-verbal processing was studied by Reynolds and Jeeves (1978) using a face-recognition task. Although college students and 13–14 year olds showed a significant left field superiority in reaction time, a group of 7–8 year olds showed a non-significant right hemifield advantage. All subjects, who were female, showed a right sided advantage on a similar task using letters as stimuli but this was statistically significant only for the two older groups. While the results of this study might be taken to imply that hemispheric specialisation for face recognition develops with age it is again possible that it is not so much the cerebral loci of processes concerned in face recognition that change over time, as that a particular strategy used to perform the task, which may indeed be referable to a particular hemisphere, becomes more firmly established.

Reynolds and Jeeves did not employ any fixation control in their study, which may be related to their failure to find a LVF superiority in their youngest subjects. By contrast, Marcel and Rajan (1975), Young and Ellis (1976), Broman (1978) and Young and Bion (1980a, 1981) have all reported significant LVF superiorities for face recognition in children aged 7 and upwards without showing any developmental increase in the magnitude of the effect. On the other hand, the failure of Reynolds and Jeeves to find a significant hemifield difference in females is consistent with the findings of Young and Bion (1979). These workers asked subjects to report how many dots were present in a random array. They found a significant left hemifield advantage in each group of 5, 7 and 11 year old boys, as did Kimura (1966) with male adults, but no asymmetry in girls of the same age.

Despite the interpretative difficulties arising from use of the tachistoscopic technique it is probably fair to conclude from the opposite field superiorities that have been obtained for verbal and non-verbal material that complementary specialisation of function in the left and right hemispheres is evident in children from at least the age of 6 years. However, the data reviewed provide no grounds for believing that the degree of such specialisation, once established, increases with age. The emergence of a tachistoscopic laterality effect at a given age, without any developmental increase, should be regarded, on this view, as reflecting an asymmetrical loss of plasticity rather than progressive lateralisation of function.

Although the left hemisphere may be specialised from birth for certain language-related functions, such as phonemic discrimination, it is conceivable that both hemispheres are involved in building up the child's early word-dictionary or lexicon. Beyond a certain age only the left hemisphere may be concerned in acquiring a vocabulary. If so, one might expect to find no tachistoscopic hemifield asymmetry for those words that are learned early in life but a right hemifield advantage for words learned at a later age. This idea has been tested by Young and his co-workers who found no effect of age-of-acquisition, whether the stimuli employed represented different ages at which children learn to speak (Ellis and Young, 1977; Young and Bion, 1980b) or to read words (Young, Bion and Ellis, 1982). However, in unpublished studies carried out in the writer's laboratory a right hemifield superiority in reaction time and accuracy of recognition was found for words acquired later in childhood but no asymmetry was found for early-acquired words (Beaton, Sykes, King and Jones, 1982). These results could not be attributed to the influence of variables such as concreteness and imageability which co-vary with age-of-acquisition. They therefore imply that the age at which words are learned may influence the degree of visual field asymmetry as predicted by the hypothesis that both hemispheres are involved in the early stages of language learning. Inspection of the data presented by Young and his colleagues suggests that their failure to find an age-of-acquisition effect can be attributed perhaps to insufficient differentiation between their early and late acquired words in terms of the ages at which these words are learned. All their early-acquired words are generally learned below 5 years of age while their later words are learned above the age of 7 years (Gilhooly and Hay, 1977). By contrast, the early acquired words used in the writer's laboratory had a lower mean age of acquisition than the stimuli employed by Young, and the later-acquired words used in Beaton's studies are generally acquired after the age of 10 years (Gilhooly and Gilhooly, 1980).

## LATERALITY AND SECOND LANGUAGE ACQUISITION

The idea that age-of-acquisition of words affects tachistoscopic visual field asymmetry receives some support from studies of bilingual subjects.

It has been known for some time that cerebral lesions may not affect the bilingual patient equally in both languages. In particular, the language learned earlier is often, but not always, the first to recover (Paradis, 1977). Rather naively, it has been suggested by some that the two languages of a bilingual person are represented in opposite hemispheres. Testing of a single bilingual split-brain subject has indicated that this is not universally the case (Dimond, 1978). Nonetheless, there is a sufficient number of positive findings to warrant examination of the notion that the two halves of the brain are differentially involved in cases of bilingualism. The evidence in favour of this proposition comes from both clinical and normal subjects and has been summarised by Albert and Obler (1978). On the clinical side they quote the results of an investigation by Nair and Virmani (1973) who found that 10 out of 12 right-handed Indian patients with left sided-hemiplegia had some dysphasic symptoms. This is a much higher incidence than one would expect, and Nair and Virmani argue that it may be due to the high frequency of polyglotism in India. Unfortunately, Nair and Virmani do not indicate whether their 12 dextral hemiplegics were, in fact, at least bilingual although 33 out of 50 aphasics in their purportedly random sample of hemiplegics were said to be so. There are also other problems in interpreting Nair and Virmani's data but their study provides an interesting starting point for a more sophisticated approach to the problem.

Experimental investigations of field asymmetry in bilingual normal subjects raise a number of methodological issues, which are reviewed by Obler, Zattore, Galloway and Vaid (1982). Despite certain problems, two conclusions emerge. Firstly, different languages may yield visual field effects of different magnitude; the direction of asymmetry may also differ. Secondly, the visual field effect even for the subjects' main language is often reduced in comparison with monolingual control subjects (Albert and Obler, 1978).

The lack of equivalent hemifield asymmetry for the two languages may be explained by supposing that languages learned at different ages will be recognised or learnt by different strategies, or perhaps by the same strategies but to differing extents. Albert and Obler (1977) suggest that the strategy used to acquire a second language will not only influence the degree of cerebral lateralisation for that language but will alter the 'standard' pattern of lateralisation of the first language. Exactly how this is achieved they do not say. There is, however, some evidence in favour of the first part of their thesis, that language acquisition at different ages involves different strategies.

Silverberg, Bentin, Gazier, Obler and Albert (1979) found that at all ages tested native Hebrew speaking children showed a RVF superiority in the time taken to react to target words. Those in their second year of learning English, however, showed a LVF superiority in responding to English words. This LVF superiority was smaller among those who had been learning English for 4 years and reversed to a RVF superiority among those who had been learning English for 6 years. In a follow-up study, Silverberg, Gordon, Pollack and Bentin (1980) tested Israelis during their second and third grades of schooling. During the second grade they showed a LVF superiority for English words while in the third grade they showed a RVF superiority for the same words. The authors argue that their results suggest a shift from a right towards a left hemisphere-based strategy with increasing experience in a language. This would be consistent with the results of other experiments which have found a shift away from a LVF advantage towards a RVF superiority when subjects are specifically trained to recognise those words which will subsequently be presented to them (Bentin, 1981; Endo, Shimuzu and Nakamura, 1981) and suggests that the degree of familiarity with the material is an important variable in asymmetry of tachistoscopic recognition (Bryden and Allard, 1976).

Support for the view that acquisition of a second language after the age of 6 years involves right as well as left hemisphere-based strategies comes from the results of an experiment using the dual-task technique. Sussman, Franklin and Simon (1982) found that only the performance of the right hand was impaired in monolingual and bilingual subjects who acquired both languages before they were 6 years old. Among bilingual subjects who acquired their second language after the age of 6 years, however, only the performance of the right hand deteriorated during concurrent vocalisation of the first-learned language, whereas both hands were impaired if the subject spoke in the second-acquired language.

The possibility that speaking a language learned after the age of 6 years requires the participation of both hemispheres does not necessarily conflict with the finding of a tachistoscopic RVF superiority for the same language. Speaking a language clearly involves different cerebral mechanisms than visual recognition of words, to some extent at least. The data of Sussman et al. (1982) suggest that differentially lateralised mechanisms are involved in language acquisition before and after the age of 6 years. This corresponds well with the end of the period of functional plasticity identified by Krashen (1973) and suggests the possibility that after this age the individual is forced to use right hemisphere processes that are not available at an earlier age.

## ENVIRONMENTAL INFLUENCES ON BRAIN LATERALISATION

The overwhelming bias in favour of left hemisphere speech processes in the majority of dextrals suggests that this specialisation is wired-in to the nervous system. There are instances in neuro-biology where pre-wired systems can be modified by environmental input (e.g. Schlaer, 1971) or deprivation (Hirsch and Spinelli, 1970; Blakemore, 1978). Can lateralisation of brain function be influenced by experience? Certainly it is clear that pathology early in life can reverse the direction of hemispheric specialisation (see Chapter 7), but what of non-pathological influences?

The suggestion has been made that the disadvantages associated with low socio-economic status may retard the process of lateralisation (Geffner and Hochberg, 1971; see also: Borowy and Goebel, 1976; Stark, Genese, Lambert and Seitz, 1977) although the findings on which this suggestion was based have not been replicated for either the visual (Dorman and Geffner, 1974; Geffner and Dorman, 1976) or auditory (Davidoff, Done and Scully, 1981) modality. Neither was lack of experience with the printed word in adult illiterates found to influence tachistoscopic hemifield asymmetry for simple words in a study by Davidoff, Beaton, Done and Booth (1982). It does not follow, however, that more extreme deprivation will be without effect. The girl Genie, found at the age of 13 and until then completely isolated from all normal social and cultural contact, gave a dichotic left ear advantage, with other indications of right hemisphere specialisation, for verbal stimuli when tested after she had acquired language (Curtiss, 1977). The suggestion has therefore been made that her isolated existence prevented the normal development of speech in the left hemisphere. The implication, presumably, is that after a 'critical period', speech, if it develops at all, is as likely to be subserved by the one hemisphere as the other. Certainly Genie represents a highly unusual case. Other circumstances of deprivation rather less extreme, though sometimes more prolonged, are those encountered by the deaf. Consequently, a number of experiments have been carried out in an attempt to discover whether auditory deprivation influences hemispheric lateralisation.

## HEMISPHERIC SPECIALISATION IN THE DEAF

McKeever, Hoemann, Florian and Van Deventer (1976) tested congenitally deaf females on recognition of words and drawings of the manual alphabet and American sign language (ASL). For hearing subjects words gave a RVF superiority whether presented unilaterally or bilaterally; deaf subjects showed a significant RVF superiority for unilaterally but not bilaterally presented words. Non-word stimuli were only presented bilaterally. These stimuli showed a significant LVF effect for hearing subjects and no significant asymmetry for the deaf. It is not obvious how to interpret the findings for deaf subjects obtained with bilateral stimulus presentation, but the unilateral RVF advantage for written words might be taken to imply that visual linguistic processes are lateralised in the brain of the deaf as in the hearing population.

In contrast to the results of McKeever et al. bilaterally presented words were found to give a RVF superiority in both deaf and hearing subjects by Manning, Goble, Markman and La Breche (1977). Photographs of signs from the American Sign Language showed no significant asymmetry. However, if a word was presented in one visual hemifield simultaneously with a sign in the opposite hemifield, then the deaf showed a RVF advantage for both types of stimuli. Possibly this represents a shift towards verbal processing of the signs. Other authors have found LVF superiority for ASL stimuli (Scholes and Fischler, 1979; Ross, Pergament and Anisfield, 1979; Kelly and Tomlinson-Keasey, 1981).

Neville and Bellugi (1978) found congenitally deaf subjects without functional speech to show a RVF superiority when signs were presented in a single half-field, but again no asymmetry was found with bilateral presentation. The asymmetry in the unilateral condition is consistent with an earlier report (Neville, 1977) of enhanced evoked potentials over the left hemisphere in deaf subjects who knew ASL, in response to line drawings of common objects. Subjects without knowledge of ASL showed no asymmetry while hearing control subjects gave responses of greater amplitude over the right hemisphere. Phippard (1977) also distinguished between subjects who were able to communicate only by signing and those who could only communicate orally. She found that signers showed no hemifield asymmetry on either a verbal or a non-verbal task while the speakers showed a LVF advantage for both tasks. Hearing subjects showed a LVF advantage for the non-verbal stimuli and a RVF advantage for the verbal stimuli. As with Neville's (1977) results the pattern of results might be taken to suggest that knowledge of sign language induces comparatively more linguistic, left hemisphere processing for ostensibly non-verbal stimuli. Nonetheless, meaningful signs may be processed by the deaf differently from non-meaningful signs. In a discriminative reaction time task both hearing and deaf subjects whose first language was ASL showed a RVF advantage to arabic numbers and a LVF superiority to non-ASL signs (Poizner and Lane, 1979).

One problem in interpreting the results of divided visual field studies using ASL stimuli is that, in the real world of the deaf, signs are not static elements in the language but are dynamic in the sense that the hands move while signing. Poizner, Battison and Lane (1979) found that whereas deaf subjects showed a LVF superiority in response to static signs, they showed no visual field asymmetry, and by inference a shift towards linguistic processing, when the stimuli consisted of three frames of a film showing the initial, intermediate and final positions of the hands. The implication is that photographs or drawings depicting a single position may not elicit linguistic processing.

Interpretation of the literature on asymmetry in the deaf is made difficult by the fact that experimenters have not controlled their subjects' coding strategies (linguistic versus spatial) by constraining the kinds of task they have been asked to perform. Ross et al. (1979) make the useful suggestion that future research should use semantic categorisation tasks in order to force semantic processing if the aim is to discover something about the lateralisation of linguistic processes.

Another difficulty is that data have usually been analysed in terms of group averages, which could easily obscure asymmetries of opposite direction in different subjects. A tendency for deaf subjects to separate into those who characteristically prefer a linguistic strategy and those who use non-verbal strategies might well relate to other variables which are not always reported, such as age of onset of deafness, extent of speaking competence, degree of skill in using sign language and so on. Where such variables are taken into account (Phippard, 1977; Neville, 1977) differences between subjects have been found. Scholes and Fischler (1979), for example, found a difference in mean asymmetry scores between deaf subjects who were linguistically skilled, as determined by various language tests, and those who were unskilled. Indeed, it is likely that the pattern of tachistoscopic asymmetry shown by hearing control subjects, at least with signs as stimuli, will also differ between those who are well acquainted with ASL and those who are naive in this respect. If the performance of control subjects is used to evaluate the influence of hearing loss on tachistoscopic asymmetry, then logically the control subjects should be matched as closely as possible to the deaf subjects on knowledge of ASL. This has not always been done.

Despite inadequate attention being directed to the methodological points outlined above (see also Neville and Bellugi, 1978) there is a fairly general consensus of results showing that the deaf show reduced and/or reversed half-field asymmetry for verbal stimuli in comparison with hearing subjects (McKeever et al., 1976; Manning et al., 1977; Phippard, 1977; Scholes and Fischler, 1979; Kelly and Tomlinson-Keasey, 1981). This is at least suggestive of a reduced reliance on the left hemisphere of the deaf for processing written English. With regard to signs, the evidence is more equivocal. But in so far as the studies reviewed have been concerned with perception rather than production of signs, the lack of a clear trend towards RVF superiority is not necessarily in conflict with the literature on signing in deaf 'aphasics', which indicates left hemisphere involvement in the production of signing, as in the production of words by normal hearing individuals.


## SUMMARY

The findings reviewed in this chapter leave little doubt that complementary hemispheric specialisation of function can be detected in the early years of life. The issue that remains unresolved is whether or not lateralisation develops gradually, in the sense that functions initially executed by both hemispheres gradually become concentrated in one hemisphere. In contradistinction to this idea is the view that functions are segregated to left and right halves of the brain from their initial emergence in the early life time of the child. What changes, according to this view, is the capacity of one hemisphere to substitute for the other. That is, cerebral plasticity of function declines. Along with this goes an increasing tendency to acquire new skills asymmetrically.

On balance, the bulk of studies reported in the literature do not show any developmental increase in the magnitude of the laterality effects obtained. This suggests that the onus of proof rests with those who believe in the concept of progressive lateralisation rather than with those who believe that lateralisation is fixed at or near birth. Not all aspects of language may be lateralised to the same extent, however, and it is at least conceivable that lateralisation of speech is progressive while receptive aspects of language are more bilaterally represented throughout the lifetime of an individual. The idea that speech lateralises progressively is derived from observation of children with early brain damage and difficulties in speaking. The experimental literature on the other hand arguably taps the processes of auditory and written language perception.

It should be noted that progressive lateralisation may develop either from an initially undifferentiated base, in which both hemispheres participate in all functions, or from an asymmetrical base in which predominantly one hemisphere subserves particular functions. Thus the question of whether the *degree* of lateralisation changes with age is distinct from the question of whether cerebral specialisation as such exists at or shortly after birth.

# 8

# Sex differences in asymmetry

During the past decade or so it has become increasingly common for researchers to analyse their data for sex-related effects. This chapter reviews evidence in favour of the notion that males and females differ in their degree of hemispheric specialisation of function. Such a difference, it has been argued, could account for certain of the putative differences between the sexes in cognitive ability. The review is not intended to be a résumé of sex differences in general, only of reports that the sexes differ in some aspect of lateral asymmetry. The findings from studies using adult subjects are reviewed first; the results of experiments with children are considered separately.

## CLINICAL INVESTIGATIONS

Credit for being the first to analyse neuropsychological test scores in relation to the sex of the patient belongs to Lansdell. In a number of investigations in the 1960s and '70s, he reported sex-by-side of lesion effects. Left temporal lobe lesions impaired the ability of men but not women to interpret proverbs (Lansdell, 1961) and left thalamotomy (removal of the left side of the thalamus) affected the performance of males but not females on a word-association test (Lansdell, 1973). Among women, temporal lobe surgery to the 'dominant' hemisphere, as defined by the Amytal test, produced a fall in scores on the Graves Design Test, a test of aesthetic or artistic aptitude, while surgery to the non-dominant hemisphere produced an increase in score. Among male patients the opposite was the case (Lansdell, 1962). These effects, however, did not persist beyond the immediate post-operative period and the data were based on relatively small numbers of patients.

The Wechsler-Bellevue Intelligence Scale has often been used to assess intellectual changes as a result of brain damage. There is some evidence that in adults damage to the left half of the brain impairs verbal IQ leaving performance IQ relatively intact, while right sided lesions impair performance but not verbal scores (Anderson, 1951; Reitan, 1955; Goldstein, 1974). However, recent evidence suggests that this relation may hold more for men than for women.

McGlone (1977; 1978) reported that only male patients showed a significant lateralised effect of brain damage, left lesions impairing verbal IQ relative to performance IQ and right lesions having the converse effect. Female patients did not show selective deficits. Stimulated by this finding, Inglis and Lawson (1981) considered that some of the discrepancy in the previous literature could be accounted for by taking the patients' sex into account. On the basis of McGlone's results they postulated that those studies showing a positive laterality effect would have an excess of male compared to female patients while studies failing to show this effect, or only partially doing so, would have a greater female-to-male ratio. They further predicted that a re-examination by sex of the negative and equivocal outcomes would reveal positive results in the case of male patients only. Examination of the literature confirmed their first prediction; analysis of data presented by Meyer and Jones (1957) confirmed the second.

The question of whether the sexes differ in the incidence of overt language difficulties following unilateral cerebral damage has not yet been satisfactorily answered. McGlone (1977) found that in a series of left brain damaged patients 14 out of 29 males were aphasic but only 2 out of 16 females (chi-square significant). A non-significant trend towards more severe aphasia in men than women was noted by McGlone and Kertesz (1973). On the other hand Kertesz and Sheppard (1981), studying a larger group of patients, found no difference in the proportion rendered aphasic by a left hemisphere insult; nor did the sexes differ in scores obtained after right-sided damage on a battery of tests designed to detect language difficulties.

The source of the discrepancy in the findings of McGlone (1977, 1978) and Kertesz and Sheppard (1981) is unclear. The latter studied only stroke patients while McGlone's sample consisted of both stroke and tumour cases, but her results held for both categories of patient. In support of Kertesz and Sheppard, examination of records for 260 males and 130 females by Micelli, Caltagirone, Gainotti, Masullo, Silveri and Villa (1981) revealed no sex difference in the incidence, severity or type of aphasia produced by different kinds of brain damage. A failure to find a difference between the sexes in frequency of aphasia was also reported by De Renzi, Faglioni and Ferrari (1980) and by Basso, Capitani, Laiacona and Luzatti (1980).

The reason for the difference between the sexes found in the study by McGlone is uncertain. One suggestion has been that among her patients more males than females had vascular damage affecting posterior compared to anterior regions of the brain, the dividing line between the two being the Sylvian or central fissure. Kimura (1980) found in a small series of patients that all twelve aphasic patients with posterior damage were male while among 6 aphasics with anterior damage, three were female. She therefore suggested that McGlone's results could be due to a sex difference in intra-hemispheric organisation, combined with the known greater frequency of restricted posterior than anterior cerebral damage due to vascular accident.

Another possibility which might account for the sex difference noted by McGlone (1977) is that among her vascular cases males and females differed in the site of occlusion. Kaste and Waltimo (1976) found for left sided lesions that relatively more males than females suffered occlusion in the trunk of the middle artery rather than in its branches. This could mean that ischaemic damage (due to lack of oxygen) tends to be greater in males; if so, this might account for the increased incidence of aphasia in men found by McGlone. At present, such a suggestion remains speculative and has in fact been rejected by McGlone herself (1980). It should be noted also that a sex difference in the degree or extent of damage to cortical tissue in the territory served by the middle cerebral artery cannot explain the fact that McGlone's tumour group showed similar sex-related deficits to those of her vascular patients. Even among the tumour cases, however, different origins of the neoplasm may have led to differing severity of damage in males and females (Kinsbourne, 1980).

The findings of McGlone, if confirmed by future investigators and found not to be related to some artefact of pathology, suggest that certain verbal functions may be differentially organised in the two sexes. Yet the fact that there is no difference between males and females on verbal tasks following right-sided brain lesions implies that the basic pattern of unilateral hemispheric representation of expressive language functions applies to both sexes. McGlone's data with regard to incidence of aphasia might then be interpreted as implying a sex difference in intra-hemispheric organisation of language, at least in terms of expressive speech. On the other hand, the finding that in females unilateral damage to either side of the brain does not appear to impair verbal and performance IQ selectively, but lowers both sub-scores to an equivalent degree, raises the possibility that some aspects of verbal function are bilaterally represented in females.

## NEUROANATOMICAL ASYMMETRY

In the light of the clinical data it is of interest to consider whether there are indications of a sex difference in the literature concerning anatomical differences between the hemispheres. Further, since anatomical asymmetries have been shown to relate to handedness it would be useful to know whether any left-right anatomical difference between the sexes relates more to differences in the handedness distributions of males and females or to some difference in their pattern of hemispheric specialisation. At present, the relevant data are sparse.

In the series of brains examined by Wada, Clarke and Hamm (1975) only 10 out of 100 adult specimens (78 of known sex) had a larger temporal plane (planum temporale) on the right and 8 had equal plana on the two sides. A predominance of the 10 brains showing a reversal of the usual asymmetry were female. Nonetheless, the large majority of females showed the usual pattern of asymmetry. There was a trend, but not statistically significant, for the left temporal plane to be larger in males than in females. Witelson and Pallie (1973), on the other hand, reported that among very young neonates (mean age 6.6 days, range 1–19 days) the left planum was on average larger than the right in females but not in males.

Indirect estimates of neuroanatomical asymmetry using radiographic techniques have consistently failed to show any significant sex-related effects (Le May and Culebras, 1972; Hochberg and Le May, 1975). Thus, overall, the evidence concerning a sex difference in neuroanatomical asymmetry is very weak.

## LABORATORY STUDIES WITH NORMALS

### Dichotic listening: verbal stimuli

A number of studies have found sex differences in the degree of ear advantage on verbal dichotic listening tasks. Remington, Krashen and Harshman (1974) found a significant REA only for males as a group and not for females although the direction of asymmetry was the same for both sexes. Harshman, Remington and Krashen (1974) looked at three studies, two already published and one of their own, and by combining data (it is not specified how) claimed to find a highly significant sex difference favouring males in magnitude of the REA. Among twins, a reduced ear asymmetry in females was also noted by Springer and Searleman (1978) and by Vaid and Lambert (1979). Similarly, Lake and Bryden (1976) found a greater proportion of males than females showing a right ear superiority, as did Briggs and Nebes (1976). However, in the latter study the sex difference was not statistically significant and in the study by Lake and Bryden the sex difference held only for subjects with sinistral relatives.

Data may be presented in the form of mean differences between left and right ears for males and females or in terms of the proportion of each sex showing an advantage for one or other ear. These two measures may not coincide. McGlone and Davidson (1973), for example, found no sex difference in mean magnitude of ear asymmetry although a considerably higher proportion of males (75.7%) than females (59.5%) showed a right ear superiority. Unfortunately, it is rare for authors to present their data in such a way that it is possible to consider together both relative frequency and mean ear asymmetry.

Against these few reports of greater asymmetry in males, most of which involved stimuli such as CV syllables, can be set a study by Dorman and Porter (1975) who reported greater asymmetry in females in dichotic recognition of single CV syllables, and a number of studies in which an analysis by sex was undertaken but no sex-related asymmetry was found (e.g. Schulman-Galambos, 1977). In addition it seems fair to suppose that where subjects' sex is not mentioned in an author's results there was no conspicuous sex difference which would otherwise have invited an analysis according to sex.

As noted elsewhere, if order of report is not controlled by an experimenter a difference in scores for left and right ears may reflect a preferential processing or reporting strategy on the part of the subject. Bryden (1979) reports that in two experiments in which possible confounding strategy effects were carefully controlled no suggestion of any sex difference in ear asymmetry was found.

Bryden's (1979) failure to find a significant sex difference with controlled order of report contrasts with the results of Piazza (1980) who also controlled for order of report but found males and females to differ in magnitude of REA. Males showed a strong, significant superiority of the right ear while the right ear advantage of females was weak and insignificant. Because male and female scores differed significantly in variance Piazza analysed her data for each sex separately. Strictly speaking, therefore, one cannot say that there was a significant interaction between sex and ear asymmetry. If there had been, then the difference in results between Bryden (1979) and Piazza (1980) would have been difficult to explain. As it is, Bryden (1979) does not provide sufficient detail to confidently suggest the cause of the discrepancy. One might hazard a guess that the larger number of subjects in Piazza's study accounted for her finding a significant sex difference. Another possibility is that although Piazza controlled for order of report, she did not analyse her data by ear of first report. Bryden, on the other hand, required his subjects to report only from one ear on any given trial. It is possible that Piazza's finding is due entirely to the data from the second-reported ear – that is, to a sex-by-ear of report interaction.

## Interaction of sex with personal and familial handedness

Since handedness interacts with ear asymmetry, the possibility of a three-way interaction between sex, handedness and asymmetry bears examination. Although Briggs and Nebes (1976) found no significant interaction between these three variables Lake and Bryden (1976) reported that, regardless of the subject's own hand preference, familial sinistrality among females increased the likelihood of a superiority of the left ear while among males the presence of left handedness in the subject's family increased the likelihood of right ear superiority. In this study, as in that of Briggs and Nebes, personal hand preference, per se, did not emerge as a significant factor. In contrast, Piazza (1980) found that whereas familial sinistrality was not a significant variable, hand preference did interact significantly with sex. Among male subjects both left and right handers showed a significant REA, which was of greater magnitude for the latter group. Among females the weak, non-significant REA was not affected by handedness.

Lake and Bryden made a deliberate effort to include subjects whose hand preference was not strongly biased to the right or left as well as subjects who did show a strong hand preference. Nonetheless, subjects were dichotomised as right or left handed, only 2 points out of the range 14–70 separating the two groups. Subjects obtaining a score of 14–51 on a 'modified version of the Crovitz and Zener (1962) questionnaire' were categorised as right handers and those obtaining scores of 43–70 were classed as left handers. It seems likely that this method of classification can account for the failure of Lake and Bryden to find a main effect of handedness, in contrast to Piazza's results. Piazza does not report her criterion for classifying her subjects' handedness but presumably she made no particular attempt to recruit subjects who fell in the middle of the handedness distribution.

That Lake and Bryden, but not Piazza, obtained a significant main effect of familial sinistrality might be due to the fact that Lake and Bryden had 18 subjects in each of their 8 subject groups (obtained by crossing sex, handedness and family sinistrality in all combinations) while Piazza had only 8 subjects per group. Moreover, Piazza analysed her data separately for each sex while Lake and Bryden performed an analysis of variance over both sexes combined. Thus Lake and Bryden's analysis was based on 144 subjects, 72 of whom had a family history of left handedness, while Piazza's analysis was based on only 64 subjects, 32 of whom had at least one sinistral relative.

Lake and Bryden's results suggest that familial sinistrality increases the likelihood of a left ear superiority in females and the likelihood of a right ear advantage in males. A differential effect of familial left handedness in males and females has also been noted by Carter-Saltzman (1979). She reports that her male subjects with sinistral relatives showed a right ear superiority whether they were themselves left or right handed. Among females with sinistral relatives however, those who preferred the left hand had a left ear supriority but those who preferred the right hand showed an advantage for the right ear.

It is worth noting that Carter-Saltzman (1979) used a different criterion to classify subjects as belonging to groups with familial sinistrality than did previous investigators. Whereas the latter classified all subjects according to the presence of one *additional* left handed member of the family, Carter-Saltzman divided subjects into families where there was one left hander *including the subject* and families where there was more than one left hander (including the subject). The point here is that under the usual system of classification right handers with one sinistral parent or sibling are classified as belonging to a familial sinistrality group, but when left handers themselves are under consideration there has to be an *additional* left hander in the family for a left handed subject to be classified as having a positive family history of left-handedness. Carter-Saltzman's criterion of familial sinistrality has the merit of applying equally to left and right handers and of distinguishing right handed subjects with two or more left handed relatives from those with only one left handed relative. The value of such a distinction is suggested by Carter-Saltzman's findings (the statistical significance of which is unstated): 1. that left handers who are the only sinistral members in a family show a similar dichotic listening asymmetry (weak REA) as right handers with one sinistral member and 2. that female right handers with two or more sinistral relatives show a somewhat larger REA than those with only one sinistral relative, the reverse being the case among male subjects.

Using conventional criteria, McKeever and Van Deventer (1977b) found that left handed females with at least one sinistral relative, and left handed males without any left handed relatives, showed a dichotic REA; conversely sinistral females without sinistral relatives and sinistral males with left handed relatives showed no significant difference between the two ears. Among right handers a positive family history of left handedness did not significantly affect the REA found in both sexes.

Of the studies discussed above, none can be interpreted in any straightforward way. Consequently, the precise significance of a positive family history of left handedness in relation to hand preference, sex and cerebral lateralisation has yet to be worked out. At the very least, the data discussed so far do suggest that these variables should be separated in future studies of laterality.

**Dichotic listening: non-verbal stimuli**

The study by Piazza (1980) discussed above is noteworthy in that she investigated both verbal and non-verbal auditory asymmetry. When environmental sounds were the stimuli she found that males showed no difference between the ears while females exhibited a significant ear-by-hand interaction: right handers had a significant LEA while left handers showed a significant REA. However, Remington, Krashen and Harshman (1974) mention that Krashen had earlier found greater asymmetry for environmental sounds among males than females, and Gordon (1978b) reports greater asymmetry among male than female musicians in recognition of dichotically presented chords. Thus the issue remains unsettled.

**Tachistoscopic half-field studies: verbal stimuli**

As Bryden (1979) remarked in his review, it is curious that although reports of sex differences in asymmetry found with the dichotic listening technique have come largely from studies involving verbal stimuli, comparable findings using tachistoscopic procedures have emerged from studies using non-verbal stimuli. Nonetheless there are a few reports of a sex difference in visual half-field asymmetry for verbal material.

Hannay and Malone (1976a) presented a single nonsense word vertically followed either immediately or after a delay of 5 or 10 seconds by a second word. The subject's task was to say whether the two 'words' were the same or different. Overall, there was a significant sex-by-half field interaction, with males showing a significant RVF superiority at the two delay intervals (with a non-significant RVF superiority at 0 secs. delay) and females showing no significant asymmetry at any delay interval. The authors interpret their data as suggesting a sex difference in retention of verbal material by the two hemispheres. In a subsequent paper by these same authors (Hannay and Malone, 1976b) it is reported that using the same procedure, right handed females without any sinistral relatives showed a significant RVF superiority at 5 seconds delay and a non-significant RVF advantage at 0 secs. and 10 secs. delay. Females with sinistral relatives had non-significant RVF advantages at all delay intervals.

If the above results can be interpreted as showing reduced visual hemifield asymmetry in females with at least one left handed relative, then they are supported by the findings of Andrews (1977) who found that the greater the degree of familial left handedness, the lower the right half-field advantage. A rank order correlation coefficient (Kendall's tau), computed between familial sinistrality and visual field asymmetry, was higher for females than for males. This possibly indicates that the influence of familial left handedness is greater in females. It is interesting, therefore, that Annett (1978) found higher heritability estimates of hand preference for mothers and their children than for fathers. She also obtained a significant correlation only between mothers and their daughters, and not between other parent-offspring or sibling combinations, in the time-difference between left and right hands on a peg-moving task. This significant correlation was found only for families in which the father was right handed and the mother left handed and not for families where both parents were dextral or both were sinistral. In short, the influence of sinistrality, according to Annett's data, may be greater in mothers than in fathers and may show up more in daughters than in sons. It should be mentioned, however, that Annett herself is cautious about viewing her isolated finding as due to anything other than a volunteer bias in her subjects.

The finding of Hannay and Malone (1976a), that females show reduced visual field asymmetry on a verbal task in comparison with males, finds support in a study by Bradshaw, Gates and Nettleton (1977) who measured reaction time to respond in a lexical decision task (deciding whether a stimulus item such as 'FALE' is a real word or not). However, the results of a later study (Bradshaw and Gates, 1978) suggest that the reduced laterality effect in females may be a transient phenomenon. Analysis of manual reaction time scores during the first half of the experiment revealed a non-significant difference between the two half-fields for females and a significant right hemifield superiority for males. During the second half of the trials, females showed a significant right field advantage comparable in magnitude to that shown by males. To explain their data Bradshaw and Gates speculate that speech mechanisms in the right hemisphere act in an 'auxiliary capacity' for females when material is initially unfamiliar. It is not clear from their paper whether this 'auxiliary capacity' is shut off when material is no longer unfamiliar (i.e. during the second half of the trials) or whether left hemisphere mechanisms dramatically 'improve' over trials, thus giving rise to an asymmetry where none previously existed. Inspection of their figures, however, reveals that performance improves for both visual fields, the right hemifield more so than the left.

It is interesting that when Bradshaw and Gates required their subjects to name words presented on the tachistoscope, rather than perform a discriminative manual response, females from the outset of the experiment showed the same right visual field superiority, though of slightly reduced magnitude, as males. This suggests that the sex difference in manual reaction time relates to decisions made at a lexical rather than articulatory level.

The finding that males show greater visual field asymmetry (cf. also Kail and Seigel, 1978; Levy and Reid, 1976) for verbal material has not been universal. Some studies have failed to find a sex-by-laterality interaction (Kershner and Jeng, 1972; Hannay and Boyer, 1978; Leehey, Carey, Diamond and Cahn, 1978; Piazza, 1980) and at least two have found greater asymmetry in right handed females than males (McKeever and Van Deventer, 1977b; Bryden, 1979). Clearly, it would be premature, given the tachostoscopic evidence presently available, to conclude that one sex shows greater laterality effects than the other where verbal stimuli are concerned.

## Tachistoscopic half-field studies: non-verbal stimuli

One of the most commonly employed, purportedly non-verbal, tasks for which data have been analysed with regard to sex differences is recognition of faces. With discriminative reaction time as the response measure Rizzolatti and Buchtel (1977) obtained a LVF superiority only for males and not for females. With accuracy of identification as the dependent variable no sex difference was found by Patterson and Bradshaw (1975), Finlay and French (1978), Leehey et al. (1977), Hannay and Rogers (1979) or Piazza (1980).

Jones (1980) required subjects to classify photographs of faces as either male or female. Response latencies for male right handers and non-familial left handers showed a RVF superiority on this task, male familial left handers showed a LVF superiority and females showed no asymmetry in any handedness sub-group. Contrariwise, in an experiment by Ladavas, Umiltá and Ricci-Britti (1980) subjects had to respond as quickly as possible to a target emotion portrayed in a photograph of a face. Response times for females, but not males, were faster for LVF presentations. This greater sensitivity of females to left hemifield presentations may reflect

lateralised processes specific to emotion (see Chapter 11). The sex difference would be consistent with the view that women are more sensitive to
non-verbal than verbal cues of emotion (Crouch, 1976) and exhibit greater
facial muscle activity than men in response to affective imagery (Schwartz,
Brown and Ahern, 1980). On the other hand, Graves, Landis and
Goodglass (1981) found for a lexical decision task involving both 'emotional' and 'non-emotional' words that males, not females, exhibited a significant field-by-word category interaction. This was due to the greater
recognition by men of 'emotional' than 'non-emotional' words in the LVF
but not in the RVF. However, inspection of the stimuli used by Graves et
al. suggests that the 'emotional' words differ from the 'neutral' words along
dimensions (e.g. ease of pronunciation, age of acquisition) other than
emotionality. Their report is also marred by the fact that although response
times were apparently recorded, no latency data are presented by the
authors. In view of a possible speed versus accuracy trade-off this is a
surprising omission.

A second type of supposedly spatial task that has given rise to significant
sex-by-laterality effects involves detection, enumeration or localisation of
dot stimuli. In one experiment Kimura (1969) obtained a significant LVF
superiority among males but not females on a dot localisation task (in
which the subject had to indicate on a response card the position where a
dot had been presented). However, the significant sex-by-field interaction
did not emerge in three subsequent experiments. Bryden (1979) also failed
to find a sex-by-hemifield interaction in any of his experiments.

McGlone and Davidson (1973) report that when subjects had to estimate
how many dots had been presented, males showed a left field superiority
while females were as likely to show an advantage for the right as for the
left field. Kimura (1967) used only female subjects but obtained overall a
significant LVF superiority in dot enumeration.

Levy and Reid (1976) used the dot localisation task with subjects
categorised according to the position of their hand while writing. On the
basis of differential field performance on this task and on a syllable
identification task Levy and Reid argued that left handers who wrote with
an inverted (hooked) hand posture had language functions represented in
the ipsilateral (i.e. left) cerebral hemisphere and visuo-spatial functions in
the contralateral hemisphere. Conversely, left handers who wrote with a
'normal' hand posture were said to have linguistic functions subserved by
the right hemisphere and visuo-spatial functions mediated by the left
hemisphere. On the dot localisation task, and to a lesser extent on the
syllable identification task, females in the right handed-normal and left
handed-inverted groups showed smaller field differences than males,
whereas females in the left hand-normal group (and a single right-inverted
female) showed field differences of similar magnitude to those of males.

Davidoff (1977) observed a left hemifield superiority in dot detection
among males but not females and Bryden (1979) quotes a study by himself

and collaborators in which detection of a faint line embedded in visual noise also yielded a clear LVF superiority in males. Females were as likely to show an advantage for one visual field as the other and the sex-by-hemifield interaction was highly significant. Sasanuma and Kobayashi (1978) also found that men but not women gave a LVF superiority in judging line orientation. Overall, there seems to be a consensus of opinion that, as a group, females show smaller perceptual laterality effects than do males on certain tachistoscopic non-verbal tasks. However, the data of McGlone and Davidson (1973) and Bryden (1979) suggest that the scores of females may well be bi-modally distributed, one peak showing a left hemifield superiority and the second peak showing a right hemifield superiority. Averaging data across all female subjects would then show up as zero mean asymmetry between the fields.

## Summary of dichotic listening and tachistoscopic half-field studies

This review of the evidence reveals that a larger number of studies report laterality effects to be greater in men than in women as compared with the number of studies reporting the reverse to be the case. It would be unwise, however, to accept this as establishing that females are less lateralised than males, although there appears to be a feeling in the literature that this is so. Birkett (1980), for example, in finding that females, but not males, showed a RVF superiority in tachistoscopic recognition of random forms, argued that this was consistent with other evidence suggesting that females are more likely than males to use verbal (i.e. left hemisphere based) encoding strategies to solve 'visuospatial' problems. He raises the point that this may be 'forced' on females by the 'fact' of their having language more bilaterally represented than men, thus pre-empting some of the neural space in the right hemisphere that otherwise would be available to subserve visuospatial processing.

The problem, of course, is that of the 'catch-all' explanation: an advantage for one hemifield or the other can be attributed either to a hemispheric specialisation effect and/or to the preferential use of a particular strategy, depending upon one's theoretical preconceptions. Certainly, differential strategy effects can help to explain some of the embarrassing inconsistency of results in the literature and have been invoked by a number of investigators to explain their findings.

Hannay (1976) found that, as a group, females showed no significant hemifield asymmetry where males had previously shown a RVF advantage. Classifying subjects according to their scores on the Block Design subtest of the WAIS, however, revealed that those with high scores, implying a well-developed visuospatial ability, showed a LVF advantage while those with low Block Design scores gave a RVF superiority. The suggestion might be made that those low in visuospatial ability preferred, for that reason, the use of a verbal strategy, leading to the RVF advantage, while those high in visuospatial skill capitalised on that fact in adopting a right-hemisphere based strategy leading to a LVF superiority.

Birkett (1978) has reported that among male subjects there was no overall asymmetry in delayed recognition of nonsense shapes but that subjects at either extreme of the laterality index (i.e. showing strong left or right hemifield superiority) were more accurate overall than subjects with no pronounced laterality effect. This he interpreted as showing that the use of a consistent strategy, be it verbal or non-verbal, was more advantageous than that of no particular strategy at all.

There is evidence that females are slower than males when subjects are required to imagine the 'image' evoked by a word but are faster when the task can only be performed on a verbal basis (Metzger and Antes, 1976). This is consistent with the view that females are better at using verbal strategies and will use them preferentially in experimental tasks. However, in the experiment by Metzger and Antes (1976) visual hemifield differences were similar for men and women regardless of the strategy that subjects were required to adopt. Thus, when strategy effects are controlled for, sex differences in lateralisation may not be apparent. This implies that where sex-by-hemifield interactions *are* found, they may be due to uncontrolled strategy differences between the sexes (see also Bryden, 1978; Segalowitz and Stewart, 1979).

## Electrophysiological studies

There are few reports of sex differences in laterality effects in the EEG literature. Ray, Morrell, Frediani and Tucker (1976) and Trotman and Hammond (1979) and Herron (1980) adduced evidence for the view that males are more lateralised than females in terms of verbal/numerical tasks on the one hand (left hemisphere) and visuospatial tasks on the other (right hemisphere). Tucker (1976) supports this view in as much as he found evidence among his male but not female subjects for right hemisphere specialisation on a spatial task requiring a synthetic approach (Mooney's Closure Test). Beaumont, Mayes and Rugg (1978) report increased interhemispheric coherence (i.e. increased similarity of waveform across the two sides of the brain) for females compared to males during performance on a spatial task. They see this as consistent with the notion that females are less lateralised than males in their cerebral organisation of function.

Other investigators have reported findings in the opposite direction, namely that females showed greater laterality effects than males (Davidson, Schwartz, Pugush and Fromfield, 1976; Davidson and Schwartz, 1976; Rebert and Mahoney, 1978). Still other researchers report no significant sex-by-laterality interactions (e.g. Molfese, 1978).

## Dual task performance

The original finding of Kinsbourne and Cook (1971) that concurrent verbalisation decreased the time for which a dowel rod was balanced on the right index finger but not on the left has been replicated by Lomas and Kimura (1976) and Johnson and Kozma (1977). Both the latter studies obtained the effect for female subjects but not for male subjects. (The sex composition of the subjects in the original study was not specified in the report.) These findings may indicate that there is a greater overlap between the neural systems underlying speech and motor control of the right hand in males than females. However, neither Briggs (1975) nor Hicks, Provenzano and Ribstein (1975) found a sex difference in their studies of dual task performance.

## Lateral eye movements

Beveridge and Hicks (1976) asked subjects a number of 'reflective' questions and noted the direction of the first lateral eye movement. Out of 16 subjects classified as right movers, 10 were male: 17 out of 22 left movers were female (chi-square significant). However, the questions designed to elicit eye movements could be considered a mixture of both verbal and spatial questions and the subjects comprised 21 right handers and 17 left handers (classified by self-report). Thus the effects of handedness, sex and question type are totally confounded. Lefevre, Stark, Lambert and Genese (1977) recorded eye movements during dichotic listening tasks. For females, the numbers of initial right and left movements for verbal and non-verbal tasks were fairly evenly balanced but for males there was an excess of left movements. The authors do not explain how this excess came about but, however caused, it contradicts the findings of Beveridge and Hicks (1976).

Ray, Georgiou and Ravizza (1979) classified subjects as left or right movers according to the direction of the largest number of glances made during interpretation of a number of proverbs. They then asked subjects to indicate, by drawing, the water level in a tilted jar – a Piagetian task. Although the subjects were university undergraduates a considerable number failed to show the level as horizontal with respect to the ground. Among females who performed the task correctly, 23 were left movers and 8 were right movers; among females who failed there were equal numbers (27) of left and right movers. The difference in proportions is significant. Among males there were equal numbers of subjects who passed and failed and equal numbers of left and right movers in each group. These data could indicate a difference in approach between the sexes to the task, females being more right hemisphere oriented.

## SEX DIFFERENCES IN HANDEDNESS DISTRIBUTION

It is commonly found in large scale analyses of hand preference that there is a higher incidence of left handedness among males than among females (Oldfield, 1971; Thompson and Marsh, 1976; Hardyck, Petrinovich and Goldman, 1976; Bryden, 1977; Loo and Schneider, 1979). The reason for this is unclear. It may stem merely from a reluctance of females to transgress sociocultural norms or it could be due to a higher proportion of pathological left handedness among males. Taylor (1969) has argued that during the second (but not the first) year of life the inception rate of febrile convulsions (calculated retrospectively among adult epileptics) is greater for males than for females. Furthermore, he believes that left-sided lesions are comparatively common in the first two years of life and comparatively rare thereafter while right sided lesions are equally common throughout the first four years. If it is permissible to extrapolate from these data for adult epileptics then the raised incidence of sinistrality in males can be explained. This explanation rests heavily upon the notion of pathological left handedness. It is of interest, therefore, that Bishop (1980a) has gathered data from which she has been able to extend the notion of pathological left handedness to school children without obvious pathology.

There is, however, an alternative explanation of the sex difference in manual asymmetry which does not rely upon arguments from brain damaged individuals being applied to the normal population. Annett (1979) reports that when differences between the hands in peg-moving times are plotted as a function of frequency, the distribution for females is shifted in the direction of relatively faster times for the right hand in comparison with the distribution for males.

The fact that the distributions for both males and females are shifted to the right of zero mean difference between left and right hands is referred to by Annett as the 'right shift'. She has argued that this 'right shift' is under the control of a single gene which induces left hemisphere representation of speech and incidentally biases hand preference towards dextrality (see Chapter 2). Annett's data imply that the 'right shift' is more effective in females than males (Annett, 1978). Annett has suggested that this may relate to the earlier development of language in girls and their relative immunity compared with boys to developmental language disorders (Annett, 1979). If the 'right shift' is more effective in females than males this might be taken to entail greater left hemisphere specialisation for language rather than decreased asymmetry as suggested by much of the literature. Clearly, the issue hangs on precisely what it is that the gene underlying the 'right shift' is thought to control. Earlier language development in girls than boys is not logically incompatible with reduced lateralisation in girls. Yet it would be strange if a gene considered to bias the degree of manual asymmetry shown by females were not thought also to influence their degree of cerebral asymmetry, given that the gene is held responsible for the direction of language lateralisation. Arguing along these lines, one would expect a more effective 'right shift' to go with increased, possibly earlier, language lateralisation in females.


## ONTOGENY AND SEX DIFFERENCES

Whatever laterality differences there may be between men and women, it is conceivable that lateralisation in younger males and females proceeds (if it changes at all, see Chapter 7) at different rates and/or from different starting points. If hemispheric asymmetry of function relates to level of cognitive ability, then the different developmental rates of boys and girls would lead us to expect differences in degree of lateralisation in boys and girls of the same age. At least as far as language is concerned, however, there is no *direct* evidence that language is represented differently in the brains of boys and girls. Hécaen (1976) does not discuss sex differences in his review of childhood aphasia; nor does Woods (1980) distinguish between male and female cases in his study of the effects of very early cerebral lesions on subsequent verbal and performance Intelligence Test scores. It can probably be assumed, therefore, that sex differences were not conspicuous in the data of these investigators. Nonetheless, the possibility remains open of a more subtle sex difference in brain laterality.


### Sex differences in infant asymmetry

In his electrophysiological studies with infants Molfese has found differentiated waveforms over the two hemispheres in response to clicks and 'linguistic' stimuli. The auditory evoked response suggests finer auditory discrimination in the right hemisphere of infant (Molfese and Nunez, 1976; Moltese and Molfese, 1979) and pre-school (Molfese and Hess, 1978) females compared to males. This raises the possibility that females are more advanced in terms of cerebral lateralisation than are males. On the other hand, Peters and Petrie (1979) collected data which hints at the opposite conclusion, at least for motor aspect of laterality. They looked at the stepping reflex in infants having a mean age of 17 days and observed that there was a significant tendency to lead with the right leg. Female neonates were rather less likely to be laterally differentiated in this respect than were males although the sex difference was not statistically significant.


### Studies with older children

*Dichotic listening*
The question of whether lateralisation does or does not develop with age is highly contentious (see Chapter 7). To attempt to examine sex differences within a developmental context is doubly problematic, the findings being inconsistent.

Some findings suggest that a dichotic right ear advantage is found earlier in boys than girls. Nagafuchi (1970) obtained a significant ear difference for boys but not girls aged 3 to 6 years and Geffner and Dorman (1976) failed to find a significant asymmetry in 4 year old girls although present in boys at that age. By contrast, other results suggest earlier lateralisation in girls. Kimura (1963b) found a significant REA in boys and girls from 5 to 8 years old except in boys of the youngest age. Similarly Piazza (1977) found a significant right ear superiority in both sexes at 5 years of age and in girls of 3 years and boys of 4 years old. Pizzamiglio and Checchine (1971) report a significant REA in girls but not boys at 5 years of age.

In addition to the above findings, apparent 'anomalies' not infrequently arise. For example, Kimura (1963b) found a right ear superiority in children of all ages from 4 to 9 years *except* 7 and 9 year old girls and Ingham (1975b) obtained a significant REA in both sexes from 3 to 5 years of age *except* 4 year old girls. As well as these 'anomalies' many studies report no sex-by-ear interaction (e.g. Knox and Kimura, 1970; Geffner and Hochberg, 1971; Satz, Bakker, Teunissen, Goebel and Van der Vlugt, 1975; Borowy and Goebel, 1976; Hynd and Obrzut, 1977; Schulman and Galambos, 1977; Kinsbourne and Hiscock, 1977; Geffen, 1978; Bakker, Hoefkens and Ver der Vlugt, 1979; Hiscock and Kinsbourne, 1980a; Eling, Marshall and Van Galen, 1981; Davidoff, Done and Scully, 1981).

Although it is generally considered more difficult to obtain significant differences between the ears using monaural stimulation, at least two experiments have done so using children as subjects. Both reported a sex-related effect. Van Duyne and Scanlon (1974) asked 5 year olds to attend to instructions presented in a particular ear (e.g. 'When the blue/red light comes on press the round/square block'). There was no main effect of sex or ear but a significant interaction emerged due to the tendency of females with left ear instructions to make more errors. The authors saw this result as supporting suggestions of earlier left hemisphere speech lateralisation in girls than in boys. Van Duyne and Sass (1979) also found girls but not boys to show a difference between the ears, favouring the right ear in interpreting logical statements.

*Tachistoscopic half-field studies*
Sex differences in visual field asymmetry among children have been rarely reported although several authors have specifically analysed their data according to sex of subject (Marcel and Rajan, 1975; Yeni-Komshian, Isenberg and Goldberg, 1975; Young and Ellis, 1976; Kershner, Thomas and Callaway, 1977; Silverberg, Bentin, Gaziel, Obler and Albert, 1979). Marcel, Katz and Smith (1974) found a relatively greater difference between the two fields in boys than girls aged 7–8 years old. However, in this study exposure durations were individually determined for each subject and boys had on average longer exposure times than girls, which may have influenced the degree of asymmetry observed.

With regard to non-verbal stimulus material, sex difference have not often been found (Marcel and Rajan, 1975; Davidoff, 1977; Piazza, 977; Young and Bion, 1981). In one study, however, Young and Bion (1980) found a LVF superiority in recognition of unfamiliar faces among males aged 7 to 13 years but no asymmetry in girls. With a smaller stimulus set both sexes gave a LVF superiority. Young and Bion (1979) also found a significant visual field difference in enumeration of dots in boys aged 5–11 years but not in girls. This sex difference was consistent with results reported by Kimura (1966) for adult males and females.

*Tactile asymmetry*
Using a procedure analogous to that of dichotic listening, Witelson (1976a) compared recognition by left and right hands of meaningless shapes in 200 school children, 25 of each sex at ages 6, 8, 10 and 12 years. She found accuracy to be significantly greater for the left hand in boys but not girls. All age groups and both sexes showed a right ear superiority on a verbal dichotic listening task. This finding was taken by Witelson to indicate relatively greater right hemisphere specialisation in boys for non-verbal tactile tasks. Rudel, Denckla and Hirsch (1977) did not find a left hand superiority in discrimination of Braille configurations by blindfold, sighted subjects until after 10 years of age, boys showing this asymmetry at an earlier age than girls. (In adults, more females than males showed an advantage for the left hand). Prior to the age of 10 years, there was a right hand advantage. Wagner (1977), cited by Harris (1980), found that learning of Braille configurations gave a 'strong' left hand superiority in female college students and in younger boys although the age-by sex-by hand interaction was not statistically significant. Thus her results are consistent to some degree with those of Rudel et al.

The findings for Braille together with those of Witelson, who used meaningless shapes, are more or less consistent in finding an earlier or stronger left hand superiority in males despite the fact that reading of Braille involves a verbal component – i.e. attaching a label to the configuration of dots.

It may be that the combination of verbal and non-verbal components of the task can explain the rather complicated age-related effects found by Rudel et al. (1977) (see also Rudel, Denkla and Salten, 1974) since differential reliance on verbal and spatial strategies at different ages may lead to differing hand superiorities. Sex effects may have a similar explanation. Cioffi and Kandel (1979) reported that the tactile identification of pairs of letters not making a word showed a right hand superiority in girls and a left hand superiority in boys of the same age. For letter pairs making up a word, however, there was a significant right hand advantage for both sexes. Unlike Witelson (1976a) and Rudel et al. (1977), left hand superiority in tactile identification was not found at an earlier age in boys than girls by Cioffi and Kandel (1979).

Although investigators have tended to view the left hand advantage as reflecting right hemisphere specialisation for non-verbal skills, it may be that the use of a dichaptic procedure (i.e. simultaneous use of both left and right hands) introduces a strategy whereby the subject prefers to focus attention in a left-right manner, analogous to scanning of written material. Partial, cued reports as advocated by Bryden (1978) for use with dichotic and tachistoscopic procedures, would be as applicable to the tactile modality and for the same reasons. The failure of Cioffi and Kandel to find an earlier left hand advantage in boys than in girls may say something about the development of a 'tactile scanning' strategy in the other studies quoted above rather than about hemispheric specialisation of function.

## THEORIES OF SEX DIFFERENCES IN LATERALISATION

It has often been asserted that men are better at spatial tasks than women and that women outdo men on verbal tasks (see Jaccoby and Jacklin, (1974) and McGee, (1979) for reviews and Fairweather (1976) for a dissenting opinion). Since these types of task depend to some extent upon opposite halves of the brain the question arises as to whether a difference in lateralisation between men and women might correlate in some way, possibly causally, with the putative difference in cognitive skills.

Buffery and Gray (1972) argued that females are lateralised to a greater degree than are males both for verbal and for spatial skills. Thus, contrary to the view currently prevailing, Buffery and Gray see females as more laterally specialised than males. They postulate that a more strictly unilateral hemispheric substrate is optional for linguistic and verbal skills but is not ideal for spatial skills, often involving integration of information across both halves of the visual field, for which a bilateral substrate is more efficient. Hence women are superior in verbal skills, men in spatial skills. Levy (1969; 1972; 1974a) claims that, as a group, left handers do worse than right handers on perceptuo-cognitive tasks mediated by the right hemisphere. She argues that this is due to the greater bilateralisation of speech among some left-handers which entails encroachment of left hemisphere mechanisms on right hemisphere territory. Thus if it were true that women do more poorly than men on visuospatial tasks, and if degree of lateralisation is relevant to the degree of ability attained, Levy would expect women to be less laterally specialised than men, at least for verbal tasks. Consistent with this expectation, Levy and Reid (1978) found females to show reduced field differences compared with males on a tachistoscopic task. The supposed inferiority of women on spatial tasks, on this view, would be due to the reduced cortical space available in the right hemisphere for executing non-verbal tasks.

The Buffery-Gray and Levy hypotheses are opposed to one another. Levy's hypothesis as it relates to left handers is discussed more fully in Chapter 10, where it is seen to falter on both logical and empirical grounds. However, this does not necessarily mean that the Buffery-Gray hypothesis should be accepted by default. Their thesis is based upon the fact that girls mature earlier than boys so that, it has sometimes been claimed, girls should show evidence of lateralisation of function at an earlier age than boys. Buffery and Gray (1972) discuss dichotic listening and other studies which show a right ear or right hemifield advantage for verbal stimuli at an earlier age in girls than in boys. However, as discussed above, the perceptual asymmetries found are inconsistent and could as easily result from strategy effects which may be entirely unrelated to any asymmetry of *cerebral* function. It is on extrapolation from this evidence obtained with children that Buffery and Gray suggest that the right hemisphere of the female adult brain is freer to subserve non-verbal functions than is the right hemisphere of the male.

This review of the literature on sex effects in asymmetry shows only a few studies supporting the Buffery-Gray hypothesis. Thus there is little empirical support for their position. Even if it were the case that girls lateralise earlier than boys, it would not necessarily follow that the degree of lateralisation finally achieved is greater in women than in men.

## BIOCHEMICAL ASPECTS OF SEXUAL DIFFERENTIATION

If there *are* sex differences in hemispheric lateralisation, the question arises as to whether the sex hormones are influential in determining the rate or level of lateralisation achieved. Harris (1978) discusses the interaction of genetic, neurological and hormonal factors influencing the expression of spatial ability, and the reader is referred to his paper for a detailed consideration of the evidence. Although he believes there is good evidence that spatial abilities are 1. related to a sex-linked genetic factor, 2. influenced by sex hormones and 3. that males and females are differentially lateralised he does not explicitly consider the relationship between these three factors.

Waber (1976) carried out a test of dichotic phoneme identification in 40 males and 40 females who were classified as early or as late maturers on the basis of physical characteristics such as body hair. Among adolescents of both sexes, late maturers showed a greater degree of asymmetry than early maturers, suggesting that neurohumoral factors influence lateralisation. If so, one might expect sex differences in performance asymmetries to become particularly marked around puberty. This does not appear to be so but researchers tend to concentrate on pre-pubertal children or on adults of college age. A longitudinally designed research programme is required to investigate this issue.

Levy and Levy (1978) state that a colleague (Reid) has found that, regardless of handedness, in 5 year old children the left hemisphere of boys and the right hemisphere of girls is the 'more developed' half of the brain. Unfortunately, they do not explain the sense in which they use this term. Be that as it may, on the basis of Reid's results Levy and Levy hypothesised that high levels of sex steroids present in males are associated with enhancement of the left side. Levy and Levy measured feet sizes in 98 females and 52 males. They found that in dextral males the right foot was larger and in dextral females the left foot was larger. Among non-right handed subjects, males had larger left feet and females larger right feet. These differences were found in subjects under 6 years old as well as in adults. Thus Levy and Levy argued that since Reid's findings for children showed no relationship of handedness to the more developed cerebral hemisphere their own results imply that the effects of sex steroids on genes are mediated through different mechanisms for head (brain) and body regions.

It may be, of course, that sex effects are influenced either directly or indirectly by different levels of neurotransmitter substance which may in turn depend upon the action of sex hormones. Glick, Schonfeld and Strumpf (1980) discuss preliminary evidence from their laboratory that male and female rats taken from the same litter differ in rotation behaviour. Since spontaneous turning behaviour in rats is related to the presence of an asymmetry in nigro-striatal dopamine asymmetry (Glick, Jerussi and Zimmerberg, 1977) it is of considerable interest that male and female mice have shown differences in degree of asymmetry in this region of the brain whenever Glick and his colleagues have looked for it. Are rats too far removed from men and women to suggest that human males and females, if they do differ in hemispheric functional lateralisation, differ also in asymmetries of brain biochemistry?

## SUMMARY

There have been suggestions in the literature that females show reduced cerebral lateralisation of function in comparison with males. However, this review of the evidence from a variety of sources reveals that a convincing case for differential lateralisation of function in males and females has yet to be made out. This is not to argue that differences between the sexes in cerebral organisation do not exist but rather that the data at present are equivocal. This applies as much to the findings of studies carried out with brain damaged patients as to experiments conducted with neurologically intact subjects. With regard to the latter, a tendency for males and females to differ in their reliance on verbal and non-verbal modes of processing could account for results showing different degrees of left-right asymmetry between the sexes.

While the proposition that males and females differ in cerebral lateralisation has only equivocal support there is more consistent evidence that the sexes differ in manual preference. The handedness distribution for females is shifted in a dextral direction relative to that for males. Annett's theory attributes this to a more effective expression of the 'right shift' gene in females. This might be thought to entail greater lateralisation of language in the female brain. If so, this contrasts with the view currently prevailing in the literature.

# 9

# Hemispheric asymmetry and reading disability

There have been suggestions that cerebral specialisation is less well developed in subjects who are severely illiterate, since aphasia resulting from right hemisphere lesions has been reported to be more common among such individuals than in the population at large (Cameron, Currier and Haerer, 1971; Wechsler, 1976). This has, however, been disputed (Damasio, Castro-Calda, Grosso and Ferro, 1976) and data from dichotic listening (Tzvaras, Kaprinis and Satzoyas, 1981) and tachistoscopic (Davidoff, Beaton, Donne and Booth, 1982) studies offer no support for the view that adult illiterates are less lateralised than competent readers. Nonetheless, the idea that deficient reading skills in children are in some way bound up with anomalies of manual or cerebral laterality is one that has persistently appeared in the literature during the past forty years or so.

Among children who fail to learn to read and write at the normal age there are some whose failure may be attributed to environmental factors, some who are handicapped by a low level of intelligence and others who have suffered organic brain damage. In addition, there are some children whose failure to acquire normal reading and writing skills may be due to none of these factors but to a specific constitutional disability (Zangwill, 1962; Critchley, 1970; Naidoo, 1972; Miles, 1978). The term specific reading disability or developmental dyslexia has been applied to the condition of this group of disabled readers, although it must be pointed out that the idea of a constitutionally determined disability is controversial. Some people believe there is no such condition; others are equally convinced that to deny the existence of specific or developmental dyslexia is to fly in the face of a compelling body of evidence.

Some authors have noted strong similarities between developmental dyslexia and certain forms of dyslexia acquired as a consequence of known brain damage (see Jorm, 1979 and rejoinder by Ellis, 1979; Aaron, Baxter and Lucenti, 1980) but the fact that children with developmental dyslexia tend to have a family history of backwardness in reading (Naidoo, 1972; Yule and Rutter, 1976) suggests that systematic rather than accidental influences are involved. A recent study found that first degree relatives of developmental dyslexics show a performance profile on a battery of neuropsychological tests that is similar to that shown by dyslexics themselves but not by normal readers (Gordon, 1980b). This supports the view that developmental dyslexia is of constitutional origin.

## ANATOMICAL INVESTIGATIONS OF DYSLEXIA

Suggestions that there may be some underlying structural abnormality of the brain in dyslexia have been made from time to time but until recently direct evidence on this point has been lacking. Such evidence as is now available is by no means compelling.

Hier, Le May, Rosenberger and Perlo (1978) examined the CAT scans of 24 adult dyslexics (aged 14–47 years, mean = 25 years) and noted a higher than expected frequency of reversals of left-right asymmetry in the parieto-occipital region. Previous research (see Chapter 6) had shown that this region is wider on the left than the right side in the majority of normal cases, with relatively few reversals of this pattern. The relatively high frequency (10 out of 24) of reversed asymmetry among dyslexics reported by Hier et al. was not confirmed by a different group of investigators using the same methods of measurement with a sample of 26 dyslexic boys (mean age = 11.7 years) only 3 of whom showed a reversal (Haslam, Dalby, Johns and Rademaker, 1981). On the other hand, an unusually high frequency of symmetric brains (11 out of the 26) were noted in this latter study. The difference in the results of the two investigations cannot be attributed to the difference in the ages of the patients studied since neonates show the same asymmetry as normal adults (see Chapter 6). There is thus some measure of agreement that dyslexia may be associated with an attenuation of the usual right-left asymmetry in the parieto-occipital region. On its own, this cannot, of course, be considered a sufficient cause of dyslexia since reversed asymmetry is also found among normal readers.

Recently, a careful cytoarchitectonic study (Galaburda and Eidelberg, 1982) was carried out on the brain of an adult (left handed) dyslexic who died at the age of 20 years as the result of an accidental fall from a great height. Given the patient's age, it can be assumed that there was no gross pathological development of the brain prior to his death. Nonetheless, it should be pointed out that he developed seizures at the age of 16 years although the findings of neurologic and EEG investigations were essentially normal. Microscopic examination of the brain post mortem revealed certain abnormalities in the posterior temporoparietal area of the left hemisphere and bilaterally in the medial geniculate and lateralis posterior nuclei of the thalamus. Since there are connections between these regions of the thalamus and temporoparietal cortex, which on the left side is implicated in language functions, it was suggested that the cytoarchitectural abnormalities played a part in the patient's dyslexia. It remains to be seen whether these findings are confirmed in other dyslexics.


## CEREBRAL DOMINANCE AND DYSLEXIA

The idea of an association between specific reading disability and incomplete cerebral dominance was first popularised by Orton (1937) and variations of his theory have been promulgated ever since.

Orton's theory was in part designed to explain the confusion between mirror image letters (e.g. b and d) experienced by children learning to read. He assumed that a letter is recorded in the 'correct' orientation in the 'dominant' hemisphere and as a mirror image in the 'non-dominant' hemisphere. If a child failed to learn to suppress the 'activity' in the non-dominant hemisphere then this would compete with the dominant hemisphere, thus leading to left-right confusion of letters, a situation for which Orton coined the term 'strephosymbolia'.

There are no grounds whatsoever for belief in Orton's theory that symbols are recorded in a mirror image fashion in the two hemispheres. It is quite clearly the case that symbols must be *perceived* in the correct orientation in both hemispheres and Orton seems to have recognised this fact in arguing that it is the *memory* of the image that is recorded in a mirror image fashion in the non-dominant hemisphere. But, as Corballis and Beale (1976) point out, how can perceptual input be veridical yet leave a memory trace that is reversed?


## LATERAL PREFERENCE AND READING

At the time when Orton proposed his theory, one cerebral hemisphere was thought to be dominant in the sense of exerting control over its partner. He claimed that failure to establish complete dominance was reflected in mixed handedness or crossed hand-eye preference (in which the preferred hand and preferred eye are on opposite sides of the body) as well as resulting in left-right confusion. Orton believed that this was why such a large number of children referred to him for reading problems had crossed hand-eye preference or were mixed handers.

The literature concerning a relationship between hand or hand/eye preference and reading retardation is thoroughly confused, some authors (e.g. Harris, 1953) reporting a higher incidence of so-called anomalous lateral preference among retarded readers, others finding no particular association (e.g. Belmont and Birch, 1965; Sparrow and Satz, 1970; Clark, 1970; Bishop, Jancey and Steel, 1979; Richardson and Firlej, 1979). Reviewers appear to have become disillusioned (e.g. Vernon, 1957) with the plethora of discordant findings. Corballis and Beale (1976) and Satz

(1976) deal only briefly with the issue; Critchley (1970) pays it rather more attention. Beaumont and Rugg (1978) admit a 'potential link between left handedness, cerebral organisation and reading disability' without specifying their own data base while Naylor (1980) relies on Benton's (1975) excellent review of the dyslexic syndrome. Benton concluded that although

> for the most part, recent studies... have reported no important differences between normal and retarded readers... many of these essentially negative studies (and this is true of the earlier literature as well) do find a weak trend in the direction of a higher frequency of deviant lateral organisation in poor readers.

There are two conspicuous reasons for the lack of agreement in findings. Firstly, the measurement of hand and eye preference has been, until recently, somewhat haphazard and inadequate and normative data on hand-eye correspondence has been lacking. Secondly, the definition of reading retardation has been vague and inconsistent.

There are no universally agreed criteria for the diagnosis of specific dyslexia and any population of poor readers is therefore likely to be heterogeneous with respect to aetiology and symptoms. A relationship between laterality and reading may only exist among certain types of disabled readers (Boder, 1973; Fried, 1979) or for certain levels of reading impairment. Zangwill (1962) wondered whether there might be two sorts of developmental dyslexia, one type occurring in 'poorly' lateralised individuals and frequently associated with delayed speech development, clumsiness and other signs, and a second type occurring in unequivocal dextrals who show no associated deficits.

Benton's (1975) review of the literature led him to conclude that 'an impressively high frequency of anomalies is sometimes reported in studies of clinically diagnosed dyslexic children' as compared with studies of children drawn from the general school population. However, the question of a possible bias in referral arises here. The climate of opinion has until recently favoured the notion that left or mixed handedness is in some way associated with reading disability. Consequently, left or mixed handed children may have been referred for specialist attention in cases where they might not have been referred had they been fully right handed.

A possible bias of another kind concerns the sex composition of the samples studied. There is general agreement that reading difficulties are more common in males than in females. If left-handedness is also more common in males (see Chapter 2) then it would hardly be surprising if in many investigations there were more males than females in the reading retarded group and hence an apparent association with left handedness. Researchers have often failed to specify the sex-breakdown of their samples.

Quite apart from inconsistent definitions of handedness, heterogeneity of the populations studied and, in some investigations, relatively small sample sizes, differences in the method of analysis of data probably constitute a source of discrepancy in the findings relating laterality and reading ability. Annett and Turner (1974) collected data on cognitive ability, reading quotients and laterality among school children and analysed their data in two ways. When children were classified in terms of their handedness no significant differences in intellectual or reading ability between different handedness groups emerged. However, when children were selected for specific reading disability, defined as a reading quotient 30 points below their vocabulary IQ, a marked and significant excess of sinistrals was found in this group. Thus, whether a relationship was found between reading and laterality depended upon the method of analysis and the definition of reading disability.

## HAND POSTURE DURING WRITING

Allen and Wellman (1980) observed 177 right handed children aged 6.5 to 12 years and noted the position of each child's hand while he or she was writing. The majority of children held their hand in the normal upright position but some inverted their hand such that they wrote with a hooked posture (see Chapter 2). Levy and Reid (1976, 1978) have suggested that the inverted posture is indicative of ipsilateral cerebral control for writing which, in dextrals, means the right hemisphere. Since Allen and Wellman noted a decline in the frequency of inversion of the writing hand with age this could be taken to imply increasing left hemisphere specialisation for written language skills, although evidence concerning the validity of hand posture as an index of language lateralisation is far from conclusive (see Chapter 5). Allen and Wellman reported that children who wrote with the normal hand posture achieved higher reading scores than those who wrote with the hooked posture. In order to control for the possibility that this merely reflects the greater maturation of those who wrote with the normal hand posture an analysis of covariance was carried out with age as the covariate. Reading scores of the children writing in the normal position were still higher than those for children adopting an inverted hand posture. The implication is that right-sided language representation is detrimental to reading.

With the advent of the dichotic listening and tachistoscopic hemifield techniques interest has largely passed from the study of handedness to the ostensibly more direct investigation of cerebral laterality in relation to reading disability.

## DIVIDED VISUAL FIELD STUDIES

### Verbal stimuli

The first published study to utilise the tachistoscopic half-field paradigm with disabled readers was that of McKeever and Huling (1970b). They presented four-letter nouns unilaterally to both normal and reading disabled subjects and found a significant RVF superiority for both groups with no group-by-hemifield interaction. Similar results were reported for 10 year old children by Keefe and Swinney (1979) and, for children aged 10–15 years, by Bouma and Legein (1977, 1980) although the latter did not analyse their data statistically.

In contrast to the above findings of no difference in hemifield asymmetry between normal and disabled readers a number of experimenters have reported finding reduced visual hemifield differences in the reading disabled group. Marcel, Katz and Smith (1974) and Marcel and Rajan (1975) used unilateral presentation of five-letter words followed by a visual pattern mask with children aged 7 to 9 years. In both experiments a RVF superiority in word recognition was obtained but the absolute difference between left and right hemifields was significantly greater for the good readers. It should be pointed out, however, that Marcel and his co-workers used a threshold procedure to establish the exposure duration used for each subject and the mean duration of exposure was much longer for the disabled than for the normal readers. This was also found to be the case by Gross, Rothenberg, Schottenfeld and Drake (1978) who used duration threshold as their dependent variable. These workers found significantly elevated thresholds in the right compared with left visual field in reading impaired children aged 10–13.5 years but not in normal control subjects of the same age.

McKeever and Van Deventer (1975) used a similar method to that of Marcel and co-workers to investigate tachistoscopic hemifield performance among rather older subjects (aged 12–19 years) referred to a university reading clinic. A small group of poor readers (N = 9) and control subjects (N = 9) showed similar visual half-field differences in vocal reaction time to numbers and letters, but overall the disabled readers were slower. In terms of recognition accuracy the poor readers showed significantly reduced asymmetry in recognition of unilaterally presented words but with bilateral word presentation both groups gave RVF superiorities which were of comparable magnitude.

Reduced hemifield asymmetry for four-letter words presented bilaterally to 19-year-old poor readers was reported by Kershner (1977, 1979) and by Pirozollo and Rayner (1979). The latter investigators presented words both unilaterally and bilaterally and in each case obtained a significant RVF superiority in good readers but not poor readers of mean age 12.5 years. The interaction between subject group and visual field was statistically significant each time. The reverse pattern of results was obtained by Yeni-Komshian, Isenberg and Goldberg (1975) who employed a rather unusual tachistoscopic procedure and vertical presentation of their stimuli. They found a RVF superiority among disabled readers but no significant asymmetry among control subjects. This latter finding must invalidate their procedure since they failed to show that their task was sensitive to the presumed asymmetry in normal readers. Their conclusion that hemispheric lateralisation is greater in the reading disabled than in the general population is therefore clearly unjustified. Witelson (1976b) using a same-different letter matching task also failed to find a significant hemifield asymmetry among control subjects, so her finding of no asymmetry in dyslexic boys is similarly of no consequence.

Two patterns of results emerge from the studies reviewed. Some experiments show reduced visual field differences in poor compared with good readers while others show no difference in hemifield asymmetry between the two groups. There is no obvious procedural difference, such as unilateral versus bilateral presentation, to explain the grouping together of the different studies, and the details provided about subjects are usually too sparse to discover whether any difference in diagnostic criteria is responsible. In any event, those studies using 'clinically disabled' readers (McKeever and Van Deventer, 1975; Witelson, 1976b; Bouma and Legein, 1977; 1980; Pirozollo and Rayner, 1979) do not stand out as in any way distinctive in their pattern of findings. Neither does consideration of the ages of subjects investigated suggest that this is a likely source of discrepancy between those studies which do, and those which do not, find a difference in asymmetry between normal and poor readers.

**Non-verbal stimuli**

As well as investigating word recognition in left and right hemifields of readers some investigators have examined non-verbal functions. Marcel and Rajan (1975) found equal LVF superiorities for face recognition in both normal and poor readers. These same two groups of subjects differed in the magnitude of asymmetry which they showed for word recognition. There was no association between the extent of hemifield asymmetry obtained for the verbal and face recognition tasks. Marcel and Rajan therefore concluded that hemispheric specialisation for verbal processing is unrelated to that for visuospatial functions.

A superiority of the left hemifield in face recognition by disabled and normal readers was also reported by Pirozollo and Rayner (1979). On a word recognition task the disabled readers revealed no hemifield asymmetry though a significant RVF advantage emerged for the control subjects. These findings again indicate a lack of association between asymmetries for verbal and non-verbal tasks.

Witelson (1977a,b) used a set of 'unfamiliar human figures' in a same-different matching task. A left hemifield superiority was found for normal but not for dyslexic boys who showed no hemifield difference. This finding helped to confirm Witelson's view that male dyslexics have atpyical bilateral cerebral representation of spatial functions (see below).

## CRITIQUE OF DIVIDED VISUAL FIELD STUDIES

The divided visual field studies have been critically reviewed by Young and Ellis (1981). These authors emphasise what they see as being four critical principles. The first involves the control of subjects' fixation. It was pointed out that the tachistoscopic half-field paradigm relies crucially on the maintenance of central fixation during stimulus exposure but that

> for a variety of possible reasons children who are poor readers may be less likely than normal readers to fixate properly when instructed and . . . some control is necessary to establish that differences between normal and poor readers are not simply due to differences in fixation.

Reviewing the same studies as the present writer, Young and Ellis concluded that only the experiments of McKeever and Huling (1970b); Olson (1973); McKeever and Van Deventer (1975); Gross, Rothenberg, Schottenfeld and Drake (1978) and Keefe and Swinney (1979) adequately controlled fixation. To these should be added the experiments of Bouma and Legein (1977, 1980) which were not mentioned by Young and Ellis. Pirozollo and Rayner (1979) also presented a central fixation digit at the same time as their word stimuli. Although these authors do not mention any more than this fact, it seems fairer to conclude that they adequately monitored fixation than that they did not. These particular studies, however, themselves reflect the inconsistency in results shown by the divided-field literature as a whole.

Ignoring the studies of Yeni-Komshian, Isenberg and Goldberg (1975) and Witelson (1976), which failed to find the expected hemifield superiority in normal readers, it is noteworthy that the remaining studies criticised by Young and Ellis for lack of fixation control all report reduced asymmetry in poor compared with normal readers. This perhaps vindicates the claim of Young and Ellis that poor readers cannot be relied upon to fixate as instructed! If poor readers have a tendency to look to one side of the fixation point as often as the other, perhaps in an attempt to anticipate the side of exposure on unilateral trials, or so as not to ignore a particular side on bilateral trials, the end result would be a lack of hemifield asymmetry. There might also be a reduced level of performance overall as a consequence of incorrect anticipations on unilateral trials, and virtually certain failure to perceive both words on bilateral trials. Both lower and normal levels of overall performance by dyslexics have been reported in the studies dismissed by Young and Ellis as methodologically unsound. As a further criticism of some of these studies it might be mentioned that the use (e.g. Marcel, Katz and Smith, 1974; Marcel and Rajan, 1975; Kershner, 1977) of absolute difference scores, rather than a laterality index, to compare different groups of subjects in terms of their asymmetry is highly questionable (see Chapter 4).

The second principle stated by Young and Ellis is that before a difference in hemifield asymmetry between normal and poor readers can be thought relevant to any hypothesised difference in cerebral organisation it must be established that the same types of stimuli were recognised by subjects in the two groups. There is evidence (see Chapters 5 and 7) that whereas abstract low-imagery words usually give rise to a RVF superiority on tachistoscopic tasks, concrete, highly imageable words yield little or no visual field asymmetry. The explanation usually offered for this is that the right hemisphere can process high-imagery, concrete words but not abstract words. The implication for research with poor and good readers is that, generally speaking, poor readers are better at coping with concrete than abstract words. In any study reporting a reduced hemifield asymmetry in poor compared with normal readers it is clearly important to ascertain whether the reduced asymmetry can be attributed to the poor readers recognising only concrete words. Unless only words of one type are used or a separate analysis is carried out of the number of abstract and concrete words recognised, the suspicion inevitably arises that this factor might have contributed to the results of the experiment. Young and Ellis conclude from their review that only the studies of McKeever and Huling (1970b) and McKeever and Van Deventer (1975) used homogeneous stimuli. Interestingly, these authors also utilised proper fixation control and found significant RVF superiorities in poor as well as good readers.

Young and Ellis's third point is that good and poor readers might differ in the extent to which they employ different cognitive strategies. Words can be identified by means of a phonological code, whereby a word is decomposed into its constituent sounds or by directly visual means. There is evidence (Cohen and Freeman, 1978; Coltheart, 1980b) that the first method is almost exclusively a left-hemisphere function whereas the second may be less clearly lateralised. If good and poor readers differ in their reliance on these two methods (see below) then a difference between the two groups in magnitude of hemifield asymmetry may relate to their preference for one code as opposed to the other, rather than to a difference in cerebral organisation of function. This possibility has not been considered in the literature to date. However, since good readers tend to rely on visual methods to a greater extent than poor readers, who rely more on a phonological code, one might expect good readers to show less lateral asymmetry than poor readers. The evidence suggests that the opposite is the case.

The final point of Young and Ellis's review concerns the control subjects with whom disabled readers are usually compared. It is usual to match disabled readers with good readers of the same chronological age. The problem is that it is not possible to distinguish whether any difference between them is due to their differing levels of reading ability per se and/or to any specific defect (dyslexia?) in the disabled group. What is required is to match disabled readers with controls of the same reading ability as well as with the usual chronological-age controls. At the time of writing the only published laterality experiment to employ both reading-age and chronological-age controls is one in which the present writer was concerned

(Davidoff, Beaton, Done and Booth, 1982). This was a study in which adults who had not learned to read as children were compared with normal literate adults of the same chronological age and with children of the same reading age. Fixation was controlled by presentation of a central digit at the same time as a 3-letter word or nonsense syllable appeared in one or other hemifield. Only trials on which the digit was correctly reported were accepted. Words and nonsense syllables were vertically aligned in an attempt to minimise the putative effect of left-right scanning processes (see Chapter 4). A superiority of the RVF was found for all stimuli and groups of subjects and the magnitude of this asymmetry did not differ when overall level of accuracy was taken into account. These results suggest that, at least for the relatively simple task and stimuli employed, neither reading age per se nor chronological age seriously affects visual hemifield asymmetry. However, the same may not be true of more complex reading tasks nor for horizontally aligned stimuli.

A point not mentioned by Young and Ellis is that it is desirable to match reading disabled and control subjects for general intellectual ability. A difficulty in using full scale IQ scores for this purpose is that disabled readers often show a considerably lower verbal than performance IQ (Naidoo, 1972). Although reading ability in normal schoolchildren was found in one study to be unrelated to discrepancy in verbal-performance IQ (Bishop and Butterworth, 1980) matching normal and disabled readers on the basis of Full Scale IQ (verbal plus performance IQ) probably does not adequately control for the cognitive skills most relevant to the task. Matching on the basis of either verbal or performance IQ alone are the alternatives.

There is a further criticism of the divided visual field studies. If the object of the exercise is to assess hemispheric specialisation for language there is something rather odd about using a technique which demands the very skills in which disabled readers are, by definition, deficient. Furthermore, tachistoscopic hemifield asymmetry may be influenced by post-perceptual scanning or order of report strategies (see Chapter 4). It is reasonable to expect that the effectiveness of such strategies, be they covert or overt, will relate to level of reading experience or proficiency. Olson (1973) found that older (10–11 years) reading disabled children showed a significant word recognition asymmetry, favouring the RVF under both unilateral and bilateral conditions of presentation. In contrast, younger disabled readers (8–9 years) did not show any significant asymmetry under either condition. This finding was seen in terms of the 'maturational lag hypothesis' (see below) but might equally as well be taken to indicate that strategies of information extraction vary with reading experience.

Even if the influence of directional scanning on hemifield asymmetry is relatively small (Bradshaw, Nettleton and Taylor, 1981a) any differential effect in normal and poor readers confounds interpretation of hemifield difference scores in terms of cerebral lateralisation of function. This is particularly relevant where differences between poor and good readers are found, if at all, only in magnitude and not direction of asymmetry. Marcel, Katz and Smith (1974) tried to take account of this possibility by examining serial-position effects in their data. They concluded that since there were no differences between their good and poor readers in left-to-right scanning, as revealed by comparative accuracy for letters occupying different serial positions in their stimulus displays, it was justifiable to infer differences in cerebral lateralisation for language between their two groups of subjects. However, Davidoff, Beaton, Done and Booth (1982) report markedly different serial position curves for normal adult readers on the one hand and semi-literate adults and children on the other. Whereas the data for semi-literates and children showed decreasing letter recall as a function of serial position, the curves for normal adult readers were generally bow-shaped, as has usually been reported to be the case for horizontally aligned letters (Coltheart, 1972). Davidoff et al. analysed the data from their experiment in terms of order of report but this could not explain the serial position effects, which were therefore attributed to the way in which information was extracted from the stimulus displays. Serial position interacted significantly with visual hemifield as it did in an experiment by Butler and Miller (1979). There is thus a clear possibility that in certain circumstances differences in tachistoscopic hemifield asymmetry found between subjects of different reading ability may relate more to strategies of information extraction than to cerebral lateralisation of function.

A final criticism of the divided visual field studies is that authors show a surprising lack of sophistication in attempting to understand the cognitive processes which are deficient in cases of reading retardation (Naylor, 1980). Given the dubious choice of the tachistoscopic procedure for determining cerebral language laterality in individuals with deficient reading skills, experimenters might argue in riposte that they are not so much interested in cerebral lateralisation in general as in hemispheric specialisation for visual-verbal language in particular, since it is in this regard that dyslexics are impaired. But even accepting the principle that specialisation for visuo-verbal processes is the proper object of study, investigators have shown little or no interest in delineating precisely those processes which are abnormally lateralised. There is a very large literature concerning word recognition and reading generally yet the concepts and techniques developed by workers in this area are hardly ever applied to laterality research. If it is thought valuable to try to identify the hemisphere or pattern of cerebral organisation 'reponsible' for certain types of reading disability, an effort to establish the processes concerned is even more worthwhile. It is not enough to be satisfied with a simple statement that words or letters are recognised more readily in this or that visual field by this or that group of subjects. One wants to know precisely what factors gave rise to the observed hemifield superiority.

## DICHOTIC LISTENING STUDIES

Sparrow and Satz (1970) compared the performance of 40 reading disabled boys (aged 9–12 years) on a dichotic digits task with that of 40 normal readers matched for performance IQ, age, sex and socio-economic status. They found a mean asymmetry favouring the right ear in both groups with no ear-by-group interaction. Similar results were reported for children in the age range 8–12 years by Bryden (1970), Yeni-Komshian, Isenberg and Goldberg (1975), Abigail and Johnson (1976), Leong (1976) and Keefe and Swinney (1979). A significant REA was not found for dyslexics by Zurif and Carson (1970) but they also failed to demonstrate a significant ear difference in their control subjects.

In contrast to the above findings some authors have found reduced ear asymmetry in dyslexic children. Satz, Rardin and Ross (1971) found a significant REA for both disabled and normal readers but the magnitude of this effect was greater for normal readers aged 11–12 years than for disabled readers of the same age. For younger children (7–8 years) there was no difference in the size of the ear asymmetry between normal and control readers. These findings were held to support Satz and Sparrow's (1970) maturational-lag hypothesis of developmental dyslexia which is discussed below.

Satz (1976) refers to a study by Darby which found a REA in both reading-disabled and control children at all ages from 5–12 years but the ear asymmetry was statistically significant only for 12-year-old control subjects. For the normal readers the trend was for the size of the REA to increase from age 7 to age 12, but for the reading-disabled group this trend, though present, was attenuated. These results point up the need to take into account developmental aspects of dichotic listening (and tachistoscopic hemifield) performance (see Chapter 7). This need was recognised by Bakker and his associates (Bakker, 1973; Bakker, Smink and Reitsma, 1973; Bakker, Teunissen and Bosch, 1976) who, in general, may be said to have found an uncertain or conflicting relationship between ear asymmetry (monaural as well as dichotic) and reading ability among younger children but evidence that at older ages (beyond 10 years or so) proficient reading is associated with a strong right ear advantage. Their data are explicated in more detail below.

Several experimenters have used children drawn from clinics for the educationally disabled. Witelson and Rabinovitch (1972) gave a dichotic digits task to children aged 8–13 years who were diagnosed as having 'auditory linguistic defects'. Although not a criterion for inclusion in the group all children were retarded in reading by 1 to 4 years. Control subjects gave a REA, the clinical group a LEA, on the task. These ear effects were not statistically significant but the group-by-ear interaction did attain significance and the mean difference between ears was significantly greater for the clinic group. Even if this is accepted as indicating a shift away from left hemisphere superiority for the task it is not clear to what extent this relates to reading disability per se rather than to the auditory deficit.

Witelson (1977a) tested children aged 6–14 years referred to a reading clinic and found a significant REA in this group as in controls on the dichotic digits task. The magnitude of asymmetry was equivalent in the two groups with no significant ear-by-age interaction. Thompson (1976) used children aged 9–12 years referred to a clinic for 'dyslexic type language difficulties' and found on a number of dichotic tasks with verbal stimuli that the clinic sample generally showed less asymmetry than control children. Among a small group of older disabled readers (12–19 years) attending a university reading clinic McKeever and Van Deventer (1975) found a significant REA for dichotically presented digits. The magnitude of asymmetry was similar to that found for control subjects.

Studies with subjects drawn from clinic populations, then, mirror the inconsistency of results obtained with school populations.


## CRITIQUE OF DICHOTIC LISTENING STUDIES

An excellent review of the dichotic listening studies was provided by Satz (1976). He concludes that most experiments have shown a REA for verbal material among disabled as well as normal readers and that those studies that have not done so are marred by procedural artefacts. What is in doubt is the magnitude of the asymmetry in each group of readers and how this relates to age. Satz does not discuss the point but it is clear that the suggestion that disabled readers should be compared with children of similar reading age as well as with those of the same chronological age (Young and Ellis, 1981) is as applicable here as it is to divided field studies.

In view of the unresolved issues with regard to interpretation of the magnitude of any ear asymmetry it is regrettable that most authors have chosen to present their experimental results in terms of mean asymmetry scores for their different groups of subjects. This is particularly unfortunate since there are indications that the variability in scores is higher among disabled than normal readers (Sparrow and Satz, 1970; Keefe and Swinney, 1979). Despite this, an analysis of the relative frequencies of individuals showing left or right ear superiorities has rarely been carried out. Sparrow and Satz (1970), however, did perform such an analysis and found a significantly higher proportion of subjects with a LEA in their reading-disabled than in their normal control group. Mean asymmetry scores did not differ between the groups. Bryden (1970) also undertook a frequency analysis but found no difference in the distribution of left and right ear superiorities among readers of different levels of ability. Whatever the outcome of such an analysis, the inferential problem concerning the interpretation to be placed on a dichotic left ear superiority (Satz, 1977) still remains.

In so far as any conclusion can be drawn at all from the studies outlined above it must be that there is no compelling reason to suppose that auditory language lateralisation is markedly different in normal and poor readers. More elaborate conclusions must await the outcome of experiments exercising tighter methodological control and more sophisticated theoretical analysis.

## ELECTROPHYSIOLOGICAL INVESTIGATIONS AND DYSLEXIA

Although a fairly large number of electrophysiological studies have been carried out with dyslexic and other learning disabled subjects (e.g. Sklar, Hanley and Simons, 1972; Symann-Lovett, Gascon, Matsumiya and Lombroso, 1977) they do not appear to have contributed greatly to our understanding of dyslexia (Benton, 1975; Beaumont and Rugg, 1978). From the point of view of the present chapter, they have largely failed to provide evidence relevant to the issue of left-right hemispheric differences (but see Fried, Tanguay, Boder, Doubleday and Greensite, 1981). For reviews of the literature the reader is referred to Benton (1975), Callaway (1975) and Beaumont and Rugg (1978).

## MODELS OF READING DISABILITY IN RELATION TO HEMISPHERIC FUNCTION

### Satz and Sparrow's developmental-lag hypothesis

Satz and Sparrow (1970) developed a theory of reading disability which postulates that lateralisation of functions to left and right hemispheres progresses with age and that the rate of lateralisation is slower in reading disabled children than in normal readers. The prediction that degree of REA would be greater for normal readers than for disabled readers of the same age was confirmed for children aged 11–12 years, but not for those aged 7–8 years, by Satz, Rardin and Ross (1971). However, the fact that the magnitude of ear asymmetry did not differ for normal and reading-disabled boys at the younger age does not necessarily count against the theory, since it is not clear that the functions of the specific task utilised (dichotic digits) should be any more lateralised in normal than disabled readers during the early years. This emphasises the point that developmental norms must be established before conclusions can be drawn on the basis of comparisons between different groups of readers at a given age. As was seen in Chapter 7 it is not universally accepted that lateralisation does increase with age.

### Bakker's balance model

Bakker, Smink and Reitsma (1973) argued that because formal reading instruction begins before lateralisation of speech is fully complete (but see Chapter 7) efficient reading at early ages, depending as it does on non-linguistic operations, is compatible with incomplete hemispheric lateralisation of function. At later ages, however, complete lateralisation is optimal for reading proficiently, which in the more mature reader relies heavily on deductions made from the linguistic context. Bakker et al. therefore predicted that ear asymmetry would be minimal among the best readers in a group of 7 year olds but maximal in the most proficient readers in a group of 8.5–9.5 year olds. These predictions were confirmed by Sadick and Ginsburg (1978) and for monaural but not for dichotic stimulation by Bakker et al. (1973). The monaural findings are compatible with Satz and Sparrow's maturation lag hypothesis. However, Bakker et al. also state that large REA scores were associated with poor reading at younger ages and with fluent reading at older ages. Although this was explained in terms of an inadequate use of non-linguistic strategies Bakker and associates subsequently revised their position.

Bakker, Teunissen and Bosch (1976) claimed that reading proficiency was associated with a REA in 5th and 6th grade boys and in 3rd grade girls and with either a LEA or REA in 3rd grade boys and in 2nd grade girls. It was therefore suggested that at the early ages of learning to read both hemispheres are 'functional' but at later stages the left hemisphere only is crucial. Some children being with a dominant right hemisphere, others a dominant left. Bakker (1979) claims that left-dominant children read rapidly but make many visual errors while right-dominant children read slowly and accurately. He argues that for efficient reading, 'left-dominants' require to utilise right hemisphere based perceptual strategies while 'right-dominants' need to adopt the faster left-hemisphere strategies which eventually predominate. There is thus some degree of compromise between the hemispheres and so Bakker refers to his theory as a Balance Model. The progression to left hemisphere based strategies, with a concomitant REA, is said to occur earlier in girls than in boys. Given the presumed slower pace of development in boys they are, according to this view, more likely than girls to remain with right hemisphere based strategies and more likely to end up as poor readers, hence the sex difference in the incidence of dyslexia. Thus Bakker and his associates support a modified version of the maturational lag hypothesis.

Tomlinson-Keasey and Kelly (1979) chose, as their measure of hemispheric specialisation, relative speed of response to left and right visual field presentations of a probe word that had to be compared in a same-different task to a previously presented word. Subjects were divided into a left hemisphere group, where responses to right hemifield presentations were 100 milliseconds or more faster than to left hemifield presentations, a right hemisphere group (LVF presentations responded to 100 ms faster) and a no specialisation group (the remainder). It was found for children in the third grade at school (mean age 8.8 years) that those in the 'no specialisation' group were superior to the other two groups in certain verbal abilities but not in reading. The 'right specialisation' group had the lowest scores in all the verbal tests. At the seventh grade (mean age = 13.3 years) the only significant difference between the groups was for mathmatical concepts; again, the 'no specialisation' group was superior.

When the division of subjects into different groups in the study by Tomlinson-Keasey and Kelly was based on response times to pictures, rather than to words, the 'no specialisation' group (third grade) came out worse than the other two groups in spelling but not in reading. In the seventh grade the 'no specialisation' group was found to be best in reading comprehension with the 'left specialisation' group poorest of all. Overall, it was concluded from 'integration of these findings with those of other investigations' that a lack of hemispheric specialisation for processing word stimuli is associated in the early years of school with better reading skills. Left hemisphere specialisation for picture processing, on the other hand, was said to go with poor reading, at least in the 13 year olds. 'Integration of these findings with those of other investigations' is indeed necessary in order to reach such conclusions. Apart from the arbitrary nature of the division of subjects into different groups of so-called specialisation, and the consequent small number of subjects (N = 6, 7 and 8 in the smallest groups), Tomlinson-Keasey and Kelly fail to present full statistical analyses of their data. From the point of view of interpretation, there is a measure of self-deception in arguing that lack of hemispheric specialisation for words is associated with better reading skills, when the data of one's own study quite clearly indicate no significant difference anywhere between the three 'specialisation' groups in levels of reading achievement!

If reading disability is conceived as being due, at least in part, to some lag in development and/or dysfunction of the left hemisphere, then one would expect to find dyslexics exhibiting a range of cognitive deficiencies related to all functions mediated by the left half of the brain, not just those related to reading and writing. There is suggestive evidence that this is the case. Holmes and McKeever (1979) presented subjects with 20 items, one at a time, and asked them to pick out those they had seen before. In a second condition subjects had also to identify those items (words and faces) seen before but this time to pick them out in the same order. Adolescent dyslexics were impaired on the latter task in comparison with college students but only for words, not for faces. It was therefore suggested that dyslexics have a specific disability in serial verbal memory, a view supported by Bakker (1970b) and Done and Miles (1978). Verbal memory defects of various kinds have also been noted by Nelson and Warrington (1980) who point out that all the deficits observed in their dyslexic subjects occurred in functions mediated by the post-rolandic regions of the left hemisphere. Defects in naming of colours (Denckla, 1972) and objects (Denckla and Rudel, 1976) as well as in recoding a visually recognised letter into its letter name (Ellis and Miles, 1978; Bouma and Legein, 1980) can also be referred to the left hemisphere. Furthermore, some but not all disabled readers show additional defects, such as finger agnosia (Sparrow and Satz, 1970; Lindgren, 1978) and/or left-right confusion (Belmont and Birch, 1965; Sparrow and Satz, 1970) which arguably can be seen as signs of left hemisphere dysfunction. (The fact that poor right-left discrimination is associated with reading failure in younger children but not older dyslexics (Benton, 1962; Sparrow and Satz, 1970), particularly in the more severe cases of reading disability (Corballis and Beale, 1976), might be held to support the maturational lag hypothesis.)

## Annett's right-shift theory

It will be remembered from Chapter 2 that Annett views lateralisation of language as being due to the presence of a specific genetic mechanism which confers some linguistic advantage on the left hemisphere and incidentally biases handedness towards dextrality. In the absence of this right-shift (RS+) factor both handedness and cerebral laterality for language occur to left and right according to chance and independently of one another. The extent of the right-shift, that is the magnitude of the effect mediated by the gene, is assumed to differ in different groups of people. The extent of the shift is reduced in twins compared with singletons (Annett, 1975) and exaggerated for females compared with males (Annett, 1979). In a recent study (Annett, 1981) dyslexic children were found to show a smaller mean difference between the hands than control subjects on a peg-moving task. That is to say, the extent of relative dextral superiority was reduced for dyslexics. This suggests a reduced right-shift effect in these subjects which is consistent with the idea of dyslexia as a deficit in the development of language-related skills.

It is important to be clear that attenuation of the right-shift factor cannot, on its own, be the *cause* of dyslexia any more than presence of the RS+ gene is the *cause* of normal language ability. Even in the total absence of the gene (RS− individuals), people learn to speak and use language. The right-shift factor should therefore be thought of not as necessary for adequate language and related skills but as providing some boost to normal language development. As Annett and Turner (1974) point out:

> In the absence of this factor language growth proceeds without the aid of some special facility enjoyed by the majority.

Thus there is no implication that all individuals who lack the right shift factor, or in whom its expression is weaker, are dyslexic. Such individuals are at greater risk but not all of them turn out to be dyslexic.

Annett's theory of a reduced 'right-shift' in dyslexics can be viewed as a modified version of the maturational lag hypothesis. Annett and Turner (1974) point out that this hypothesis must account for three things. First, that the delay depends upon a 'naturally occurring variant of cerebral function' (as opposed to accidental damage to the CNS); second, that the delay is 'fairly specific to language functions' and, third, that it is associated with a slight increase in sinistral tendencies without implying that *all* left and mixed handers are at risk. If Annett's right-shift factor is seen as conferring some advantage in the development and/or attainment of language-related skills then the absence of this factor carries certain implications. These are, one, that there is likely to be a delay in, and/or upper limit to, the acquisition of language skills and, two, since the 'right-shift' factor induces left hemisphere lateralisation of language, in the absence of this factor there would be no consistent cerebral asymmetry for language, which would then be mediated by the left or the right hemisphere according to chance. Since the 'right-shift' factor, when present also incidentally biases handedness towards dextrality, absence or weakness of the RS+ factor eliminates or reduces this bias, resulting in an increased proportion of sinistrals among that section of the population in whom the 'right-shift' factor is missing. It does not, however, mean that all sinistrals lack the RS+ factor. Annett's theory thus neatly copes with the three points which the maturational lag hypothesis is required to accommodate.

Annett believes that a reduced 'right-shift' in twins (Annett, 1975) is consistent with their delay in language development compared with singletons, and that an exaggerated shift in females (Annett, 1979) is consistent with the earlier language acquisition of girls compared with boys. Annett evidently sees the 'right-shift' factor as involving a temporal element. In as much as dyslexia is not a condition which one simply 'grows out of', but is a lasting difficulty, the absence or attenuated influence of the 'right-shift' factor, if it is to account in any way for the dyslexic's difficulties, must be considered to entail some permanent disadvantage. The fact that the handedness distributions for RS+ females and for twins are deemed to be, respectively, more and less to the right of the RS− distributions for singletons and for males (see Chapter 2) also implies that the right-shift factor entails permanent effects, since the handedness data, from which the extent of the shifts were calculated, came from mature adults. Thus unless Annett wishes to invoke a pervasive influence due to the lack of certain events occurring during a critical period of development, she would seem committed to the belief that absence or attenuation of the RS+ factor represents an unremitting biological handicap (as far as language related skills are concerned). In this sense, her theory conflicts with the maturational lag hypothesis, which implies that dyslexic children eventually 'catch up' as the lag in their cerebral development is reduced. The fact that dyslexia persists into adulthood means that 'maturational lag' cannot be the whole story. Annett's theory is an attractive alternative account.

## Witelson's model

Witelson (1976b, 1977a,b) administered a number of tests to right-handed dyslexic and normal children aged 6–14 years. The results of these, taken separately for boys and girls, were as follows.

In a dichotic digits task, dyslexic and normal boys showed an equivalent

REA, but overall level of recall was lower for the dyslexics. On a test of tactile perception designed to be analogous to the dichotic listening procedure (and hence designated a 'dichaptic' test) dyslexic boys showed a significant left hand superiority with letters as the stimuli while control subjects showed a significant right hand superiority. Using non-verbal shapes as stimuli for matching to a visual replica, dyslexics showed no difference between left and right hands, but the normal controls performed significantly better with the left hand, the group-by-hand interaction being statistically significant. The normal readers, but not the dyslexics, also showed a significant left hemifield advantage on a tachistoscopic non-verbal task.

Like their male counterparts, dyslexic and normal girls showed a significant REA on the dichotic digits test, with the dyslexics performing at an overall lower level. Neither the dyslexics nor the control readers showed any significant difference between left and right hands on either of the dichaptic tasks.

On the basis of these findings, Witelson (1977c) concluded that both male and female dyslexics have normal left lateralisation of language but, in view of their low level of accuracy on the dichotic digits test, she argued that they had some degree of dysfunction of the left hemisphere. Naylor (1980) believes that this inference is not justified since the asymmetry between left and right ear scores was similar for dyslexics and controls. However, at certain sites within the left hemisphere a unilateral lesion has been reported to depress performance for both the ipsilateral and contralateral ears (Kimura, 1961a,b; Sparks, Goodglass and Nickel, 1970) so it is at least conceivable that a less severe 'dysfunction' of the left hemisphere may do the same without markedly affecting the *difference* between the two ears.

The results on the dichaptic tasks led Witelson to propose that dyslexic boys have bilateral (as opposed to normal right hemisphere) specialisation for non-verbal spatial functions. Since the performance of dyslexic girls on these tasks was similar to that of the control girls there was no need to postulate an abnormal representation of spatial function in dyslexic girls as in dyslexic boys. Females, whether dyslexic or not, were assumed to have spatial functions represented bilaterally. However, therein lies a paradox. If dyslexics of both sexes are assumed to have spatial functions represented in both cerebral hemispheres, why should this be inimical to efficient reading by boys but not by girls? Witelson attempts to resolve the paradox by suggesting that

> dyslexia in boys is associated not only with bilateral spatial representations, but also with deficient left hemisphere functioning. Normal females have no deficient left hemisphere. (p. 39)

Since dyslexia in girls is, according to Witelson, also associated with a deficient left hemisphere, it may be asked why the representation of spatial function is thought relevant at all. The reason is that Witelson views dyslexia as being associated with a type of cerebral organisation which might be termed two right hemispheres, rather than one left and one right. She believes that dyslexics have a predilection for non-verbal cognitive processing, even where linguistic processing would be more appropriate. Witelson attributes this to the dyslexic's pattern of cerebral organisation. It is the same pattern that is typical of normal girls, but the latter do not get 'locked into' a non-verbal cognitive mode because their left hemisphere, unlike that of both male and female dyslexics, is functioning normally.

In support of Witelson's view there is some independent evidence that reading performance correlates with preferred cognitive strategy and that visuospatial strategies are associated with lower levels of performance. Caplan and Kinsbourne (1981) presented 6–12 year old children with a set of words from which the subject had to choose the odd one out. The choice could be made either on a linguistic basis or on the basis of the shapes specified by the words. The extent of each subject's preferred strategy was measured in terms of the difference between the number of choices of each type. It was found that for the group as a whole, and with age and IQ taken into account, preferred strategy correlated with reading performance. The greater the preference for a verbal mode of selection, the higher was reading performance.

Despite the elegance of Witelson's theory there are several problems concerning interpretation of the data from which it was derived. On the dichaptic non-verbal task, shapes presented to the left hand of male control subjects were reported first more often than shapes represented to the right hand. These subjects gave a left hand superiority on this task. Dyslexic boys showed no asymmetry and were as likely to report first those shapes presented to the right as to the left hand. Surprisingly, Witelson sees the order-of-report data as complementing the accuracy scores but does not consider the possibility that, for normal controls, the bias in order of report actually causes the asymmetry. This could come about if there is a drop in accuracy for the second-reported shape, as has been found for the second-reported ear in some dichotic listening studies (see Chapter 4). For female subjects, there was no asymmetry in accuracy for the two hands and no bias in order of report was observed for either dyslexic girls or controls. On the dichaptic letters task, accuracy for normal boys was significantly greater for the right than the left hand and they also tended to report the right hand shape first although not to a statistically significant degree. Dyslexic boys were significantly more accurate with the left hand and in addition showed a significant bias towards initially reporting the stimulus presented to the left hand. The correlation between order of report and inter-manual asymmetry seems too close to be comfortably ignored.

A problem also arises over the left hand superiority shown by the dyslexic boys on the dichaptic letter task. If dyslexics have normal left hemisphere lateralisation of linguistic processes, why did they not show the same right hand superiority on this task as the controls? The dyslexic boys' 'deficient' left hemisphere was not said to reduce the *degree* of asymmetry shown on the dichotic digits task but to lower their overall performance.

It is of interest that Witelson's suggestion, that visuospatial functions are represented bilaterally in certain male dyslexics, has also been made by Dalby and Gibson (1981) on the basis of results obtained with the dual task technique. Right handed boys aged 9–12 years were required to tap alternately with the first and second fingers of one hand at the same time as repeating aloud animal names (verbal task) or solving items from the Raven's Coloured Progressive Matrices Test (spatial task). The disabled readers were classified as non-specific dyslexics, dysphonetic dyslexics (who make a large number of phonetic errors in spelling) or dyseidetic dyslexics (poor visualisation of known words). The pattern of results with concurrent performance of the tapping and cognitive spatial task suggested that the first two groups of disabled readers had bilaterally represented spatial functions, in contrast to the last group and to normal control subjects, both of whom had spatial functions lateralised to the right hemisphere.

## The Corballis and Beale model

Corballis and Beale (1976) conclude from their review of the relevant literature that

> the evidence ... is at least consistent with the proposition that there is a syndrome characterised by left–right confusion, poorly established lateralisation, and reading disability. (p. 174)

They are careful to point out, however, that

> there are many disabled readers who do not appear to exhibit left–right confusion or anomalies of cerebral lateralisation, just as there are many poorly lateralised individuals who suffer no reading disability.

Corballis and Beale argue that different manifestation of laterality – handedness, eyedness, visual-verbal processing skills, auditory processing and so on – may develop at different times and at different rates. They speculate that in the absence of any inherited predisposition to be lateralised in one particular direction (*pace* Annett's right-shift factor) the direction and degree of different manifestations of laterality may be largely a matter of chance, each developing at random and more or less independently of one another. They suggest that unusually 'symmetrical' individuals comprise a high proportion of dyslexics. Corballis and Beale propose that

> If different aspects of laterality contribute to the development of reading at different stages ... a person (lacking the right-shift factor) might find himself hampered at any one of these stages. Moreover, persons in this category would be especially prone to the confusing influence of crossed lateralisation, perhaps between hand and eye, perhaps between visual and auditory lateralisation in perceptual processing. (p. 176)

Unfortunately, it is not clear from their account precisely why Corballis and Beale believe that crossed lateralisation should constitute a 'confusing influence'. In general terms, these authors believe (Corballis and Beale, 1970; 1976) that for any system to be able to tell left from right, or distinguish between mirror-images of the same object, the system itself must be asymmetrical. The gist of their argument as it applies to dyslexia seems to be that lack of a clear or consistent lateral asymmetry makes for difficulty in tasks involving directional orientation. If so, this assigns a central role in dyslexia to errors of letter-reversal and other manifestations of directional confusion which the authors themselves admit are restricted to a relatively small proportion (about one quarter) of individuals who are poor readers. Nevertheless, Corballis and Beale submit that this proportion of cases represents the incidence of 'true' dyslexia in the population. A cynic might suggest that this disposes of the problem of accounting for the remaining 75 per cent of disabled readers, simply by a definitional sleight of hand.

## The Beaumont and Rugg model

Beaumont and Rugg (1978) briefly reviewed the literature on reading disability and hemisphere dysfunction. They concluded that, on balance, published reports show dyslexics to have a normal degree of right ear advantage on verbal dichotic listening tests but a reduced degree of tachistoscopic hemifield asymmetry for letters and words. They therefore suggest that

> in dyslexics there is a normal lateralisation for auditory language processing, but a relative bilateralisation for visual language processing, which results in difficulties of integration between these two functions. In normals these functions both possess a relatively high degree of left hemisphere specialisation, and can thus be successfully integrated, but the relative bilateralisation of the visual-verbal function in dyslexics may be causally associated with their deficit. (p. 20)

Beaumont and Rugg point to the study of McKeever and Van Deventer (1975) as providing tentative support for their hypothesis. These authors found that in comparison with control subjects poor readers showed normal dichotic listening performance combined with reduced visual field asymmetry.

One implication of Beaumont and Rugg's hypothesis is that the participation of the right hemisphere in visual language processing, as opposed to the more usual left hemisphere participation, makes for a difficulty of integration between left hemisphere auditory language processing and such visual language processing as is undertaken by the right hemisphere. The underlying assumption is that interhemispheric integration is less successful than intra-hemispheric integration. Other authors have also suggested that there is a specific defect of interhemispheric integration in dyslexia. Vellutino, Steger, Harding and Phillips (1975) compared poor and normal readers on tasks of visual-auditory non-verbal learning, visual-verbal (written) and visual non-verbal learning. Poor readers were only impaired in the visual-verbal task and it was suggested that a deficit in integration of left hemisphere verbal with right hemisphere spatial functions was responsible. However, this was only an inference from the data with no direct experimental evidence in support.

To date, no direct test of integration between the hemispheres in the auditory or visual modalities has been undertaken with dyslexics, although there have been tests involving bi-manual motor performance. Badian and Wolff (1977), for example, required subjects to tap in time to a metronome either with one hand alone or alternating between the hands. Disabled readers performed as well as controls in the unimanual condition but not in the bimanual condition. This result was confirmed in a study by Klicpera, Wolff and Drake (1981) who showed that the bimanual defect was not present when subjects had to tap simultaneously rather than alternately with left and right hands. Since there is evidence that co-ordinated manual activity involves interhemispheric integration (Cohen, 1970; Kreuter, Kinsbourne and Trevarthen, 1972; Denckla, 1974; Finlayson, 1976) it was argued that the defect in dyslexics involves a deficiency in co-operation between the hemispheres.

## The Dunlop model

Dunlop and Dunlop (1974) presented a theory to explain confusion between mirror-image letters. Their theory relates to the idea that there is a small strip of the visual field on either side of the vertical midline that is represented in both left and right visual cortices. The image of a stimulus falling within this strip is therefore received in both the left and right hemispheres. Dunlop and Dunlop cite evidence suggesting that the process of interhemispheric transfer is accompanied by mirror-image inversion of the stimulus. They therefore argue that if cross-callosal transfer of a letter occurs from the right hemisphere to the left hemisphere, in order to arrive at language centres, there will be a problem in deciding which 'image' to accept. If some system does not develop such that one image consistently 'dominates' the other, mirror image confusion will result. The theory is, in essentials, similar to that of Orton and is intended to explain what the Dunlops see as a central feature of dyslexia, namely mirror-image reversal of letters. However, Fischer, Liberman and Shankweiler (1978) found that dyslexic children did not make more errors of letter reversal than normal readers, although they did make twice as many errors in a right-to-left direction (e.g. rotating b from right to left gives d) as in a left-to-right (d to b) direction. Normal control subjects showed no directional bias in their errors.

The idea of bilateral cortical representation of a central region of the visual field is one derived from direct neurophysiological evidence in the cat (Blakemore, 1969; Bishop and Henry, 1971) and the monkey (Stone, Leicester and Sherman, 1973) but equivalent data for Man is lacking. Admittedly there is suggestive evidence, particularly the phenomenon of macular sparing, that the same holds true for Man but at least two studies of motor responses to lateralised visual stimuli have not supported the idea of an overlap of the visual and temporal hemifields in Man (Haun, 1978; Harvey, 1978). Similarly, the notion that mirror-image reversal accompanies interhemispheric transfer is, according to Corballis and Beale's (1976) review of the evidence, 'equivocal at best'. Nonetheless, the Dunlops have devised an orthoptic test said to evaluate whether the left or right eye's image predominates in a subject's percept of the central region of the binocular visual field. It is not clear how they see this as relating to the bilateral cortical representation of the area abutting the vertical meridian. However, despite the generally weak theoretical underpinnings of the Dunlops' work they claim that the reference eye and preferred hand are on opposite sides of the body more frequently among dyslexic than normal readers (Dunlop, Dunlop and Fenelon, 1973; Dunlop, 1976). Recently, Bishop, Jancey and Steel (1979) were unable to confirm an association between this form of 'crossed-laterality' and poor reading.

## SUMMARY

This chapter began by reviewing Orton's theory and subsequent studies concerning manual laterality in relation to reading disability. There is considerable confusion in the literature, stemming in part from the inadequate assessments of handedness that have been adopted in many of the studies reported. Nonetheless, it is probably fair to conclude that there is a slightly higher incidence of left and mixed handedness among disabled readers than in the population at large. This is more likely to emerge when it is severely disabled readers that are under consideration rather than the entire range of individuals with reading problems.

The question arises as to whether a trend towards increased sinistrality among severely disabled readers can be explained according to the 'pathological left handedness' model. If so, the nature of the 'pathology' might also explain the difficulties with reading. It is also conceivable, in view of the tendency for reading difficulties to run in families, that a genetic factor is involved.

In recent years interest has largely centred on cerebral rather than manual laterality in relation to reading disability. Attempts to assess hemispheric asymmetry in disabled readers using the divided visual field technique have been dogged by certain methodological problems. There is also the difficulty that the tachistoscopic paradigm may not be appropriate for use with subjects whose difficulties reside precisely in dealing with visually presented verbal material. On the other hand, there may be no reason to suppose that aspects of laterality tapped by dichotic listening or other techniques are in any way unusual in cases of reading disability. This highlights a distinction between those theories that postulate a global impairment in left hemisphere function, or in the development of lateralisation, and those theories which relate to specific functions.

Whatever the nature of the theory under experimental test, it is important to consider the possibility that differences in performance asymmetry between subject groups are due to differences in reporting or processing strategies. This has been conspicuously lacking in experiments reported to date.

A majority of the studies reviewed fail to distinguish between reading retardation per se and difficulties of a genuinely dyslexic nature. While there are no universally agreed criteria, use of the term dyslexia implies the existence of certain kinds of error or areas of difficulty in learning to read and write. Backwardness in reading, on the other hand, implies a developmental lag in the acquisition of reading and writing skills. It is possible that this in turn results from some delay in maturation of the nervous system. However, in the absence of precise information as to the normal developmental pattern of cerebral lateralisation it is difficult to relate any reduced perceptual asymmetry among disabled readers to a maturational lag in hemispheric development. Furthermore, it is possible that anomalous patterns of laterality, if they exist at all in cases of reading difficulty, are found only in certain diagnostic sub-categories. Heterogeneity in the subject populations studied could account for much of the discrepancy in the literature of this field.

# 10

# Laterality of hand and brain
## sinistral versus dextral

The fact that a minority of the population prefers to use the left rather than the right hand has given rise to all manner of speculation. The Latin word for left, *sinister*, is used in English to refer to something which provokes mistrust. The French word for left, *gauche*, is also used pejoratively. 'Left-ness' it seems, carries with it undesirable connotations. However, the view that sinistrality is associated with the dark side of human nature belongs firmly to the realm of folklore. Nevertheless, it may still be the case that left and right handers differ in certain aspects of their psychological make-up. What, if any, is the scientific evidence for differences in mental functioning between left and right handers?

## LEVY'S MODEL

Levy (1969) argued that unilateral hemispheric specialisation for language is biologically adaptive. Hence, she reasoned, individuals without such specialisation should show some cognitive impairment relative to well lateralised individuals. She speculated that those people who have language represented in the right as well as the left hemisphere would be at a disadvantage in those cognitive functions which are the special responsibility of the right half of the brain, since there would be less neural tissue available to subserve those functions. Levy therefore compared a group of right handers with a group of left handers on the grounds that left handers were likely to have some degree of bilateral language representation. Since there is some evidence that scores on the verbal sub-tests of the WAIS are depressed after left hemisphere injury, and scores on the performance sub-tests are decreased after right hemisphere injury (e.g. Anderson, 1951; Reitan, 1955; but see Goldstein, 1974) she hypothesised that, as a group, the left handers would have lower performance IQs than the right handers.

The results supported this hypothesis. The two handedness groups did not differ in verbal IQ but were significantly different in mean performance IQ. More dramatically, the mean discrepancy between verbal and performance IQ for the left handers (25 points) was significantly greater than for the right handers (8 points). However, it has been shown that on the digit sub-test of the WAIS the performance of some subjects is adversely affected by the subject's hand obscuring the symbols, an artefact more common among left handers (Bonier and Hanley, 1971). Since Levy does not provide a breakdown of scores on the individual sub-tests of the WAIS it is possible, as suggested by Gilbert (1977), that the overall difference between the mean performance IQ of left and right handers is largely attributable to an artefactual difference in scores on the digit sub-test.

If less neural 'space' for non-verbal processing means a lower performance IQ then, by the same argument, more neural 'space' for verbal processing should entail a higher verbal IQ. Yet left handers, despite presumed bilateral speech representation, in some cases at least, did not have a significantly higher mean verbal IQ than right handers. At first sight this casts some doubt on the 'neural competition' hypothesis, but since Levy employed only 10 left handed subjects in her study it is unlikely, given the percentage of left handers with bilateral speech (see Chapter 5), that more than 2 of Levy's sinistral subjects did in fact, have speech represented bilaterally.

Levy's results were supported by those of Miller (1971) who compared 29 right handers with 23 mixed handers and found the latter to be inferior in terms of the NIIP Form Relations test but obtained no significant difference between the groups on NIIP Group Test 33, a test of verbal ability.

In the studies of Levy (1969) and Miller (1971) no distinction is drawn between male and female subjects. It is arguable that a sex difference in the composition of the handedness groups can account for the findings. Females are often found to perform more poorly on spatial tasks than males (Maccoby and Jacklin, 1974). An over-representation of females among left handers in these studies might then lead to an apparent effect of handedness. Although females are generally less likely to be strongly sinistral than are males (see Chapter 2) it is possible, given the small numbers employed in these experiments, that female subjects were, in fact, relatively more numerous among the left handers.

The matter of sex differences in relation to cognitive ability is highly contentious. It is no less so when considered in conjunction with handedness. Kocel (1976) reports that while left handed males were superior to right handed males in spatial ability, the reverse held true for women, but Johnson and Harley (1980) failed to find any significant main effect or interaction involving the sex variable. These authors found no difference between right, mixed and left handers on sub-tests of the WAIS but report that sinistrals were significantly superior to dextrals and mixed handers in Millhill Vocabulary scores although inferior on a test of spatial thinking.

Annett (1970b) reports finding a difference in vocabulary scores between consistent left and right handers and those of mixed hand preference; the latter had lower scores than either of the other two groups who did not differ. Mixed handers were also found (by Swanson, Kinsbourne and Horn, 1980) to be inferior to those of consistent hand preference in Thurstone's Primary Mental Abilities Test. It is possible that mixed handers include a high proportion of individuals whose hand preference has shifted from one extreme or other as a result of early brain damage, which would account for their lower scores.

Against reports of a difference between sinistrals and dextrals in certain aspects of intelligence (see also: Flick, 1966; James, Mifferd and Wieland, 1967; Berman, 1971; Gilbert, 1973; Nebes and Briggs, 1974; Hicks and Beveridge, 1978) must be set the negative results of a number of other studies (Gibson, 1973; Newcombe and Ratcliff, 1973; Fagan-Dubin, 1974; Newcombe, Ratcliff, Carrivick, Hiorns, Harrisson and Gibson, 1975; Heim and Watts, 1976; Hardyck, Petrinovich and Goldman, 1976; Fennel, Satz, Van den Abell, Bowers and Thomas, 1978; Carter-Saltzman, 1979b). It is possible that differences between the positive and negative studies relate to the subjects sampled although Fennel et al. argued that the weight of evidence against Levy's hypothesis is such that it should finally be laid to rest (for critique of Fennel et al. see Berenbaum and Harshman, 1980).

Those studies in which positive findings emerged were carried out with highly selected groups of subjects, namely members of university communities, who are unrepresentative of the population at large. However, Teng, Lee, Yang and Chang (1979) compared preference for hand, eye and foot in university students and normal school children (mean age 11 years) and found no differences between the two samples in the distribution of lateral preference. It was therefore concluded that there is no association between laterality and level of scholastic achievement, which implies that the use of highly educated subjects should not bias the findings on IQ in relation to laterality. By contrast, Hicks and Dusek (1980) found a lower proportion of extreme right handers among 578 intellectually gifted schoolchildren, defined as those having Stanford-Binet IQs over 132, in comparison with 391 less 'gifted' peers. This result is something of an anomaly since it is unclear, on theoretical grounds, why extreme dextrals, rather than sinistrals, should be under-represented in those with high IQs.

The negative findings concerning a relationship between laterality and intelligence have generally involved larger numbers of subjects than the studies reporting positive findings. It is nonetheless possible that some of the negative findings may have proved positive had closer attention been paid to certain variables thought to be related in some way or other to bilateralisation of language in sinistrals, such as familial sinistrality and hand posture.

Bradshaw, Nettleton and Taylor (1981b) found only familial sinistrals to have significantly lower WAIS performance IQs relative to dextrals. Similarly, Eme, Stone and Izral (1978) found only familial left handed children to be significantly inferior (1-tail test) to dextrals on Block Design and Object Assembly sub-tests of the WISC. However, Briggs, Nebes and Kinsbourne (1976) reported that the presence of a left handed parent or sibling in the (left handed) subject's family lowered WAIS full scale scores without differentially affecting verbal and performance sub-scales.

With regard to hand posture, Gregory and Paul (1980) reported that left handers with an inverted writing position performed significantly more poorly on WAIS tests of Picture Completion and Vocabulary than either right handers or left handers with an upright writing posture. This is the opposite of what Levy might expect, since it is those left handers who write with an upright posture that are thought by her to have right hemisphere speech. In line with Levy's views, however, McGlone and Davidson (1973) found that only those left handers showing a dichotic left ear advantage (indicative of right hemisphere speech?) were impaired relative to right handers on the Spatial Relations test of Thurstone's Primary Mental Abilities Test. Similar results were not found by McKeever and Van Deventer (1977c).

Levy's hypothesis has been investigated using dependent variables other than IQ scores. Nebes (1971b) reported that mixed and left handers were inferior to right handers on a visuotactile matching task on which split-brain patients showed a clear right hemisphere superiority (Nebes, 1971a). The task involved feeling with one hand an arc of a circle and matching this to the complete circle to which it belonged among a number of alternatives presented visually. Other investigators have used this same task but have failed to replicate the original findings (Hardyck, 1977; Kutas, McCarthy and Donchin, 1975). In an experiment by Thomas and Campos (1978) in which the Nebes arc-circle test was extended to other geometric forms (squares and trapezoids), the overall performance of subjects who were said to be strongly left or strongly right-handed was superior to those whose hand preference was less extreme. This curvilinear relationship was statistically significant, as revealed by multiple regression analysis, suggesting to the authors that it is the degree, rather than the direction, of handedness that is important. As pointed out earlier, the reason for both lowered scores and inconsistent hand preference could be early brain damage.

The majority of the studies reviewed above concern visuospatial ability. Byrne (1974) wondered whether non-visual functions subserved by the right hemisphere would suffer in the way hypothesised by Levy (1969) for visuospatial skills. Rather than compare left and right handers on musical ability, Byrne chose to examine handedness among musicians on the grounds that if musical functions mediated by the right hemisphere are impaired by bilateral language representation then there should be a shortage of left handers among competent musicians. In the event Byrne found no preponderance of right handers among musicians compared with control subjects. This is in agreement with the findings of Oldfield (1969).

In contrast to Levy's view that sinistrality is associated with certain *non-*

*verbal* deficits, some experimenters have found left handers to show a *verbal* deficit in comparison with right handers. Subtle damage to the left hemisphere, but sufficient to cause a shift towards pathological left handedness in some individuals, could explain such findings. Bradshaw and Taylor (1979) found sinistrals, especially non-familial sinistrals, to be significantly slower than dextrals in vocal identification of laterally presented words, and Bradshaw, Gates and Nettleton (1977) reported familial sinistrals to be significantly slower than dextrals in a lexical decision task. However, Sherman, Kulhavy and Burns (1976) found sinistrals to remember highly concrete words better than dextrals. No such difference emerged for abstract words. Since damage to the right temporal lobe impairs learning of concrete but not abstract words (Jones-Gottman and Milner, 1978) the result of Sherman et al. may be interpreted as suggesting that the presence of language bilaterally (or subtle damage to the left hemisphere) does not impair functions mediated by the right temporal lobe.

Among university students Cohen and Freeman (1978) found differences between left and right handers in various aspects of reading. Left handers were slower, but no more efficient in comprehension, than right handers and were more likely to be *relatively* inefficient readers (though still reading perfectly adequately). It was suggested that the slower reading of left handers was due to their greater tendency to include a phonemic stage in reading (that is, processing words in terms of their sounds prior to analysis of meaning). Further, the distribution of left handed subjects' scores for reading proficiency showed two peaks compared with the unimodal distribution for right handers. This suggested to Cohen and Freeman that the left handers in their experiment were of two types, corresponding to left and right hemisphere speech specialisation. They therefore carried out a verbal dichotic listening test with their left handers and found that those who showed a right ear advantage were, on average, significantly faster readers than those showing an advantage for the left ear. The correlation between ear asymmetry and reading 'speed' was statistically significant. Cohen and Freeman next went on to obtain evidence in right handers that phonemic or phonological encoding of words is a specialised function of the left hemisphere, but they do not appear to have looked for differences in extent of phonological encoding between their sinistral subjects with left and right ear advantages.

On a variety of tasks it is right handers that have been reported not to do as well as left handers. Cohen (1972) and Hermann and Van Dyke (1978) found that left handers were significantly faster at judging two visually presented patterns as same or different while Craig (1980) found left handers to be better in tapping out rhythms heard in both ears. In the latter experiment a control task ruled out the possibility that this effect was due to left and right handers using their preferred (and hence opposite) hands for tapping. This implies that the difference found between left and right handers resulted from a central processing difference rather than output factors. Diverse other observations also suggest that left and right handers do not differ simply in terms of speed of motor responses (Kappauf and Yeatman, 1970; Olson and Laxar, 1974; Beaumont, 1974; Deutsch, 1978; Pirozollo and Rayner, 1980). For example, the auditory pitch illusion (Deutsch, 1974) is reported with greater frequency among right than left handers (Craig, 1979a) and left-right confusion is more common among (female) left handers (Harris and Gitterman, 1978). Further, the fact that a family history of sinistrality has been found to relate to performance on such tasks as the Seashore test of musical abilities (Byrne and Sinclair, 1979) as well as such simple sensory functions as the detection of tactile pressure (Fennell, Satz and Wise, 1967) implies that differences between the sinistral and dextral brain go further than motor functions and language.

## HARDYCK'S MODEL

An attempt has been made by Hardyck (1977b) to account for differences between the left and right handed within the same theoretical framework, rather than regarding sinistrals simply as exceptions to whatever rules are held to apply to dextrals. In his view, there is a continuum of cerebral organisation, ranging from extreme lateralisation of function at one end to extreme bilateralisation at the other. The highly lateralised brain consists of two hemispheres, each with their highly specialised functions (although to some extent all functions are represented, however unequally, at the two sides). At the opposite extreme, bilateralisation of function means that all functions are represented more or less equally at the two sides. Between these two extremes there are intermediate degrees of lateralisation.

Hardyck believes that the highly lateralised brain is characteristic of right handers without any left handers in their pedigree, that extreme bilateralisation is characteristic of left handers with a family history of sinistrality, and that right handers with, and left handers without, sinistral relatives fall somewhere in between. He argues that this conceptualisation allows him to make predictions, not only concerning the relative magnitude of interhemispheric differences that will be observed for these three groupings in experiments utilising lateralised presentation, but also of the relative level of performance that will be achieved. In order to make such predictions, however, Hardyck developed certain postulates based largely on the assumption that interhemispheric transfer of information is 'limited' in the highly lateralised brain, but that bilateral organisation is accompanied by 'a high degree of interhemispheric transfer'. There is no experimental evidence relevant to this assumption, which seems to have arisen from the logical difficulty of specifying the nature of the information flow between two hemispheres which, as it were, do not speak the same language. (Hardyck rejects the idea that impulses representing raw sensory data are what transfer. If this were so, he maintains, we should not be able to detect differences in performance – other than reaction time – as a function of the hemifield, or side of stimulation. However, this only follows if callosal transmission is achieved with one hundred per cent fidelity. Under conditions of tachistoscopic presentation and the like, this is unlikely to be the case.)

Hardyck tested his model against data in the literature and reports a close correspondence, in certain instances, between his predictions and the obtained results. In other instances relevant data were lacking. In fact, data concerning the distinction between familial and non-familial sinistrals were lacking in virtually all instances, so his model was effectively tested only as regards left and right handers. As such, it was only a re-statement of the proposition that left handers as a group show reduced perceptual asymmetries in comparison with right handers. Nonetheless, as experimenters increasingly take account of familial handedness as a variable in their studies it will become possible to test the full power of Hardyck's formulation.

Although the majority of reported findings show left handers (as a group) to yield reduced tachistoscopic hemifield differences in comparison with right handers, certain results, reported by Beaumont (1974), show the reverse pattern. In as much as Hardyck's account makes no provision for such reversals it is disconcerting that he does not mention them despite referring to Beaumont's article.

## BEAUMONT'S MODEL

Beaumont (1974) summarises the results of a number of experiments he carried out in collaboration with Dimond. Right and non-right handers were compared in terms of the degree of visual hemifield asymmetry they showed on a variety of visual tasks. Non-right handers showed larger hemifield differences on tasks involving speed of arithmetical calculation, Stroop-type interference, paired-associate learning of digits and symbols, and word association. Right handers showed larger between-field differences of tests of colour-naming, matching of digits and matching of letters between and within the Greek and English alphabets. Beaumont viewed those tasks on which dextrals showed larger hemifield differences as relatively 'high-level' cognitive tasks and those tasks on which non-right handers showed larger between-field differences as relatively 'low-level'. It should be noted, however, that Beaumont provides no theoretical basis for this distinction between different 'levels' of task.

Beaumont argued from his findings that non-right handers are less lateralised for non-verbal visuospatial and perceptual mechanisms as well as for language. He took this idea a stage further in proposing that non-dextrals have a more diffuse intra-hemispheric organisation of function as well as being less lateralised in terms of interhemispheric organisation. He conceptualises the sinistral brain as made up of a large number of small but well-connected 'functional units'. The brain of the right hander, by contrast, is said to be characterised by a smaller number of relatively disparate, less well-connected 'units'. Beaumont considers this conceptuali-sation fits well with the fact that right handers showed smaller hemifield differences than non-right handers on the relatively 'low level' tasks and larger hemifield differences on the 'high-level' tasks.

## HANDEDNESS AND OCCUPATIONAL CHOICE

In view of the reports of differences in ability between sinistrals and dextrals it is of interest that some investigators have suggested that there is an association between handedness and choice of occupation. Although generally unstated, the tacit assumption underlying studies of handedness and occupational choice is presumably that left handers are likely to rely more heavily on the right than the left hemisphere.

Bogen, de Zure, Tenhouten and Marsh (1972) coined the term 'hemi-sphericity' to refer to the tendency for people to rely more on one half of the brain than the other. They argued that in so far as different cultures favour different modes of thought, this concept of hemisphericity could be applied to the different patterns of cognitive performance shown by people of differing cultural backgrounds. Middle-class American whites were found to perform better on the similarities sub-test of the WAIS than on the Street Completion test while Hopi Indians showed the reverse pattern. The similarities sub-test stresses verbal-logical skills while the Street test emphasises visual pattern recognition. Bogen et al. suggested that their findings reflect relatively greater development of left hemisphere mediated abilities in American whites and of right hemisphere skills in Hopis. Since there is evidence that left or right hemisphere processes can be differen-tially activated (e.g. Galin and Ornstein, 1972) it is not unreasonable to suppose that individuals tend to rely more on one half of the brain than the other. The question at issue is whether left handers characteristically employ right hemisphere modes of thought to a greater degree than do right handers.

Peterson and Lansky (1974) claimed to have found a higher than expected proportion of left handers in a sample of student architects. Their estimate of the incidence of left handedness in the general population was based on a consensus of reports in the literature, while their own ad hoc definition of left handedness in their student sample was not obviously related to the same criteria. Since the incidence of sinistrality in the population depends upon the criteria of left handedness adopted it is doubtful that Peterson and Lansky's claim should be taken seriously. In a more recent experiment, however, the same authors found that 251 of 405 right handed architectural students entering college in 1970 eventually graduated compared with 58 out of 79 left handers (Peterson and Lansky 1977). The difference in proportions is statistically significant. As left and right handers were defined in terms of the same set of criteria this finding is rather more convincing than their report of a higher-than-expected frequency of left handers among architects. It might be noted, incidentally, that if Levy (1969) is correct in her theory that sinistrals are impaired relative to dextrals in visuospatial skills then sinistrals should be under-represented among architects.

Differences in cerebral blood flow between student architects and students of English were inferred from the results of an experiment by Dabbs (1980). The architects were said to have greater flow on the right side of the brain, English students on the left side. A greater blood flow on one side suggests increased metabolism, indicative of greater neural activity on that side. The findings of Dabbs therefore support the results of the handedness survey in suggesting that architects preferentially employ a right hemisphere mode of thought.

Rather than compare the incidence of left and right handers in a given sample of people, Jones and Bell (1980) examined the total distribution of handedness scores among psychology and engineering students. The engineers were significantly more dextral in their hand preference than were the psychologists. The latter did not differ from the distribution reported by Oldfield (1971) for undergraduates generally. The total distribution of responses to a handedness questionnaire was also considered by Mebert and Michel (1980) who found a significant shift away from dextrality in the responses of art students compared with other college majors.

The implied inference from demonstrations of an excess of sinistral tendencies among members of certain occupational groups is that such individuals are more likely to utilise right hemisphere processes than are members of other more left hemisphere 'dominated' occupations. There is little to justify this inference. However, one or two attempts have been made using the lateral eye movement paradigm to relate hemisphere 'activation' to occupational characteristics.

Using only a single occupational group, Harnard (1972) found among mathematicians that those who characteristically looked towards the left professed to use more visual imagery in their work, and were rated by peers as being more creative, than right lookers. In a between-groups study Combs, Hoblick, Czarnecki and Kamler (1977) reported that a group of students in language related fields of study showed a larger ratio of right to left lateral eye movements in response to verbal and non-verbal questions than did students of the visual arts or of a variety of other subjects. Galin and Ornstein (1974) compared lawyers and ceramicists but found a difference only in vertical, not lateral, eye movements.

Electrophysiological techniques were used by Doktor and Bloom (1971). These authors presented verbal-analytic or spatial-intuitive problems to 8 management executives and 6 operational researchers. Using asymmetry of alpha-blocking in the EEG as their dependent variable they found that operational researchers, but not the management executives, showed a shift towards greater left hemisphere involvement during the verbal-analytic questions. Arndt and Berger (1978) presented faces and letters in a tachistoscopic reaction time task to lawyers, psychologists and sculptors but found no difference in the pattern of visual hemifield asymmetry shown by those three groups of people.

To summarise the findings on occupational choice and 'hemisphericity,' there is some weak evidence that different occupational groups may differ in the extent to which they rely preferentially on the skills of one hemisphere rather than the other. In the event that left hemisphere functions fail to 'predominate' there may be a tendency to choose those occupations which benefit in some way from a visual mode of thought. The findings suggest that this is sometimes associated with a hand preference that is other than fully dextral. While this may arise from purely natural circumstances, the possibility should be borne in mind that both the trend towards sinistrality and the tendency to utilise right hemisphere skills may arise from very early insult to the left half of the brain. The ranks of architects, artists and the like may, in other words, be swelled by the presence of pathological sinistrals (see Chapter 2).

Not all cognitive differences between left and right handers relate to level of intellectual functioning.

## THE DIMENSION OF FIELD DEPENDENCE-INDEPENDENCE

Field dependence refers to the tendency of an individual to be affected in his judgements by the surroundings of a stimulus display. The Rod-and-Frame Test employs an apparatus consisting of a rod which is free to rotate about a centre pin. The board to which the rod is attached has a square frame and this frame, too, can be rotated. The board is maintained in an upright position in the fronto-parallel plane and the task of the subject is to set the rod to the vertical position. The rod and the frame are luminous, the frame is tilted away from the vertical position and the task is carried out in the dark. Some subjects are strongly influenced in their setting of the rod by the frame being offset from the vertical. Such subjects are termed 'field dependent'. Others are not so affected by the tilt of the frame and set the rod to the vertical regardless of the tilt of the frame. These subjects are referred to as 'field independent' (Witkin, Dyk, Faterson, Goodenough and Karp, 1962). In another test, the results of which correlate with those of the Rod-and-Frame test, a subject is asked to pick out a particular figure, say a triangle, from a stimulus pattern in which this target is embedded. In this Embedded Figures Test subjects who are field independent are readily able to pick out the target item, whereas field dependent subjects are influenced by the presence of the distractor shapes.

In a study by O'Connor and Shaw (1978) handedness, field dependence and EEG coherence were found to be significantly associated, suggesting that the dimension of field dependence-independence may be related in some way to cerebral asymmetry of function. Garrick (1978) has reviewed studies comparing field dependent and independent subjects on a variety of tasks and comments that

> it is readily apparent that there are many similarities between . . . characteristics (of the field-independent individual) and the characteristic mode of functioning of the right hemisphere and between field-dependent characteristics and the left hemisphere mode of functioning. (p. 65)

Silverman, Adevai and McGough (1966) found that left handers as a group performed more poorly than right handers on both the Rod-and-Frame Test and the Embedded Figures Test, though the difference was significant only for the former. Left handers, in other words, were more field dependent, a result replicated by O'Connor and Shaw (1978). So-called ambidextrous subjects were also found by Pizzamiglio (1974) to be more field dependent than right handers.

In an attempt to relate cerebral lateralisation and field dependence more directly Pizzamiglio (1974) considered dichotic ear asymmetry regardless of subjects' handedness. He found that those with a strong preference for one ear, be it left or right, were less field dependent than those with a weak ear advantage. Oltman and Capobianco (1967) reported subjects with less well-established eye dominance to be more dependent than those with a clear dominance of one eye. Similarly, Hoffman and Kagan (1977) found that individuals whose lateral eye movements were consistently in one direction, either left or right, were less dependent than those whose direction of eye movement was less consistent.

Thus far, then, there is reasonable agreement that lack of a strong hand or eye preference, dichotic ear advantage or directional consistency in lateral eye movements is associated with reduced field independence. Whether this entails the conclusion that field independence is related to strong lateralisation at the cerebral level is less clear.

Zoccolotti and Oltman (1978) found, in a study using only male subjects, that those classified as field independent showed a mean RVF superiority in reaction time on a tachistoscopic letter discrimination task and a mean LVF superiority in responding to faces. Field dependent subjects showed no asymmetry for either task. Unfortunately, the interaction between visual hemifield and subject group was not tested statistically. This omission was rectified in a later study (Pizzamiglio and Zoccolotti, 1981) in which the earlier findings were replicated for subjects of both sexes. Thus a clear visual field asymmetry was associated with field independence. In another experiment, degree of field independence was found to correlate signifi-cantly with the extent of left hemifield advantage in a face matching task, but no relationship was found for field dependent subjects (Oltman, Ehrlichman and Cox, 1977). These findings suggest that field independent subjects have more clearly differentiated hemispheric specialisation of function than field dependent individuals. An alternative possibility is that field independent subjects are more consistent in their strategies of tachisto-scopic processing or report.

## HANDEDNESS AND PERSONALITY

The dimension of field dependence-independence is correlated with certain aspects of social behaviour, field dependent subjects being more influenced by their social environment than field-independent subjects (Witkin, Dyk, Faterson, Goodenough and Karp, 1962). In a similar vein Hicks and Pellegrini (1978a) reported that in comparison with subjects of mixed hand preference those who were strongly left handed or strongly right handed were more 'externally' controlled, as defined by Rotter's (1966) locus of control questionnaire. Individuals having an external locus of control are said to see their destiny as being controlled by Fate or some other extrinsic agency, while those with an internal locus of control consider themselves to be responsible for their own predicament.

Intriguingly, it has been claimed (Hicks, Pellegrini and Hawkins 1979) that mixed handers tend to sleep for shorter durations than either left or right handers. Harburg, Feldstein and Papsdorf (1978) found that proportionately more left handers than right handers smoke 10 or more cigarettes a day, and Mascie-Taylor (1981) reported right handers to be less neurotic than either left or mixed handers. With regard to drinking, Bakan (1973) claimed a higher than expected incidence of sinistrals, defined according to the hand used for writing, among a group of 47 alcoholics. However, it is more likely that factors other than cerebral laterality mediate such relationships. Bakan himself has argued (Bakan, 1971) that first born and only children are more frequently left handed than are those of later birth rank (but see Chapter 2). Schacter (1959) has found that first born and only children are more susceptible to the arousal of anxiety in certain situations than are later born children. Conceivably, alcohol abuse and high levels of nicotine intake occur in response to anxiety.

## HANDEDNESS AND ANXIETY

Orme (1970) reported that among a group of girls attending an assessment centre 23 left handers showed greater emotional instability than did 277 right handers. Although Orme's investigation was criticised on statistical and methodological grounds by Hicks and Pellegrini (1978b) they too found that 23 left handers and 12 mixed handers were significantly more anxious than 35 right handed college students. More recently, Wienrich, Wells and McManus (1982) have reported that strongly left handed and strongly right handed subjects, classified according to Annett's (1970a) inventory, are more anxious than those whose hand preference is less extreme.

The issue of handedness and anxiety was taken up by Beaton and Moseley (1984). University students were asked to complete anonymously the trait scale of the Spielberger Anxiety questionnaire and also Annett's Handedness inventory. No relationship was found between anxiety and handedness.

Beaton and Moseley suggest, as do Weinrich et al., that the finding of Hicks and Pellegrini – greater anxiety among left and mixed handers – may have been due to a disproportionate number of females among the left and mixed handers studied by Hicks and Pellegrini. Females are often, but not universally, found to give higher anxiety scores on questionnaires. The results of Weinrich et al., as the authors themselves point out, may be due to a tendency on the part of some subjects to avoid giving or endorsing extreme responses to questionnaire items. This would have the effect of showing such subjects to have anxiety scores intermediate between high and low levels on the Taylor Manifest Anxiety questionnaire, the instrument used by Weinrich et al.

When the Annett Handedness questionnaire is scored according to the method suggested by Briggs and Nebes (1975) any tendency to give middle-of-the-road responses results in subjects being classified as intermediate in strength of hand preference. As Weinrich et al. adopted the Briggs-Nebes scoring procedure it is conceivable that their results, showing a U-shaped relationship between hand preference and anxiety, reflect nothing more than a bias in the way in which some subjects filled in both their handedness and anxiety questionnaires. Beaton and Moseley, by contrast, classified subjects' handedness into eight preference categories as recommended by Annett (1970a). Since inclusion of a subject in a particular category depends upon the pattern of responses he gives to the items of Annett's questionnaire, and not on a numerical score, there is no opportunity for a particular hand preference category to be artefactually linked to a particular range of anxiety scores. Beaton and Moseley's results, indicating no systematic relationship between anxiety and handedness, therefore suggest that the findings of Weinrich et al. are indeed the consequence of a methodological bias.

## SUMMARY

This chapter has considered suggestions that left and right handers differ in cognitive ability and in certain personality characteristics.

Evidence of a difference in ability seems more likely to be found in studies of samples drawn from university and similar populations. It is also possible that some of the negative findings might have proved positive had attention been paid to such factors as hand posture and familial sinistrality.

Handedness has been found in some studies to relate to occupational choice. The concept of 'hemisphericity' may be invoked to explain such findings. Differences between left and right handers have also been reported in relation to the dimension of field dependence-independence and other aspects of personality.

# 11

# Emotionality

One of the most intriguing ideas to have gained ground in recent years is that the two halves of the brain differ in the contribution that they make to our emotional life (for review see Tucker, 1981). This raises the possibility that normal emotional experience depends upon the relative balance of excitation and inhibition between the two cerebral hemispheres. It also suggests the further possibility that certain pathological states, such as depression or hyper-mania, result from abnormal hypo- or hyper-activation of systems lateralised to one or other side of the brain. Evidence concerning laterality effects in relation to psychopathology is discussed in Chapter 12. The present chapter focuses on the notion that different emotional responses may characterise left and right halves of the normal brain.

## CLINICAL OBSERVATIONS

The earliest suggestions of differential hemispheric involvement in emotion can be found in the observations of clinicians such as Goldstein, who noted that following damage to the left side of the brain patients commonly showed an emotional reaction which he described as 'catastrophic' (Goldstein, 1939). This was a response of exaggerated despair to the situation in which the patient found himself. The opposite reaction of 'indifference', often seen among cases of right hemisphere damage, was also described in the neurological literature of this period (Hécaen, Ajuriaguerra and Massonet, 1951; Denny-Brown, Meyer and Horenstein, 1952). These indications of an association between laterality of brain lesion and different types of emotional reaction were later supported by observations made on candidates for neurosurgery who were undergoing the Wada test.

Terzian and Cecotto (1959) reported that after injection of sodium amytal into the left carotid artery patients often exhibited a 'depressive catastrophic' reaction while injection on the right side was more likely to precipitate a 'euphoric-maniacal' reaction. Perria, Rosadini and Rossi

(1961) agreed with these observations, commenting that in reactions of a depressive type after injection on the side dominant for speech 'the patient cries and says he will never recover, that his family will go to ruin' and so on. After non-dominant injection, on the other hand, 'he is optimistic about his future, makes jokes and often breaks into laughter'. These findings are consistent with the results of other amytal studies (but see Milner, 1967 and Kolb and Milner, 1981b) and of systematic investigations carried out among groups of unilateral left and right brain damaged patients (Gainotti, 1969; 1972).

It is tempting to look on the 'depressive-catastrophic' reaction following left sided lesions or amytal injection as a reflection of the patient's concern with his motor or speech disabilities. The reaction of 'euphoria' or 'indifference' to right hemisphere insult, on the other hand, might be accounted for by the more frequent occurrence of unilateral neglect (Hécaen and Albert, 1978) with injury to this side of the brain. This may take the form of denying that there is any impairment of the limbs on the side contralateral to the lesion. Not being 'aware' of any hemiplegia and not suffering from dysphasia, is it surprising that some patients with right hemisphere damage do not show the same emotional reaction as their fellows with damage to the opposite side of the brain? Yet this notion can be rejected for several reasons.

Firstly, Perria et al. noted that the onset of an emotional reaction was not coincident with the other symptoms caused by barbiturate injection but occurred approximately four to six minutes after introduction of the drug into the carotid artery, that is, some time after the paralysis and arrest of speech have worn off. Nor was an emotional reaction seen to invariably follow the occurrence of these symptoms. Conversely, reactions were sometimes observed even after doses of barbiturate too low to produce dysphasia or hemiplegia (Rossi and Rosadini, 1967).

Secondly, although Gainotti (1972) found that reactions of a 'catastrophic' type were more common among aphasic than among non-aphasic patients with lesions of the left hemisphere, reactions were still seen among non-aphasics. Also, different features of the 'catastrophic' reaction were associated with different varieties of aphasia, which implies that the precise form which a reaction takes depends upon the locus of lesion within the left hemisphere. The pattern of aphasics' responses suggested to Gainotti that the determinants were neurological as well as psychological (see also Robinson and Benson, 1981).

With regard to 'indifference' reactions consequent upon right hemisphere damage, the data of Gainotti's study show that although these were more common among patients exhibiting unilateral neglect, this was neither a necessary nor sufficient condition for the occurrence of such reactions. Nor could the presence of neglect account for certain components of the indifferent or euphoric response, such as the tendency towards making fatuous jokes.

There is a third reason for rejecting the idea that different emotional reactions to left and right hemisphere damage can be attributed to the presence of aphasia in the former case, and of neglect in the latter. This is that evidence of differential involvement of the two halves of the brain in emotion is emerging from work with normal subjects who are, by definition, neither dysphasic nor suffering from any other intellectual impairment. Such investigations are discussed later in this chapter.

An ever-present problem in interpreting the effects of unilateral brain damage is that it is rarely possible to say whether the damage has impaired or destroyed a particular mechanism or has released one area of the brain from the influence of other regions. Sackeim, Greenberg, Weiman, Gur, Hungerbuhler and Geschwind (1982) analysed 90 case reports in which the patient suffered either from uncontrollable laughing or from pathological crying. The former was associated predominantly with right-sided space occupying lesions, the latter with lesions of the left hemisphere. Literature reports of hemispherectomy cases were analysed, too, which suggested that removal of the right hemisphere was associated with euphoria. This implies that the pathological laughter observed in cases of restricted right hemisphere damage – and perhaps other instances of increased positive affect – do not arise from the release of inhibition (disinhibition) of cortical mechanisms on the right side. (Release of ipsilateral subcortical mechanisms remains a possibility). Sackeim et al. further examined reports of epileptic patients having outbursts of pathological laughter and noted that (unlike the cases with space-occupying lesions) this was associated with left sided epileptic foci. They therefore argued that, taken overall, their analyses suggest that 'uncontrollable outbursts of laughing more often result from disinhibition of, or excitation within, the left side of the brain'.

Flor-Henry (1969; 1974; 1976) has argued that among temporal lobe epileptics diagnosed as psychotic, patients with an epileptic focus in the left hemisphere exhibit schizophrenic-like symptoms while those having a right hemisphere focus show symptoms similar to those of the affective psychoses. Although some of Flor-Henry's data and interpretations can be criticised, (see Chapter 12) he marshals a considerable body of evidence to link schizophrenia with damage to the temporal-limbic system of the left hemisphere and, rather less impressively, the affective psychoses, including depressive and manic states, with dysfunction of the corresponding regions on the right. Similarly, Bear (1979), working with temporal lobe epileptics unselected for the presence of psychotic symptoms, concluded that those with a left sided focus are generally more concerned with intellectual and philosophical issues relating to their own place in the grand scheme of things, while right temporal epileptics are 'emotive', showing considerable emotional lability. Thus there is some agreement that seizures originating in the left hemisphere may be associated with disturbances of thought, and right sided foci with alterations of mood.

## EXPERIMENTAL INVESTIGATIONS WITH BRAIN DAMAGED PATIENTS

A shortcoming of many clinical investigations is reliance on the subjective impression of the investigator. It is reassuring to note, therefore, that Gasparrini, Satz, Heilman and Coolidge (1978) and Black (1975) have confirmed, using the Minnesota Multiphasic Personality Inventory, that left brain damage is associated with greater depression than comparable damage to the right side of the brain. Right hemisphere damage, on the other hand, has been associated with other symptoms specifically related to emotion.

Wechsler (1972) read two very short stories to his brain injured patients. One story was entirely factual, whereas the other was supposedly 'affective in tone'. The patients' task was to recall each story immediately after hearing it. For the purposes of analysis, each story was abritrarily divided into twenty segments and responses were scored both for quantitative and qualitative content. Brain injured patients as a whole made proportionately more qualitative errors in recall of both narratives than either psychiatric controls or neurological patients with peripheral nerve damage. However, patients with left hemisphere damage made comparatively fewer qualitative errors in recalling the affective story than patients with right sided damage, although there was no difference between the groups in qualitative recall of the neutral story. Since the left lesion group included aphasic patients, the greater number and proportion of qualitative errors by the right brain damaged group in recalling the affective story is particularly striking. Wechsler therefore concluded that the right half of the brain normally plays a dominant role in the recall of affectively charged material. Unfortunately, it is not clear from his paper what statistical analysis Wechsler carried out on his data and this selective deficit in recall therefore requires to be confirmed.

Other authors have suggested that emotional stimuli may not be perceived as well, far less recalled, by patients with right compared with left sided damage. Heilman, Scholes and Watson (1975) required 6 aphasic left lesion patients and 6 patients with right hemisphere disease to point to one of 4 faces (expressing sadness, happiness, anger or indifference) as a means of testing their ability to appreciate the emotional tone of various sentences which were read out to them. Patients with right sided disease were impaired relative to those with left lesions only in appreciating the emotion concerned but not in understanding the factual content of the sentences. This finding was not attributable to an inability on the part of right hemisphere patients to recognise faces, a disability often associated with right sided damage, since prior to testing all patients could point correctly to the sad, happy, angry or indifferent face on demand.

Tucker, Watson and Heilman (1977) asked brain injured patients to judge whether the emotional tones of particular pairs of spoken sentences were the same or different. Again, those with right sided lesions were more impaired than those with damage on the left. More recently, Cicone,

Wapner and Gardner (1980) demonstrated that left brain damage led to impaired appreciation of the emotional significance of linguistically presented material, whereas right sided damage impaired appreciation of the emotion carried by both linguistically and pictorially presented information.

Using the electrodermal skin response as their measure of arousal, Morrow, Vrtunski, Bart, Kim and Boller (1981) found right hemisphere damaged patients to show virtually no arousal in response to stimulating photographic slides. Patients with damage confined to the left half of the brain were more aroused, but less so than control subjects.

The above results suggest that the ability to perceive the emotional tone of auditorily or visually presented material is dependent upon the integrity of the right cerebral hemisphere. However, Schlanger, Schlanger and Gerstmann (1976) found no difference among left and right brain damaged groups in distinguishing between sadness, happiness and anger when tested using a similar procedure to that employed by Tucker et al. As the latter utilised four response categories but Schlanger and his colleagues employed only three it might be thought that the test used by the latter was insufficiently sensitive to reveal a left-right difference. Yet Schlanger et al. obtained a significant difference in their experiment between left hemisphere damaged aphasics who were classified into two groups according to their verbal ability. This suggests that the relevant factor is not a lack of sensitivity in the method of testing. What is more likely, perhaps, is that the discrepancy between the results of Schlanger et al., who found no difference between the effects of left and right brain damage, and those of Tucker et al. and Heilman et al., who did find a significant difference, is related to the site of the lesions in the right hemisphere of the patients studied by these different investigators. All right lesion patients in the Tucker and Heilman studies had damage encroaching on the parietal region and all showed unilateral neglect. By contrast, in the experiment by Schlanger et al. only 3 out of 20 patients had damage involving the parietal lobe and none showed symptoms of unilateral neglect.

If localisation of damage in the right parietal area turns out to be the crucial factor in determining whether any asymmetry in perception of emotional tone is observed between left and right brain damaged patients this will be rather surprising. In view of the assumed anatomical substrate of emotion as consisting principally of sub-cortical structures, frontal cortex, and the limbic areas of the temporal lobe (Strongman, 1973), it might be predicted that hemispheric asymmetry in emotional response would be revealed by damage to the temporal rather than parietal region of the brain.

## STUDIES WITH NORMAL SUBJECTS

If it is true that the perception of emotionally intoned sentences requires an intact right hemisphere, then it should be possible to demonstrate a comparable left ear effect in normal subjects. Safer and Leventhal (1977) independently varied the tone of expression and the affective content of sentences in a factorial design and found that 29 out of 36 subjects judged the sentences (as positive, negative or neutral) according to the tone of the sentence when it was presented to the left ear, but 21 out of the 36 judged it according to the content when listening with the right ear. Bryden, Ley and Sugarman (1982) presented 7-note tonal sequences dichotically and found that subjects judging the emotional quality of the sound were more in accord with previous ratings by a panel of (binaural) listeners for the left ear than for the right ear stimulus. A problem in interpreting these results in terms of emotional asymmetry, however, is that some evidence implicates the right hemisphere in perceiving certain aspects of vocal sounds, both verbal and non-verbal (Haggard and Parkinson, 1971; Carmon and Nachson, 1973b; Zurif, 1974). Thus if subjects hear stimuli at the left ear they may attend to those attributes which, quite concidentally, are important for discriminating between different emotions. In short, the results tell us little about the involvement of the right hemisphere in specifically emotional material. The same point can be made with regard to the clinical studies quoted above, as was, in fact, acknowledged by Heilman et al. (1975).

The present writer attempted to assess the role of the right hemisphere in emotion using a different approach. Rather than presenting stimuli to be discriminated from each other, subjects were asked to rate stimuli on each of three nine-point scales in the hope that any difference between the hemispheres would show up as a difference in the ratings assigned to the same material heard at the left and right ears. The stimuli consisted of poems and passages of orchestral music played via headphones to the subject's left or right ear with white noise in the opposite ear. The intention was that in this competitive situation suppression of the uncrossed by the crossed pathways would result in lateralisation of the input from each ear to the contralateral hemisphere (see Chapter 4). Subjects were asked to listen to three poems and three musical passages and to rate each item on the dimensions 'pleasant-unpleasant', 'soothing-irritating' and 'cheerful-depressing'. Pilot subjects had previously rated one item of each class of material as pleasant, one item as unpleasant and the remaining item as neutral. The experimental items were selected from a larger number of items, on the basis that they were relatively unfamiliar to university students and showed close inter-rater agreement on the pilot rating scales.

The results of the study (Beaton, 1979b) showed that, by and large, all items were judged to be more pleasant, and the music as more soothing, by the group hearing the stimuli at the left ear in comparison with those for whom the material was presented to the right ear. These ear differences were statistically significant and could not be attributed to differential familiarity with the items by subjects of the two groups. Since the results were in the same direction for both the poems and the music the effect cannot have been due to a relative dominance of one hemisphere in perceiving a particular class of material. The poems, being verbal, should have been processed predominantly by the left hemisphere while the music should, perhaps, have engaged primarily the right hemisphere (see Chapter 4). Since both types of stimulus were 'appreciated' more at the left ear this suggests a super-ordinate role for the right hemisphere in mediating responses to emotionally arousing stimuli.

It has sometimes been claimed that the right hemisphere is the emotional half of the brain, the left hemisphere being emotionally flat in comparison. Several experimenters have reported that subjects make proportionately more lateral eye movements towards the left, implying activation of the right hemisphere (see Chapter 4), in response to emotionally arousing questions compared with neutral questions (Schwartz, Davidson and Maer, 1975; Tucker, Roth, Arneson and Buckingham, 1977). In support of this view Sperry, Zaidel and Zaidel (1979) state that when significant stimuli were presented to the right hemisphere of two split-brain subjects the emotional responses evoked were 'if anything ... somewhat more intense and less restrained than those from the left'. Sperry et al. add the rider, however, that this may have reflected 'the added mental stress of having to use the mute hemisphere'.

If it were the case that the right hemisphere is emotional while the left is unresponsive it might have been expected in Beaton's (1979b) experiment that, relative to right ear responses, pleasant items would have been rated as more 'pleasant' and unpleasant items as more 'unpleasant' when heard at the left ear. Since no such 'exaggeration' effect occurred it seems unlikely that stimuli are invested with emotional properties solely at the behest of the right hemisphere.

The findings of Beaton (1979b) do not support the results of an experiment by Dimond, Farrington and Johnson (1976). These investigators fitted subjects with specially constructed contact lenses which, when worn in conjunction with partially blacked out spectacles, meant that the subjects could see only with a restricted portion of the retina. By suitable positioning of the clear areas of spectacle in relation to the contact lenses, information could be channelled to any desired hemi-retina and hence to either left or right side of the brain. The subjects watched each of three films while wearing the lenses: one film was of travelling on Lake Lucerne, one was a humorous cartoon and the third was a film of a rather gory surgical operation. After seeing each film, the subjects had to rate it on each of several rating scales. It was found that all the films were rated as significantly more 'unpleasant', and the cartoon and operation as more 'horrific', when viewed by the right compared with the left hemisphere. This result was seen by the experimenters as consistent with the findings that injection of sodium amytal into the left carotid artery and natural pathology of the left hemisphere are often followed by a depressed reaction on the part of the patient. Presumably Dimond et al. see dysfunction of the left hemisphere as leading to a subsequent emotional 'supremacy' of the right hemisphere. It would, however, be just as reasonable to argue that a lesion in one hemisphere releases that half of the brain from the restraining influence of its fellow.

It should be pointed out, in connection with Dimond's claim that the right hemisphere is more 'depressive', that although a significant difference was obtained between the groups of left and right hemisphere subjects on the dimension 'unpleasant' no corresponding difference was obtained on ratings of how 'pleasant' the films were. This inconsistency necessarily diminishes the force of the claim that the right half of the brain is more 'depressive' than the left. On the other hand, Dimond and Farrington (1977) reported finding a significant difference in heart rate changes, a physiological measure of arousal, between the groups of left and right hemisphere subjects, which supports the general notion that there is some degree of hemispheric emotional asymmetry.

Most of the experimental investigations so far mentioned in this chapter have involved the presentation of material via the auditory modality. Yet there is a group of studies which seem to suggest a prominent role for the right hemisphere in the visual perception and memory of a special class of emotionally significant material.

Suberi and McKeever (1977) gave subjects photographs of faces to remember which subsequently had to be discriminated from other faces presented tachistoscopically. In agreement with many other studies on facial discrimination the results on this task showed a significant left field superiority in reaction time. The important point in the present context, however, was that the left field superiority was significantly greater for those subjects who memorised target faces expressing either happiness or sadness than for subjects who had to remember neutral faces. The kind of non-target face presented did not affect this visual field asymmetry. Since the effect of memorising emotional rather than neutral faces was to increase response latencies only for right field presentations, and not for both visual fields, Suberi and McKeever argued that their result cannot be attributed to an overall increased difficulty in remembering emotional faces. Since response times to emotional and neutral faces presented in the left visual field were more or less identical, Suberi and McKeever concluded that the mechanisms mediating memory for emotional faces are not additive with the mechanisms mediating memory for neutral faces. These workers therefore felt that their results pointed to a particular involvement of the right hemisphere in the storage of emotional material.

Similar findings to those of Suberi and McKeever were reported by Ley and Bryden (1979) and by McKeever and Dixon (1981). These latter workers also found that recognition of faces was faster in the left than the right visual hemifield, but only if the faces were rated as emotional. In addition, a LVF superiority emerged only when subjects engaged in emotional imagery in order to remember the faces and not when imagery was neutral. McKeever and Dixon argued that the effects of emotion and of facial recognition per se were dissociable. Hansch and Pirozollo (1980), however, did not find any difference in response times to neutral and emotional faces, and concluded that it is not possible to differentiate between right hemisphere superiority for faces and a special involvement of the right hemisphere in emotion.

Evidence of a somewhat different kind which points to right hemisphere involvement in emotion was put forward by Sackeim and his collaborators (Sackeim and Gur, 1978; Sackeim, Gur and Saucy, 1980; but see critique by Ekman, 1980 and reply by Sackeim and Gur, 1980). It has often been found that in right handers the left side of the face is perceived as more emotionally expressive than the right side, although the finding may hold more for deliberately posed than for spontaneous expressions (Ekman, Hager and Friesen, 1981; but see Cacioppo and Petty, 1981). This impression is gained both from looking at real faces (e.g. Chaurasia and Goswami, 1975; Moscovitch and Olds, 1982) and from experiments where subjects are presented with particular kinds of photographs (see Ekman and Oster, 1979 for a review). Sackeim and Gur (1978), for example, took photographs of faces and split them vertically down the centre. By appropriate means they were able to make up a second set of photographs of a complete face made up entirely from either the true left or the true right side of an original face. These composite photographs were then shown to subjects who judged that those constituted entirely from the left side of a face were more intense in emotional expression than the right side composites. This was the case even though the latter were judged to be more similar to the original undoctored photographs, perhaps because the right side of the face is often larger (Ekman, 1980; Koff, Borod and White, 1981). On the other hand the left side of the face has greater 'mobility' (Koff, Borod and White, 1981; Alford and Finnegan-Alford, 1981) which could account for its greater intensity of expression.

The fact that the left side of the face is usually considered more emotionally expressive than the right side (Campbell, 1978; Sackeim and Gur, 1978; Sackeim, Gur and Saucy, 1980; Borod and Caron, 1980; Heller and Levy, 1981) suggests that the right half of the brain is intimately involved in the expression as well as the perception of emotion. This view is supported by the finding that left-side composite photographs are rated as more expressive only if taken of patients with left hemisphere brain damage but does not hold for photographs taken of those with right cerebral lesions (Bruyer, 1981). Futher, when neurologically intact subjects are asked to guess which of several emotionally arousing pictures are being looked at by brain damaged patients, guesses are significantly more accurate if the patients have left hemisphere damage than if they have right hemisphere lesions (Buck and Duffy, 1980). Thus, although control of the facial muscles is bilaterally organised in the brain, these findings suggest a dominance of the right hemisphere for emotional expression which influences facial gesture through the contralateral pathways. It is of considerable interest, therefore, that Ross and Mesulam (1979) observed that two patients with presumed right sided, anterior lesions were unable in their everyday life to exhibit emotion through facial gestures, although apparently experiencing the emotions which they could not express. They were also unable to convey emotion through the appropriate intonation of their speech. This supports the findings of Tucker, Watson and Heilman (1977), who asked brain damaged patients to say various sentences in different tones of voice. Tape recordings of the patients' efforts were given to a panel of judges who had to say whether each sentence sounded happy, sad, angry or indifferent. The efforts of patients with right hemisphere lesions were judged to accord with the experimenter's instructions significantly less often than those of left brain injured patients.

There have been suggestions that so-called positive and negative emotions may give rise to opposite left-right asymmetries (Reuter-Lorenz and Davidson, 1981). In one experiment subjects were told about events in their past which they had previously described to the experimenter. These events were classified as either positive (calm, pleasant) or negative (sadness, fear, anger). EEG recordings showed changes which were interpreted as showing relatively greater involvement of the left hemisphere during negative emotions and relatively less involvement during the re-creation of positive emotional states (Harman and Ray, 1977). By contrast, Ahern and Schwartz (1979) found that questions inducing positive emotions (happiness; excitement) elicited more right than left lateral eye movements, whereas negative emotions (sadness; fear) elicited more left movements. It was also reported (Schwartz, Ahern and Brown, 1979) that measurement of muscle movement, made in spontaneous response to these questions, showed greater right sided activity of the zygomatic muscles (which elevate the cheeks, as in smiling) for these positive emotions whereas there was more left sided activity for the negative emotions. The corrugator muscles (which raise the eyebrows) showed no significant laterality effect. However, when subjects were asked to deliberately produce facial expressions, there was significantly greater left sided activity in this muscle group, among female but not male subjects, while the zygomatic group showed no laterality effect. The sex-by-side interaction was statistically significant. In contrast to the above results Alford and Finnegan-Alford (1981) found that males showed a greater ability than females in voluntarily raising their left eyebrow. For both sexes this was easier than raising the right eyebrow.

Safer (1981) argued that females are better able than males to integrate verbal and emotional processing codes while Davidson and Schwartz (1976) suggested that females show greater lateralisation of emotional arousal to the right hemisphere. This latter claim, in particular, requires to be viewed with caution in as much as the statistical analyses supporting it concentrated on selected comparisons within the data from a number of different experimental conditions. Inspection of the entire set of data suggests that greater activation of the right hemisphere among females than males was not only associated with conditions of emotional arousal.

## THEORETICAL INTERPRETATIONS OF LATERAL ASYMMETRY IN EMOTION

The experiments reviewed above strongly suggest some degree of cerebral lateralisation of emotion. Although some of the individual studies require to be replicated and interpretation of certain findings is problematic, data from a variety of sources point consistently in the direction of there being a difference of one sort or another between left and right sides of the brain. The nature of such a difference remains obscure. Hécaen and Angelergues (1963) suggested that sensory data represented in the left hemisphere 'undergo complex, conceptual elaboration by means of language while sensory data in the right hemisphere retain their immediacy and rich affective value'. This might be termed an 'attenuation' hypothesis: emotional feelings are diluted in the left hemisphere by verbal encoding.

A different explanation of the particular contribution of the right hemisphere in emotion was put forward, in a footnote, by Davidson and Schwartz (1976). These authors drew on the views of Semmes (1968) who argued that somatosensory functions were organised in a different way at the two sides of the brain. Her argument was based on research into the effects of penetrating brain injuries, which revealed that somatosensory impairment was correlated fairly precisely with locus of lesion in the left hemisphere but appeared to be relatively non-specific as regards lesion location in the right hemisphere. Semmes concluded that somatosensory functions were mapped onto the cortex of the left hemisphere in a fairly precise manner but were more diffusely organised in the right hemisphere. Furthermore, Semmes and her colleagues found evidence for a similar organisation along hemispheric lines of a more cognitive, higher-level function, namely spatial orientation. She went on to propose that functions involving spatial relations in different modalities – visual, kinaesthetic and vestibular – might have overlapping neural representations in the right hemisphere such that integration of such sub-systems leads to a single, supra-modal representation of space. In the left hemisphere, however, the more concentrated, focal representation of function would not favour such multi-modal integration. Taking this as their point of departure, Davidson and Schwartz suggested that, not only might different neural sub-systems having to do with the same psychological function overlap in the right hemisphere, but so also might the systems subserving widely different functions. They proposed that

> Since emotion typically involves the integration of heteromodal or unlike inputs (e.g. cognitive and visceral), it would appear . . . that the right hemisphere is uniquely suited to perform such a task.

This extrapolation from the data of Semmes and her co-workers is, of course, purely speculative.

Schacter (1975) has described experiments in which subjects are injected with either adrenaline or a placebo according to a double-blind design and are subsequently exposed to different environmental conditions. The adrenaline causes physiological changes which are identical for the different environmental conditions of the experiment. Nonetheless it is found that subjects attribute such physiological changes to different emotions depending on the environmental circumstances. This view of emotion, that we label autonomic activity according to the cognitions available to us, might offer an explanation of how the same event can be interpreted differently by left and right sides of the brain.

Because of the presence of language in one hemisphere but not in the other, the cognitive experiences of each half-brain may differ to some degree. Memories of the past may be encoded in different ways which will affect one's construction of the present. The memories of one hemisphere's registration system may not always be available to the hemisphere on the opposite side.

Risse and Gazzaniga (1976) placed various objects in the left hand of a patient undergoing the intra-carotid Sodium Amytal Test. After the effects of the drug had worn off, the patient was asked to say what objects she had felt during administration of the test. She was totally unable to report what the objects were, but could readily retrieve the same objects with the left hand. This result suggests that engrams encoded in the absence of verbalisation do not subsequently become available to the verbal system. It is conceivable, therefore, that subtle differences in the memories of left and right hemispheres might lead to a slightly different evaluation of the same event.

Although not specifically concerned with the question of emotion, certain experiments carried out with two split-brain subjects are relevant in the present context. Sperry, Gazzaniga and their collaborators wished to discover whether the right hemisphere could be accorded the same conscious status as its partner on the left (Le Doux, Wilson and Gazzaniga, 1979; Sperry, Zaidel and Zaidel, 1979). In short, their aim was to find out whether a split brain entails a split mind or whether the property of 'mind' belongs only to the left, speaking hemisphere. By presenting stimuli to a single hemisphere and asking the patient to respond by pointing to one of a number of alternative choices, it was shown that the right hemisphere of these patients did indeed have its own sense of identity, could look to the future, feel emotion and generally exhibit conscious awareness. What is interesting for present purposes is that, if a stimulus evoked some emotional reaction when presented to the right hemisphere, it was almost always the case that the left hemisphere could verbally indicate that it, too, was aware of the emotion involved even though it did not know what triggered that emotion.

Because the left hemisphere was able to accurately gauge the emotional tone of the right hemisphere's response it has been suggested (Le Doux, Wilson and Gazzaniga, 1979) that the theory of emotion put forward by Schacter must be in error. This theory holds that bodily changes such as increased heart rate, sweating and vasodilation which occur during emotional arousal are non-specific, that is they do not differ very greatly in different emotional states. Various emotions are experienced as different because these bodily changes are interpreted in the light of different environmental circumstances according to which we feel love, anger, fear or whatever (Schacter, 1975). Emotion is, according to this view, as much a cognitive as a physiological event. The reason why the split-brain researchers believe this view to be mistaken is that the left hemisphere can appreciate the quality of the emotion generated in response to a stimulus presented in the left visual field, even though knowledge of the stimulus is confined to the right hemisphere. If Schacter is right, they argue, how is it that the appropriate emotion is recognised by the left hemisphere in the absence of the relevant cognition?

The split-brain findings are not necessarily damning for Schacter's theory. Reading the original reports it appears that the left hemisphere's awareness of the right hemisphere's emotional reaction is at a fairly primitive level. Certainly the right hemisphere's emotion is recognised as 'good' or 'bad' but the evidence does not suggest that the left hemisphere is capable of any more sophisticated appreciation than this. It is not inconceivable that there is in the brain some mechanism whereby a feeling of 'goodness' or 'badness' can be registered quite independently of how a particular emotion originally arose. Thus Schacter's idea as to how emotions are labelled in the first place is not inconsistent with the possibility that a specific emotion is evaluated as desirable or not. Of course, it is implicit in this suggestion that certain neural events which follow the labelling of bodily arousal in terms of a specific emotion must necessarily differ, in some way, according to whether such an emotion is registered as good or bad. Perhaps it is this general evaluation which transfers in the callosum-sectioned patient. Le Doux et al. themselves suggest how this might come about.

Much recent research on the biochemistry of the brain suggests that specific substances may be associated with changes in emotional state. These substances, known as enkephalins, have been identified in an area of the limbic system known as the amygdala (Snyder, 1977). It has been established that the anterior commissure, which was not sectioned in the two split-brain patients used in the experiment described above, connects the amygdala on the two sides of the brain. It is possible, therefore, that mechanisms located in the amygdala are involved through the action of enkephalins in experiencing an event as 'good' or 'bad' and that the anterior commissure provides a link for the transfer of such feeling between one half-brain and the other.

Robinson (1979) has suggested that the lateralisation of emotional response following brain injury may depend upon an asymmetrical response of catecholaminergic neurones to cortical injury. Recently, evidence has been obtained to suggest that subcortical catecholaminergic pathways are asymmetrical (Oke, Keller, Mefford and Adams, 1978). Since ascending subcortical catecholaminergic pathways branch out within the cerebral cortex, asymmetry within these pathways could cause lesions of one hemisphere to retrogressively affect catecholaminergic neurons differently from lesions in the opposite hemisphere (Robinson, 1979). (This explanation cannot account, of course, for differences between left and right hemisphere emotional responses within the intact brain.)

Glick, Weaver and Meibach (1980) implanted electrodes in the lateral hypothalamus of laboratory rats. The animals were trained to press a lever to produce a brief electric current through the electrode, thereby stimulating this region of their brains. By having electrodes implanted in the lateral hypothalamus at left and right sides it was possible to find out whether there was any difference in self-stimulation at the two sides. Glick et al. reported that the intensity of current that was required for the rats to engage in self-stimulation differed for the left and right sides of the hypothalamus of each rat. In attempting to relate their findings to the human situation Glick et al. proposed:

> One might hypothesise ... that the side of the brain having lower reward thresholds would also have higher aversion thresholds. An individual with mood-related material engendering mixed affect might have one side of the brain more receptive to perceiving it as pleasant and the other side as more likely to perceive it as unpleasant. In this way quantitative differences in neuronal mechanisms (for example rate of firing) could be translated into apparently qualitative specialisation with one side of the brain appearing to be specialised for high mood and the other for low mood. (p. 1094)

Needless to say, this is a highly speculative suggestion.

## SUMMARY

Data from brain-damaged and normal subjects have been reviewed which suggest that each side of the brain may make a different contribution to emotions. Neuropsychological studies show that damage to the left hemisphere is often followed by feelings of depression, while damage to the right hemisphere is associated with euphoria or feelings of indifference. Comparable findings, obtained using the sodium amytal technique, suggest that these different reactions are unlikely to be simply a consequence of the different psychological sequelae of left and right sided lesions. They may instead be due to a lateral difference in neurotransmitter substances or neuronal arborisation within the hemispheres.

A particular role for the right hemisphere in emotion is suggested by findings which show that damage to this side of the brain may affect the patient's ability to express (but not necessarily to experience) emotion. Right sided lesions also impair the perception and recall of emotionally charged material presented in the auditory or visual modality. This effect appears to be independent of whether the information is conveyed by verbal or non-verbal means.

Results from experiments carried out with neurologically normal subjects have been interpreted as supporting the view that the right hemisphere is particularly implicated in emotion. However, it has not always been clear that the results obtained can be attributed specifically to a right hemisphere specialisation for emotion, as opposed to specialisation for other non-verbal processes.

# 12

# Asymmetry and psychopathology

The findings reviewed in the previous chapter point to a particular, though not exclusive, involvement of the right half of the brain in matters of mood or feeling. It is therefore of some interest that psychiatrists have occasionally noted something similar. Halliday, Browne and Kreeger (1968) reported that electro-convulsive shock (ECS) applied to the right side of the head was slightly better than bilateral ECS in relieving clinical depression, while left sided ECS was slightly worse. These differences, though not statistically significant, are supported in part by the findings of Cronin, Bodley, Potts, Mather, Gardner and Tobin (1970) who found unilateral left ECS to be significantly less effective than bilateral ECS or right sided ECS which did not differ in therapeutic efficacy. Deglin and Nikolaneko (1975) obtained similar findings in comparing unilateral left and right sided ECS.

These are, however, all isolated findings since most authors have found no differential effect of left and right sided electrode placements in relieving depression.

Current interest in the topic of hemispheric asymmetry in the psychiatrically disabled (for reviews see Shimkunas, 1978; Wexler, 1980; Gruzelier, 1981; Colbourn, 1982; Walker and McGuire, 1982) owes much to the work of Flor-Henry (1969; 1974) who marshalled considerable evidence from diverse sources to suggest that certain psychopathological syndromes were associated with dysfunction of lateralised cerebral systems. He argued (Flor-Henry, 1974) that the evidence implied 'a disturbance of dominant hemispheric function in schizophrenia and psychopathy ... whilst non-dominant hemispheric dysfunction appears to be correlated with depressive symptomatology'.

The notion that schizophrenia is related to a disturbance of left hemisphere mechanisms is intuitively appealing, given the well-documented anomalies of language behaviour shown by many schizophrenics

(Andreasen and Grove, 1979). The fact that language problems may be particularly conspicuous in schizophrenia, however, does not necessarily mean that right hemisphere functions are unaffected.

## NEUROPSYCHOLOGICAL STUDIES

The argument put forward by Flor-Henry has been much quoted in the literature. It is therefore worth looking a little more closely at this before passing to the work of others. Flor-Henry builds his case largely on two different kinds of evidence. Firstly, he cites psychometric studies which, he claims, show that psychopathy and schizophrenia are associated with high scores on performance, compared with verbal, sub-tests of intelligence while depression is associated with the reverse pattern. Secondly, Flor-Henry maintains, in psychoses associated with temporal lobe epilepsy schizophrenic-like symptoms are found most commonly in conjunction with left hemisphere damage and manic-depressive symptoms are found in conjunction with right-sided epileptic foci. For both strands of the argument, the evidence is equivocal. Regarding the psychometric data, Alpert and Martz (1977) have pointed out that Matarazzo in his revision of Wechsler's handbook states that 'no Wechsler sub-test pattern or profile . . . has been reported which reliably differentiates "schizophrenic" patients from either normal individuals, or from patients described clinically as falling into other psychiatric diagnostic categories'. Thus although Flor-Henry's own (1974) studies bear out (to some extent) his claim many other investigations have failed to do so.

The data for the second part of Flor-Henry's thesis come from temporal lobe epileptics. In his 1969 paper, he compared 50 cases of temporal lobe epilepsy who showed psychotic symptoms with 49 cases of temporal lobe epilepsy without such symptoms. In the former group 19 patients had a focus in the left hemisphere, 9 patients had a focus in the right hemisphere and 22 patients had bilateral foci. Among non-psychotic epileptics there were 13 cases with a left hemisphere focus, 25 with a right-sided focus and 11 with bilateral foci. A chi-square analysis shows this difference in distributions to be statistically significant. However, as Alpert and Marz pointed out, among the non-psychotics there was a relatively large number of cases with right-sided foci and it is largely to this that the significant chi-square value can be attributed. Since other authors have reported a more or less equal incidence of left and right temporal foci, Alpert and Martz tested the distribution of lateralised lesions among Flor-Henry's psychotics against an expected distribution of equal numbers of left and right foci among non-psychotic epileptics (bilateral cases were omitted). This analysis yielded a value of chi-square which was not significant at the 5 per cent level of confidence. Shukla and Katiyar (1980) also failed to find any relationship between laterality and psychosis among temporal lobe epileptics.

In more recent papers Flor-Henry (1976; 1979a) has repeated his arguments. Based on psychometric profiles, 114 patients were classified as showing primarily left or right hemisphere dysfunction. The sample consisted of 54 schizophrenics and 60 patients with affective psychoses unselected for the presence of epilepsy. Although Flor-Henry's presentation of his data is confusing, it appears that severe schizophrenics are characterised largely by presumed left-sided cerebral dysfunction while affectives are impaired mainly on tasks associated with the right hemisphere. However, the group labelled 'affective psychoses' includes patients classified as schizoaffective who, it appears, may well have exhibited 'florid symptoms' of schizophrenia but had, *at some time or another*, exhibited in addition definite episodes of depression or mania. It is not at all clear why such patients should have been categorized among the 'affective psychoses' any more than with the schizophrenics. Unfortunately, from the data presented by Flor-Henry it is not possible to eliminate 'schizoaffectives' from the figures provided so as to see whether a significant relationship between affective psychoses and laterality of lesion remains.

In a second study reported in Flor-Henry's (1976) and (1979a) papers, schizophrenics, manic-depressives and normal controls were compared in terms of EEG power spectra. Compared with normals, manic depressives were said to show more power, averaged across different conditions, in the 20-30 Hz band of the EEG in both left and right temporal regions. Flor-Henry gives statistical significance levels of less than 10 per cent for the left temporal region and less than 5 per cent for the right temporal region. However, quite apart from the fact that averaging data across conditions may obscure potentially valuable information which could qualify any general statement of laterality effects, it does not necessarily follow from Flor-Henry's significance levels that there is a significant difference between manic depressives and normals *only* for the right side of the brain (which is what a lateralised deficit hypothesis claims). The result for the left temporal region did not show a significant difference between normals and depressives but the fact that there was a significant difference for the right temporal region does not establish that there is a left-right difference. To assume that it does is to commit a fundamental error of statistical inference. Moreover, as only one-tail probability levels were quoted, the 5 per cent level of significance for the right temporal region is not particularly impressive. Thus from these data, the case for a special right hemisphere involvement in the affective psychoses is not particularly strong, as Flor-Henry himself admits. Nor has the research of other investigators (e.g. Goldstein, Filskov, Weaver and Ives, 1977; Kronfol, Hamsher, Digre and Wazir, 1978) provided any really convincing support. Nonetheless, there is no doubting the heuristic value of Flor-Henry's work and he is to be commended for the impetus he has given to the study of hemispheric asymmetries in this field.

## PSYCHOPHYSIOLOGICAL STUDIES OF HEMISPHERIC ASYMMETRY

Several lines of converging evidence (reviewed by Flor-Henry, 1976; Gruzelier and Venables, 1974) implicate the temporal-limbic structures of the brain in the symptomatology of schizophrenia. As well as the evidence already discussed, other work points specifically to involvement of the temporal-limbic system on the left side. For example, on an auditory temporal discrimination task schizophrenics performed more poorly than normals and in particular omitted to make responses more often to stimuli presented to the right ear compared with the left ear (Gruzelier and Hammond, 1976). Gruzelier and his co-workers, in looking for possible electrophysiological indices of lateralised nervous system disturbances in the major psychoses, decided to concentrate upon a physiological measure known to be disrupted by lesions of the limbic system in the monkey. The measure they chose was the change in level of arousal produced when a subject, animal or human, is presented with a novel stimulus.

Arousal can be measured in terms of the galvanic skin response, a measurable change in the electrical conductivity of the skin. Increased sweating on the palms, for example, reduces the resistance of the skin to the passage of a minute electric current and shows up as an increased voltage difference between electrodes connected to the hand and some reference point. On being presented with a novel stimulus normal subjects show initially increased levels of electrodermal activity (the orientation response) but quickly habituate, that is, their responses return to normal. Small differences in the level of activity recorded initially from the two hands are not uncommon. Some subjects show higher left,than right hand response amplitude while others show the reverse asymmetry (Gruzelier and Venables, 1974). The direction and extent of such lateral asymmetry is dependent upon a variety of factors (see e.g. Ketterer and Smith, 1977; Myslobodsky and Rattok, 1977; Diekhoff, Garland, Dansereau and Walker, 1978; Lacroix and Comper, 1979).

Gruzelier (1979) reports that among a group of endogenously depressed patients electrodermal response amplitudes were consistently higher on the left hand while among institutionalized schizophrenics they were higher on the right. It is not easy to interpret Gruzelier's results in terms of a specific disturbance of one or other cerebral hemisphere, since normal subjects may show a reversal from greater levels of electrodermal activity on the left hand in a low state of arousal to greater right hand amplitudes in a higher state of arousal (Gruzelier, 1973). However, the opposite direction of asymmetry for the two clinical groups suggests some perturbation in the balance of lateralised cerebral processes. This balance may be restored by appropriate medication since significant left-right response asymmetries were not seen in a number of schizophrenic patients whose symptoms had regressed under drug therapy. (Nor, incidentally, were response asymmetries observed in patients whose depression was diagnosed as reactive rather than endogenous.)

If the reduced level of response on the left hand of (non-remitted) schizophrenics is associated with left temporal lobe pathology, as predicted by Flor-Henry's hypothesis, then it would be expected that complete left temporal lobectomy should produce a decrease in skin conductance level on the left hand. This prediction was not confirmed in a study by Toone, Cooke and Lader (1979) carried out with 10 patients who had undergone surgery at least ten years previously. Alternatively, if autonomic functions are primarily mediated through contralateral pathways, as some have argued (e.g. Myslobodsky and Rattok, 1977), then a lower level of response should have been found for the right hand, but this was not found either. The precise significance of asymmetries in electrodermal activity therefore awaits clarification.


## EYE MOVEMENT STUDIES

Schweitzer and his colleagues have studied lateral eye movements in response to 'reflective' questions and report that schizophrenics made more lateral eye movement than normal controls (Schweitzer and Chacko, 1980). A greater proportion of these were towards the right than towards the left (Schweitzer, Becker and Welsh, 1978; Schweitzer, 1979; see also Gur, 1978; 1979). Similar results were obtained by Myslobodsky, Mintz and Tomer, (1979) who also found a preponderance of leftward movements in a sample of depressed patients. If a greater trend towards eye movement in one direction can be equated with a greater degree of activation of the contralateral cerebral hemisphere (see Chapter 4) and if more eye movements mean greater arousal, then these results imply over-activation of the left hemisphere in schizophrenia (Gur, 1978; 1979). This might be held to support the view that schizophrenia involves some left hemisphere dysfunction since over-activation might arise as a result of damage to inhibitory systems.


## SCHIZOPHRENIA: THE DISCONNECTION HYPOTHESIS

An alternative to the view that schizophrenia is associated with pathology of the left temporal-limbic system arose as a result of a report by Beaumont and Dimond (1973) which suggested some degree of functional disconnection between the hemispheres in long-standing schizophrenia. The starting point for Beaumont and Dimond's experiment was the report by Rosenthal and Bigelow (1972) that the only abnormality found on post-mortem inspection of the brains of a group of schizophrenics was an increase in the width of the corpus callosum relative to normal brains, a finding recently replicated by Bigelow, Nasrallah and Rauscher (1983). Beaumont and

Dimond carried out a tachistoscopic study in which subjects had to match stimuli in a same/different task. It was found that schizophrenic patients performed worse than control subjects in a condition in which one member of the pair of stimuli was directed to the left hemisphere and the other member of the pair was projected to the right hemisphere. Beaumont and Dimond therefore suggested that their data pointed to an impairment of inter-hemispheric transfer in schizophrenia. However, they also report that schizophrenics perform significantly more poorly than normal control subjects or other psychiatric patients on matching letters presented to the left hemisphere. In addition, the performance of schizophrenics was significantly worse than that of the psychiatric controls in matching shapes or digits presented to the right hemisphere. Thus the conclusion that there is a defect of inter-hemispheric transfer seems unwarranted, since poor performance within either hemisphere could account for poor matching across hemispheres. Nonetheless, since the original observation by Beaumont and Dimond a number of other workers have tested the 'disconnection hypothesis'.

Tress and Kugler (1979) tested schizophrenics and normal control subjects on the movement after-effect. This occurs after viewing, say, a rotating disc for some time after which the disc, when stationary, appears to rotate in the opposite direction. If the rotating disc is viewed with one eye the stationary disc still appears to rotate in the opposite direction even when viewed by the other eye. This inter-ocular transfer implies that the movement after-effect is, at least in part, dependent upon processes occurring after the convergence in the brain of the neural pathways from the left and right eyes. Tress et al. found that schizophrenics showed a greater degree of inter-ocular transfer than normals and they therefore proposed that there is an increase in callosal pathways in the schizophrenic brain. Dixon and Jeeves (1970) had earlier reported a complete absence of inter-ocular transfer in 2 out of 3 cases of callosal agenesis and a very much reduced degree of transfer in a third case. This suggested that the callosum is in some way involved in the phenomenon of inter-ocular transfer and provided the rationale for the experiment carried out by Tress and Kugler.

There is no prima facie reason on anatomic grounds why the corpus callosum should affect inter-ocular transfer, which is usually interpreted as being mediated by neurones receiving input from corresponding parts of the left and right eyes (Blake, Overton and Lema-Stern, 1981). Such cortical binocular units are presumed to be absent or deficient in individuals lacking stereoscopic vision, who, for the most part, fail to show inter-ocular transfer of various after-effects (for review see Aslin and Banks, 1978). More recently, Jeeves (1979) has reported that inter-ocular transfer is absent or reduced in 2 further patients with congenital absence of the corpus callosum. Since these patients are unable to correctly discriminate depth along the vertical midline, an ability shown by Mitchell and Blakemore (1970) to depend upon an intact corpus callosum, Jeeves thinks it likely that the inter-ocular transfer deficit in his acallosal subjects is linked to their defect of midline stereopsis. Certainly there is evidence that the callosum is involved in the region of visual space spanning approximately 2 degrees either side of the vertical meridian (Berlucchi, Gaazaniga and Rizzolatti, 1967; Berlucchi, 1972; but see Innocenti and Frost, 1979). Jeeves (1979) suggests that

> stimulation falling on that part of the retina 2°–3° either side of the vertical meridian has a major role to play in the inter-ocular transfer of the movement after-effect. One way to test this could be to study normals for whom the monocular stimulation is absent from the strip 2°–3° either side of the vertical meridian and test for inter-ocular transfer.

The present writer has collected data from 12 normal subjects, 10 of whom showed the usual degree of inter-ocular transfer of the movement after-effect despite stimulation falling just beyond 3° from the vertical meridian. Thus it does not seem that there is anything crucial about the central 2°–3° either side of the midline.

Although results from the visual modality do not readily support the hypothesis of a defect of inter-hemispheric integration in schizophrenia, the findings of experiments carried out in the tactile modality are a little more encouraging. Both Green (1978) and Carr (1980) from Dimond's laboratory have reported a deficit in patients' ability to compare information across left and right hands although within-hand comparisons on each side were relatively intact.

Jones and Miller (1981) recorded the contralateral and ipsilateral evoked response to stimulation of the index finger and found no overall latency difference in a group of 12 schizophrenic patients. Among normal control subjects the ipsilateral evoked response lagged behind the contralateral response. The ipsilateral evoked response in normals was considered to depend upon conduction of impulses across the callosum from the contralateral sensory cortex (see e.g. Green and Hamilton, 1976). The reduced latency of the ipsilateral evoked response in schizophrenics was therefore seen as due to the development of ipsilateral pathways from the brain stem to compensate for the lack of a normally conducting corpus callosum. However, this study by Jones and Miller was criticised on methodological and theoretical grounds by Connolly (1982) and by Shagass, Josiassen, Roemer, Straumanis and Slepner (1983). The latter group of workers failed to replicate the finding of a reduced latency of the ipsilateral evoked response in schizophrenia. Since in the study by Jones and Miller there was no difference between the pattern of response shown by four patients who had never received neuroleptic medication and the pattern shown by the remaining group of patients, it is unlikely that differences in medication can account for the discrepancy in results between Jones and Miller (1981) and Shagass et al. (1983).

Weller and Kugler (1979) and Kugler and Henley (1979) examined performance on a number of manual tasks, including three shown to be impaired in commissurotomised patients. In brief, where significant deficits for the schizophrenic patients were found they were in the direction of poorer right hand performance. Although the authors interpreted their results in terms of both a callosal transfer deficit and a lateralised left hemisphere impairment, certain aspects of their data appear to have been incompletely, and hence inappropriately, analysed. In addition, their findings may be explained more parsimoniously by factors unrelated to inter-hemispheric transfer (Gruzelier, 1979).

Dimond, Scammell, Bryce, Huws and Gray (1979) claimed that approximately one third of their sample of chronic, medicated schizophrenics showed an inability to name correctly a variety of everyday objects placed in their left hand, though objects placed in the right hand were named correctly. Such left hand anomia occurs reliably in split-brain patients and thus the results of Dimond et al. might be held to support the idea that some, but by no means all, schizophrenics show defective inter-hemispheric transfer. Unfortunately, the data in the paper by Dimond and his collaborators were poorly presented and the suspicion arises that the statistical analyses carried out were inappropriate. However, if the finding proves robust it draws attention to the fact that not all patients may show the same defects.

Although the 'disconnection hypothesis' has stimulated considerable interest, it is by no means clear that the disconnection symptoms reported are, in fact, related to schizophrenic pathology per se. In all the studies reported, the patients were chronic schizophrenics virtually all of whom were taking phenothiazine medication. Even if taken temporarily off drugs, the effect of prior medication can be very persistent. It is possible, therefore, that the medication, rather than any structural abnormality intrinsic to schizophrenia (or certain sub-types of schizophrenia), is responsible for the disconnection effects observed. The same applies to the finding by Rosenthal and Bigelow (1972) of an enlarged callosum in the brains of 'schizophrenics', a possibility acknowledged by the authors.

A second point is that lack of inter-manual transfer or cross-transfer between the visual fields could be due to an attentional, rather than a structural deficit. Perhaps schizophrenics are unable to focus their attention to more than one side of the mid-line simultaneously, or have difficulty allocating attention to each side of lateral space in turn. This could itself be a consequence of some defect in the callosal transfer system, given evidence which suggests that the callosum is intimately involved in the distribution of attention between the two sides of the body (Kreuter, Kinsbourne and Trevarthen, 1972; Ellenberg and Sperry, 1980).

Whatever criticsms may be levelled against the original formulations, the 'left hemisphere disturbance' and 'defective inter-hemispheric transfer' hypotheses concerning schizophrenia have, either explicitly or implicitly, guided much recent research. It is convenient to review this work according to the stimulus modality employed.

## AUDITORY TASKS AND SCHIZOPHRENIA

According to the disconnection hypothesis, schizophrenics should show a large right ear advantage in dichotic listening, that is, the left ear response should extinguish, as has been found in split-brain patients (see Chapter 3). If, on the other hand, schizophrenia is associated with pathology of the left hemisphere, poor recall from the right ear might be expected. These opposing predictions were tested by Gruzelier and Hammond (1980). The results offered no support for either hypothesis. No overall difference was found for either ear in comparison with normal controls. However, when patients were classified in terms of degree of arousal and presence of paranoia a difference between the ears was found for a small group (n = 5) of high arousal paranoid patients due to a poor score for the left ear. Although this is what the 'defective inter-hemispheric transfer' hypothesis predicts, Gruzelier and Hammond point out that the large difference between ears disappeared when attention was directed to the non-preferred ear. This suggests not so much a faulty structural pathway in these patients as an initial bias towards processing stimuli presented to the right ear.

Colbourn and Lishman (1979) reported that male schizophrenics failed to show the right ear advantage seen among normal controls on a dichotic task. Left ear performance of the schizophrenics was similar to that of the control subjects and thus provided no evidence for a 'disconnection' hypothesis. Inspection of the data revealed a bimodal distribution of ear-difference scores, 5 out of 10 male patients showing a significant left ear advantage (as measured by the phi coefficient) and 2 showing a significant advantage for the right ear. There is thus a suggestion of a left hemisphere deficit among some schizophrenics, since the frequency of a significant left ear advantage among normal and psychiatric control subjects was not as high as in the schizophrenic group. (Among manic-depressive patients there was nothing to suggest a right hemisphere deficit).

Caudrey and Kirk (1979) also found evidence suggestive of left hemi-sphere deficits in schizophrenics compared with normal or depressed patients, in so far as schizophrenics were relatively impaired on a task involving identification of phonemes as compared with a musical task. Male, but not female, manic-depressed patients were relatively poor on the music task compared with the verbal task. The two groups of patients, schizophrenic and (manic) depressed, also showed opposite ear advantages on the phoneme-task, at least when high levels of noise were presented to the ear not receiving the verbal stimuli. Schizophrenics showed a right ear advantage, depressives an advantage for the left ear. However, with a low level of contralateral noise both schizophrenics and depressives showed right ear advantages comparable to those of non-psychiatric control subjects. Thus there was a shift to a left ear advantage among depressives in the presence of a high level of contralateral noise. This might suggest that at high intensity levels of noise the left hemisphere in depressives is less adept than normal at 'unscrambling' relatively meaningless verbal stimuli from 'background' noise. Again this fails to support the view that it is the right hemisphere that is perturbed in depression.

Gruzelier and Hammond (1979) found no difference between schizophrenics and control subjects in degree or direction of dichotic ear asymmetry when strings of competing digits were presented to the two ears. This pattern of performance in schizophrenics was uninfluenced by drug administration or by severity of symptoms. However, if the intensity of the stimuli presented to each ear was different a discrepancy between the performance of patients and control subjects did emerge.

When subjects were required to recall the lower intensity (weaker) digits from one ear before the louder digits from the opposite ear, both groups performed similarly in the condition in which the louder digits were presented to the left ear.

When the weaker signal was in the left ear and had to be recalled before the louder right ear signal, schizophrenics showed poor performance in both ears compared with controls. Moreover, this deficit was greater for the first item presented in each ear than for the third item. Gruzelier and Hammond therefore interpreted their results to mean that schizophrenics were unable to inhibit left hemisphere processes. This was said to reflect a cognitive failure, involving semantic encoding and retrieval of items, rather than a deficient sensory pathway.

If two clicks are sounded in quick succession to each other they may be heard as one. The minimum interval required for the two clicks to be heard separately defines the two-click threshold. The presence of a cortical lesion lengthens this interval, especially if the first click is presented to one ear and a second, longer click is presented to the ear contralateral to the lesion. Yozawitz, Bruder, Sutton, Sharpe, Gurland, Fleiss and Costa (1979) therefore reasoned that resolution of two clicks should be poorest when the long click is presented to the ear opposite a dysfunctional hemisphere. These workers tested 9 patients with diagnoses of affective psychosis, 10 schizophrenics, 10 control patients and 2 patients with right temporal lobe lesions. The patients in the affective group showed the greatest difficulty in resolving the two clicks when the long click was presented to the left ear, as did the temporal lesion patients. Performance of the affective group differed significantly from that of the normal and schizophrenic groups combined. Since the 'affective' patients had been initially diagnosed as schizophrenic, the affectives and schizophrenics were receiving equivalent medication at the time of the study and therefore this cannot be the cause of the different patterns of performance in the two groups. The data thus suggest that there was some degree of right hemisphere functional disturbance in the affective patients, but there was little to suggest a left hemisphere deficit in schizophrenia.

## The effect of changes in clinical state

The possibility that changing clinical state may be correlated with changing laterality patterns of performance has been noted by some authors. Wexler and Heninger (1979), using a dichotic syllable test, found no difference in mean laterality scores between 8 schizophrenic, 6 schizo-affective, 12 depressive patients and 23 normal control subjects. Ratings of psychotic thought and behaviour were highest when laterality scores were lowest, reflecting a minimum degree of asymmetry. On the other hand, when symptoms were most improved there was, for all clinical groups, an associated increase in laterality scores. As this change was unrelated to medication levels and to overall accuracy on the task the finding was thought to support the view that during acute psychotic illness there is a breakdown in the inter-hemispheric inhibition that normally maintains a laterality balance. However, given that there was no control over order of report it is just as likely that the results reflect an ear or order-of-report strategy change on the part of the patients.

Gruzelier and Hammond (1979) found a right hand advantage in reaction times to the words 'left' and 'right' presented to either the left ear, the right ear or binaurally among those schizophrenics who responded to medication. Where symptoms were refractory to drug treatment there was a left hand advantage. Thus active symptoms of schizophrenia appeared associated with a relative impairment of the left hemisphere in response organisation whereas patients whose symptoms were in remission showed a relative superiority of the left hemisphere. In addition, responses to binaurally presented instructions were significantly slower than with monaural input, suggesting a possible impairment in inter-hemispheric integration. Green, Glass and O'Callaghan (1979) noted a similar pheno-menon in that their schizophrenic patients showed poorer comprehension of stimuli presented binaurally than monaurally. They also showed a significantly greater laterality effect than control subjects in the monaural condition due to particularly poor performance at the left ear.

## The influence of particular symptoms

Among normal subjects, as the duration of a signal is decreased, its intensity may have to be increased in order for it to be detected. Among *hallucinating* schizophrenics Bazhin, Wasserman and Tonkonogh (1978) found an asymmetrical increase in threshold for the right ear which was not found among schizophrenic patients free of hallucinations. Since the effect of a cortical lesion is to produce an increase in threshold for the contralateral ear, the authors saw their result as supporting the idea that schizophrenia involves left temporal lobe pathology. However, the absence of effect in non-hallucinating patients implies that the 'pathology' is of a dynamic rather than a structural nature. Indeed, it has been reported that over a period of two years, during which a young male schizophrenic patient was tested several times on inter-manual transfer tasks and monaural tests of speech comprehension, defects on transfer and asym-metry between the ears were found only when the patient had been 'hearing voices' during the preceding seven days (Green, Glass and O'Callaghan, 1979).

The importance of considering particular schizophrenic symptoms is shown in a study by Alpert, Rubinstein and Kesselman (1976) who presented information monaurally against a background of binaural noise. The information was either semantically well integrated (SWI) such as 'the cows ate the grass' or poorly integrated semantically (SPI) such as 'the artist found the badge'. The subject's task was to repeat the stimulus sentence. There were 10 normal subjects, 21 schizophrenics with hallucinations and 11 non-hallucinating schizophrenics. Normals and hallucinators showed higher right ear recall for SWI than for SPI information; the non-hallucinators showed this effect for the left ear, suggesting some degree of defective left hemisphere function.

Lishman, Toone, Colbourn, McMeehan and Mance (1978) gave a dichotic digit test to recently recovered manic-depressive and schizophrenic patients. Inter-ear differences were larger in the psychotic patients than in normal control patients, although, considering the patient groups and each sex separately, only male schizophrenics differed significantly from male control subjects in terms of a laterality quotient. Schizophrenics did poorly on the left ear but better than normals on the right ear. Thus in these recovered schizophrenics there is no evidence of left hemisphere dysfunction.

In a dichotic listening study by Lerner, Nachson and Carmon (1977) paranoid schizophrenics were found to show a greater difference between the ears in favour of the right side than either non-paranoid patients or normal control subjects. When the data were examined for the number of shifts from ear to ear in order of report the rank order of the groups reversed. Thus paranoid patients who showed a large right ear advantage tended to report from the right ear first more consistently than did the other two groups. Similar findings were obtained by Gruzelier and Hammond (1976; 1978) who also noted that in some circumstances paranoid patients showed a larger right ear advantage than non-paranoid cases. These data may be interpreted in terms of a relative over-activation of the left hemisphere among paranoid compared with non-paranoid schizophrenics.

Gruzelier (1981) discusses psycho-physiological findings which indicate heightened left hemisphere arousal in paranoid schizophrenics. More recently he has found that asymmetry of the galvanic skin response of left and right hands is related to specific features of schizophrenia (Gruzelier and Manchanda, 1982). A higher right than left hand response tends to go with a cluster of symptoms suggestive of over-activation of the left hemisphere. Patients showing electrodermal asymmetry in the opposite direction tend to show symptoms suggesting a reduction in left hemisphere activity. The symptom-clusters identified overlap with, but are not identical to, the clinical distinction between paranoid and non-paranoid schizophrenia.

This brief review of auditory studies reveals some support for the hypothesis that there is a disturbance of left hemisphere functions in schizophrenia. However, the absence of positive findings in some studies, combined with evidence of a shift in the balance of lateralised cerebral processes as a function of diagnostic category, clinical state or task requirements, suggests that any dysfunction of the left hemisphere has a dynamic rather than a structural basis. The often lower performance levels of schizophrenics relative to controls for verbal material is consistent with a left hemisphere deficit even in the presence of a greater REA among schizophrenics (e.g. Caudrey and Kirk, 1979; Green et al., 1979) if it is assumed that schizophrenics have difficulty in switching attention from the right ear.

## TACHISTOSCOPIC STUDIES AND SCHIZOPHRENIA

In a widely cited study Gur (1978) presented three-letter nonsense syllables followed by a visual mask to schizophrenics and normal control subjects. Whereas the latter showed the expected RVF superiority, schizophrenics gave a significant LVF superiority. Since an advantage for the LVF was found in both groups of subjects for a dot localisation test this suggests a specific left hemisphere impairment with verbal stimuli among schizophrenics.

Hillsberg (1979) presented pairs of arrows to either one cerebral hemisphere or a single member of the pair simultaneously to the two hemispheres, a task first employed by Dimond (1970a). It was argued that a callosal deficit would lead to poor performance in the bilateral condition while a left hemisphere deficit would lead to poor performance in both the bilateral and the RVF condition. The results showed that four out of five medicated schizophrenics were slower in responding to RVF than to LVF presentations, while the normal control group showed no visual hemifield asymmetry. It turned out that the one patient whose results were not in line with those of the other four patients had shown 'strong affective components in his behaviour towards the end of his participation' and Hillsberg therefore considered that 'the diagnosis of schizo-affective psychosis might have been more correct'.

With regard to the bilateral condition, no evidence that schizophrenics performed especially poorly on these trials was obtained. All subjects, psychotic and normal control, were slower on bilateral trials. However, both Dimond (1970a) and Davis and Schmit (1971) reported faster response times to stimuli 'shared' between the hemispheres in comparison with the same information directed to a single hemisphere. The performance of the control subjects in the bilateral condition of Hillsberg's study, therefore, may be considered atypical, which may have minimised any real difference between patients and controls.

Pic'l, Magaro and Wade (1979) also failed to find any visual-field difference between schizophrenics and normal control subjects (or between paranoid and non-paranoid patients), both groups showing a RVF superiority on tachistoscopic letter recognition. However, Eaton (1979) reports that medicated, acute schizophrenics were significantly slower than normal subjects on a tachistoscopic letter-name matching task but not on tests of digit or form matching. Like normal subjects the schizophrenics showed a RVF superiority on the letter-matching task and a LVF superiority in matching forms. Unfortunately, the full data are not readily available for scrutiny but if the schizophrenic reaction time deficit is confined to verbally mediated tasks this would be presumptive evidence of a left hemisphere deficit. Eaton also provides data which Gruzelier (1979) takes as showing that the effect of neuroleptic medication is to significantly improve left hemisphere performance. If so, this supports the findings of Hammond and Gruzelier (1978) that administration of chlorpromazine improves right ear performance on a temporal discrimination task. Both sets of findings are consistent with those of Serafetinides (1973) who showed that chlorpromazine produced an increase in voltage of the EEG over the left side of the head relative to control recordings taken during periods of placebo administration.

Further evidence in favour of the 'left hemisphere deficit' hypothesis comes from a study by Connolly, Gruzelier, Kleinman and Hirsch (1979) who found reaction times of schizophrenics in a so-called lexical task to be faster in the LVF than in the RVF, unlike non-psychotic control subjects who showed the expected superiority of the RVF. However, the lack of a RVF superiority among schizophrenics was also shown by depressive patients and therefore cannot be considered characteristic only of schizophrenia.

There seems to be a measure of agreement from the studies carried out using tachistoscopic recognition or reaction time procedures that there is some impairment of left hemisphere processes in schizophrenia, although different methods of treating results may lead to different interpretations of the data. Colbourn and Lishman (1979), for example, found a RVF impairment in schizophrenics' word recognition performance when the raw data were analysed in the 'conventional' way, but when a laterality ratio (the phi coefficient – see Chapter 4) which takes into account overall level of accuracy was computed the left hemisphere 'deficit' disappeared. Of course, this sort of effect is not restricted to study of clinical groups but it does caution against too ready a description of a particular result as reflecting a purely structural impairment.

## ELECTROPHYSIOLOGICAL STUDIES

Electrophysiological studies of inter-hemispheric functional relations have been increasingly carried out among psychiatric patients in recent years. Unfortunately, the increasing sophistication of the methodology employed (see e.g. Perris, Knorring and Monakhov, 1979) has not generally been matched by equal sophistication in theorising about the data obtained. The effects that have been reported remain isolated findings, often unreplicable, the meaning of which, in psychological terms, is largely unknown. The fact is, we still do not know what conclusions can be drawn from a demonstration that, for example, two groups of subjects differ in terms of waveform stability, inter-hemispheric coherence or whatever. In a summary of EEG studies reported at a conference and subsequently published in a book under his editorship (with Flor-Henry) Gruzelier (1979) had this to say:

> It is clear from all the studies that functional cerebral organisation in psychosis differs from that found in the comparison groups, usually right-handed normals, ... beyond this the evidence as to lateralized and inter-hemispheric dysfunction seems to be in almost total conflict as to the form that these might take (p. 660)

Given this conflict, only very brief mention will be made here of some of the findings reported at the conference.

From studies comparing resting EEG records (in which the subject simply relaxes, supposedly thinking of nothing in particular!) little can be concluded, since the results of three studies were in total disagreement. Shaw, Brooks, Colter and O'Connor (1979) found more power on the left side of the brain among controls but not patients while Weller and Montague (1979) found such asymmetry among patients but not controls. When the EEG was recorded while subjects performed cognitive tasks purporting to reflect left hemisphere functions, Flor-Henry et al. (1979) obtained results implying comparatively less left hemisphere involvement in such tasks among schizophrenics while the findings of Shaw et al. suggested the contrary.

Buchsbaum, Carpenter, Fedio, Goodwin, Murphy and Post (1979) recorded the evoked visual potential at contralateral and ipsilateral sides of the head in relation to the hemifield of stimulation. They found the increase in amplitude with attention to be equivalent in normals and schizophrenics for stimuli transmitted along the direct visual pathway to a given hemisphere, e.g. from LVF to right hemisphere. For information transmitted from the RVF to the right hemisphere, that is, across the midline commissures, the amplitude increase was much less marked for schizophrenics. This result is consistent with the notion of left hemisphere pathology in schizophrenia, rather than with the inter-hemispheric deficit hypothesis, since the amplitude difference observed between schizophrenics and normals was not found for the LVF-left hemisphere route. This view is reinforced by the finding that left temporal lobectomy patients (and, to a lesser extent, right temporal patients) also showed enhanced potentials for the direct compared with indirect route.

Tress, Kugler and Caudrey (1979), examining 12 schizophrenics and 7 normals, recorded evoked potentials to tactile stimuli applied to the forearm. A longer latency in the left hemisphere to stimulation of the right arm most discriminated patients from controls. Shagass, Roemer, Straumanis and Amadeo (1979) also found differences in the left hemisphere between schizophrenic and depressed patients on the one hand and control subjects on the other.

These latter studies employing evoked potential techniques are consistent, then, with the idea that the left hemisphere in schizophrenics is in some sense perturbed in comparison with normals. Since these studies did not use verbal stimuli the results imply that the 'perturbation' is not related only to verbal processing but manifests itself in other operations of the left hemisphere.

## RÉSUMÉ

Although data are not entirely consistent there is converging evidence from studies using different methodologies that the left hemisphere is perturbed in schizophrenia, at least for certain categories of patient and at certain stages of illness. The evidence includes decreased amplitude (Buchsbaum et al. 1979) and increased latency (Tress et al. 1979) of the evoked potential response, decreased accuracy (Gur, 1978) and increased reaction time (Hillsberg, 1979; Connolly et al., 1979) to right visual hemifield stimuli presented tachistoscopically, increased right ear auditory detection threshold (Bazhin et al. 1978) and decreased right ear advantage (Colbourne and Lishman, 1979) in dichotic listening. Consistent with the idea of a left hemisphere deficit, Golden, Graber, Coffman, Berg, Newlin and Bloch (1981) reported that chronic schizophrenic patients showed a significant loss of density in the left hemisphere relative to controls, as indicated by CAT scans. Since there were no apparent correlations between density loss and medication level it is unlikely that this finding is an artefact of drug treatment. Although certain aspects of the behavioural evidence (Bazhin et al. 1978; Gruzelier and Hammond, 1979; Wexler and Heninger, 1979; Gruzelier, 1981) imply that the nature of any left hemisphere impairment in schizophrenia is 'dynamic', this is not necessarily incompatible with a structural impairment. For example, a structural defect might influence, though not entirely determine, the degree of arousal or attention within a hemisphere. Thus the finding of an exaggerated dichotic right ear advantage but decreased recall in schizophrenics compared with normals (Lishman et al., 1978; Green et al., 1979; Gruzelier and Hammond, 1980) is not inconsistent with the notion of a left hemisphere structural impairment, if the effect of the latter is, under some circumstances, to overactivate the left hemisphere, making it difficult to switch attention from the right ear.

There is only slight evidence (Yozawitz, et al., 1979) to support the idea that the right hemisphere is specifically impaired in cases of depression, while some data suggest a left hemisphere impairment (Caudrey and Kirk, 1979; Connolly, et al., 1979; Shagass et al., 1979).

## LATERALITY OF MOTOR PERFORMANCE

Anomalies of lateral motor performance among psychiatric patients have been reported by a number of authors. Such anomalies are interesting in so far as they may, in some cases, be indicative of lateralised cerebral dysfunction.

Walker and Birch (1970) observed poorly developed preference for hand and eye usage in a high proportion of children at risk for schizophrenia, and Oddy and Lobstein (1972) reported an increased incidence of 'crossed dominance' between preferred hand and eye among adult schizophrenics. While such 'crossed dominance' (and comparable inconsistencies of preferred hand, ear and foot) is very common in the normal population (see Chapter 2) and must be considered to have no pathogonomic significance, it does not necessarily follow that a raised incidence of such cases among clinical patients is of no theoretical interest. It is quite possible that in some cases crossed laterality, indeterminate hand preference or whatever, is entirely without significance, while in other cases the same effect may have a different cause. In other words, crossed laterality or left handedness in an individual not otherwise predisposed to show this pattern or direction of preference might have pathological implications.

Dvirskii (1976) reported an excess of left handers among schizophrenics compared with normal adults. Among 660 male schizophrenics, 5.25 per cent were said to be left handed in comparison with 3.02 per cent of 2190 controls. The figures for females were 9.24 per cent among 610 schizophrenics compared with 5.07 per cent of 2150 normal controls. It is unfortunate in view of the commendably large number of patients and controls that Dvirskii's paper does not provide explicit criteria for the designation of left handedness. It is merely stated that, 'To determine left-handedness attention was paid to preferential use of the left hand' during a number of activities. It may be that the excess of left handers among the schizophrenic subjects can be attributed to some sort of bias. Taylor, Dalton and Fleminger (1980) state that among their schizophrenics, 'A number... claimed increased use of the left hand' which was not substantiated at interview. If Dvirskii asked patients for self-reports as part of, or wholly as, the method of determining handedness the increased sinistrality he reported may be entirely artefactual. The same may be true of the finding of increased use of the left hand among 84 schizophrenics compared with 82 hospital employees reported by Nasrallah, Keelor, Van Schroder and Whitters (1981). However, a tendency towards greater left-handedness among 200 schizophrenics compared with 200 normals was also found by Gur (1977) who assessed sensorimotor preference by means of a questionnaire combining items concerning hand and foot preference and three objective tests of eye dominance. In contrast to the results of Oddy and Lobstein (1972) no increase of crossed hand-eye preference was found in schizophrenics. A subset of 30 male and 30 female schizophrenics was subsequently asked to perform each of the tasks specified in the question-naire. The validity of the instrument as a test of handedness in schizo-phrenics was found to be comparable to that for normals (Raczkowski, Kalat and Nebes, 1974). Thus the increased left sidedness found by Gur is unlikely to be attributable to unreliable reporting by schizophrenics.

In contrast to the above results Wahl (1976) found no excess of sinistrality among schizophrenics but his final groups consisted of only 21 schizophrenics, 19 non-schizophrenic hospital patients and 18 staff members. These figures are too low to give reliable estimates.

Lishman and McMeekan (1976) used Annett's (1970a) questionnaire to assess handedness in 65 male and 65 female psychotics, whose responses were compared with those given by a group of control subjects and by Annett's samples of normal subjects. A greater proportion of mixed and left handers was found among psychotics, especially the young, and mainly in manic-depressive and schizo-affective patients compared with schizoph-renics. This greater incidence of sinistrality and mixed hand usage was more marked for male than for female psychotics. Contradictory results were published by Fleminger, Dalton and Standage (1977a) who also used Annett's questionnaire. These workers had 357 males and 443 females in each of their clinical and normal control groups. Unlike Lishman and McMeekan, Fleminger's group found a higher proportion of fully dextral patients among schizophrenic and affective patients, particularly females, than among normal control subjects. There was no significant difference in the distribution of handedness between the two clinical sub-groups. Since Fleminger et al. classified as dextral only subjects who answered 'right' to all 12 items of the questionnaire, their findings of an excess number of right handers cannot be put down to a more lax criterion of right handedness than was adopted by Lishman and McMeekan, who classified their patients in terms of the groupings identified by Annett.

The issue of handedness among psychiatric patients was recently re-examined by Taylor, Dalton and Fleminger (1980). Using a new series of patients but the same classification scheme for handedness as Fleminger et al. (1977a) they found a similar result, namely an under-representation of left and mixed handers among psychotics (this time composed entirely of schizophrenics).

Taylor, Dalton, Fleminger and Lishman (1982) looked more closely at the discrepancy between the findings of Lishman and McMeekan (1976) and Fleminger et al. (1977a). Data from the Fleminger study were re-classified according to the criteria used by Lishman and McMeekan. The effect of this reclassification was to abolish most of the apparent differences in handedness patterns between the total psychotic samples in the two investigations. The remaining discrepancy was that there was still an excess of young, left handed males in the Lishman series. This excess was due entirely to 6 patients. The schizophrenics from the two series were remarkably similar in their pattern of handedness. In both samples there was a non-significant trend towards excess sinistrality among males and a trend in the opposite direction among females. This suggests that future studies should take account of a possible sex difference in the pattern of handedness shown by schizophrenics.

Differences in diagnosis and length of illness also suggest themselves as factors to which attention should be paid in future investigations. Dvirskii (1976) claimed that the excess of sinistrals found in schizophrenics was due very largely to patients whose illness was continuous rather than episodic or recurrent. Age, too, though not an obvious source of discrepancy in the studies published hitherto, must be considered relevant in view of the suggestion of changing patterns of handedness with age (Fleminger, Dalton and Standage, 1977b). In addition, the results of studies quoted elsewhere in this chapter suggest that it may be worthwhile to consider the possibility that medication influences hand usage. It is also conceivable that preferential use of one hand alters with a patient's changing clinical state.

Flor-Henry (1979b) discusses three cases of depression and manic-depressive illness in which a change in the patient's clinical state was associated with a shift in hand usage. One woman was said to be fully sinistral during manic episodes and fully dextral when free of symptoms. In another case a male patient was ambidextrous normally but used his right hand preferentially for all skills, except writing, when he became depressed.

## Twins, handedness and schizophrenia

Much of the impetus for considering laterality effects in psychopathology stems from the search for biological foundations of psychiatric disease. An investigation into the biological basis of schizophrenia was undertaken by Boklage (1977). In comparing monozygotic and dizygotic twins, in whom at least one twin was schizophrenic, he discovered that the incidence of non-right handedness was much higher among monozygotic than among dizygotic twins, and higher among the former than has been reported to be the case for normal monozygotic twins. The frequency of non-right handedness among 'schizophrenic' dizygotic twins was similar to that for normal dizygotic twins, which is higher than among non-twin siblings (Springer and Searleman, 1980). In the monozygotic sample, there were 12 pairs of twins where both were fully dextral and in 11 of these cases both twins were, or had been, diagnosed as schizophrenic. Out of 16 pairs where one or both twins were non-right handed, only 4 pairs were concordant for schizophrenia. Further, the severity of psychosis was less in the non-right handed twins than in the fully dextral cases. Among the twin pairs discordant for schizophrenia the illness was no more likely to affect the dextral than the non-dextral member of the pair.

The difference between the relative frequencies of non-dextrality in monozygotic and dizygotic twins (selected for the presence of schizophrenia in at least one twin) suggested to Boklage that the embryonic abnormality that leads to monozygotic twinning is in some intimate way related to schizophrenia on the one hand and, on the other, to brain laterality (inferred from the high incidence of non-dextrality). The anomaly of cellular differentiation leading, in the pithy words of Taylor (1977), to the result that 'what should be the left side of Henry becomes George' also, in the view of Boklage, leads to anomalous cerebral laterality, which protects the individual from the worst ravages of the so-called 'schizophrenic syndrome'. In the same way, Boklage notes, personal or familial sinistrality carries with it an improved prognosis for recovery from aphasia, which is not to say either that the mechanism of 'protection' is the same with regard to aphasia and schizophrenia or to suggest that the two conditions are related. Recently Boklage, Elston and Potter (1979) have adduced evidence from measurement of teeth in twins to support the view that monozygotic twinning is associated with developmental anomalies of bilateral symmetry in body structure. At first blush, teeth seem a long way from brains, but as Boklage et al. point out,

> since the tooth buds derive from cells almost immediately adjacent to the forebrain end of the neural plate we may study twinning-related developmental influences very close to the source of our primary concern. (p. 88)

The above findings are likely to stimulate an increased interest generally in the embryogenesis of cerebral laterality in relation to psychiatric illness. Owing to the difficulty of finding sufficient numbers of twin pairs, especially where at least one twin is schizophrenic, progress will inevitably be slow. However, one study has already failed to corroborate the finding of an increased incidence of non-right handers among 'schizophrenic' monozygotic pairs, although supporting Boklage in finding a decreased severity of symptoms in the schizophrenic member (Luchins, Pollen and Wyatt, 1980). It should also be borne in mind that Springer and Searleman (1978) found no difference in mean laterality scores on a dichotic listening task between monozygotic twins, dizygotic twins and singletons. Within the twin groups, the correlation between ear asymmetry scores for each member of the pair was similar for monozygotic and dizygotic twins. Among normal twins, therefore, there is no evidence from this study of a genetic component in the magnitude of ear asymmetry and, by implication, cerebral laterality. The findings of Boklage with regard to handedness might, therefore, be viewed as secondary to the schizophrenia, rather than indicating a common biological origin for handedness and schizophrenia.


## CHILDHOOD AUTISM

Taylor (1977) wondered whether we can 'really afford to allow conceptual discontuity between childhood autism and schizophrenia'. As with schizophrenia, some researchers have investigated autistics for evidence of unusual patterns of laterality (for review see McCann, 1981).

A number of studies have looked at handedness. Only one (Colby and Parkinson, 1977) reported a significant difference between autistic children and control subjects. Out of 20 autistics, 13 were non-right handed in comparison with 3 out of 25 control children. However, it is possible that these findings reflect an earlier developmental stage in autistics since studies controlling for level of cognitive functioning as well as for age have not revealed any difference in hand preference between autistic and control groups (Boucher, 1977; Barry and James, 1978).

No left handers and only 6 mixed handers were observed in a group of 23 autistics studied by Prior and Bradshaw (1979). This contrasts with the excess of non-right handers reported by Colby and Parkinson. The discrepancy may be due to sampling differences or to different degrees of mental retardation, which has been shown in at least one study to relate to the incidence of left handedness. The greater the severity of retardation, the higher was the incidence of left handedness (Hicks and Barton, 1975).

There have been suggestions that autism is characterised by an unusually high proportion of cases with right hemisphere speech representation. Prior and Bradshaw (1979) found significantly more autistics than normal subjects showing a left ear advantage on a verbal dichotic listening task. Dawson, Warrenburg and Fuller (1982) deduced from EEG recordings that 7 out of 10 autistics had language in the right hemisphere. In addition, the pattern of cognitive skills of the autistics revealed a profile suggestive of left hemisphere impairment. This may explain the relatively large number of subjects with putative right hemisphere speech.

As well as a high number of reversals of the usual pattern of hemispheric speech representation, it has been reported that autistics show an unusually high frequency of anatomic asymmetry. Using computerised tomography to measure brain size, Hier, Le May and Rosenberger (1979) found that a greater proportion of autistics showed a wider parieto-occipital region on the right side than either of two control groups (mentally handicapped and neurological patients without space-occupying cerebral lesions). However, these findings were not replicated by Damasio, Maurer, Damasio and Chui (1980) or by Tsai, Jacoby, Stewart and Reisler (1982). The latter authors suggest several methodological reasons why Hier et al. obtained the results that they did. D'Angelo (1981) criticised Hier et al. for including cases not showing any asymmetry in the parieto-occipital region with those cases in which this area was larger on the left. When symmetrical cases were treated as a separate group, no significant difference emerged between Hier's autistic patients and the control groups (D'Angelo, 1981).

The idea that autism is characterised by anomalies of manual or cerebral laterality is interesting in view of the aetiological implications. However, adequate supporting evidence for this idea has still to be provided.


## HYSTERICAL PHENOMENA

In a paper considering the implications for psychiatry of cerebral asymmetry of function Galin (1974) pointed out the similarity of the right hemisphere's mode of functioning to the Freudian notion of 'primary process thinking'. Briefly, this concept embodies the idea of mental events that are non-verbal, that operate by analogy rather than according to the principles of logic, and are unconstrained by temporal limitations. Galin suggested that since the right hemisphere is unable to verbalise, tensions and anxieties cannot be worked-through by means of language. They might, therefore, find expression through non-linguistic means. In short, the right hemisphere might be the source of many hysterical symptoms – the physical expression of sub-conscious mental conflicts.

Risse and Gazzaniga (1976) asked a patient to feel objects with her left hand while the left hemisphere was 'anaesthetised' by the action of sodium amylobarbitone. After the effect of the drug had worn off, the patient was asked to say what she had felt with her left hand. She was unable to do so. Subsequently, she was asked to feel for the objects again with the left hand: she succeeded in retrieving the correct objects from among a number of alternatives. These findings suggest that the memory of the objects did not spontaneously transfer to the left hemisphere after the effect of the drug had worn off. Thus it is conceivable that, in the absence of verbal encoding, events may escape registration in the conscious mind and be manifest through non-linguistic modes of expression. The right hemisphere would employ such channels and therefore one might expect to find hysterical symptoms in particular association with the right side of the brain, and hence the left side of the body.

Galin's idea that the right hemisphere may generate hysterical symptoms more than the left hemisphere receives support from studies reporting a higher incidence of left than right sided hysterical or psychosomatic symptoms among psychiatric patients (Stern, 1977; Galin, Diamond and Braff, 1977; Merskey and Watson, 1979). However, conflicting results (Bishop, Mobley and Farr, 1978) have been obtained.

An experimental test of the idea that there is a special relationship between the right cerebral hemisphere and unconscious mental processes was undertaken by Fleminger, McClure and Dalton (1980). These workers required subjects to listen to a tape suggesting that they had a tingling sensation in the hand. The subjects were then asked to say in which, if either, hand the sensation was felt. More reports were obtained of left than right hand sensations. Frumkin, Ripley and Cox (1978) had earlier found a significant reduction during hypnosis in the extent of a dichotic right ear advantage, an effect that has been replicated with a split-brain patient (McKeever, Larrabee and Sullivan, 1981). These findings raise the possibility that hypnotic suggestion is associated with greater activation of the right than the left hemisphere.

The 'phantom limb' phenomenon refers to the experience reported by some people of feeling pain in a part of a limb that has been amputated. It has been suggested from time to time that this represents an unconscious denial that the limb has been removed. This view has been disputed (see Weinstein, 1969) but it is interesting in the present context that reports of a phantom breast following mastectomy occur earlier after surgery to the left compared with the right breast (Weinstein, 1969). (It is also intriguing to note (MacManus, 1977) that left-sided breast tumours are more common than right-sided tumours.) However, the phantom limb phenomenon was found to occur more frequently after right than left limb amputations by Shukla, Sahu, Tripathi and Gupta (1982).

The right hemisphere may well be the seat of unconscious processes but on present evidence the case remains *sub judice*.

## SUMMARY

This chapter has reviewed evidence that certain psychopathological syndromes are associated with unusual patterns of cerebral organisation. In particular, it has been argued that schizophrenia is associated with disturbance of left hemisphere functions, and the affective psychoses with perturbation of right hemisphere functions. Although the evidence in favour of the second part of this proposition fails to convince, there is a measure of agreement that schizophrenia is associated with some dysfunction of the left hemisphere. Whilst a functional deficit can co-exist with a physical abnormality, psychophysiological research, and studies using the dichotic listening technique, suggest that any left hemisphere dysfunction is as likely to be of a dynamic as of a structural nature.

In addition to the 'left hemisphere-deficit' hypothesis, evidence was discussed relating to the claim that schizophrenia is accompanied by a defect in the transmission of information between the cerebral hemispheres. The strongest support comes from studies of inter-manual transfer, but overall the evidence is far from compelling.

There have been suggestions that schizophrenia and childhood autism are associated with anomalies of motor as well as cerebral laterality. The relevant literature reports are conflicting. It is possible that closer attention to such factors as the clinical state of individual patients and the presence or absence of particular symptoms will in future resolve some of the inconsistencies. However, in so far as an excess of left sided tendencies among schizophrenics has often been reported, this might be considered consistent with the left hemisphere deficit hypothesis.

Finally, it has been suggested that the right hemisphere is the seat of unconscious processes. The implications of this view were considered with particular reference to hysterical phenomena. There is some evidence that conversion symptoms occur more frequently on the left than the right side of the body. While this may be so, it would be unwise on present evidence to conclude that the left hemisphere has no part to play in the genesis of psychosomatic symptoms.

# 13

# Channel capacity, attention and arousal

Asymmetry of function between the left and right cerebral hemispheres should not be allowed to obscure the basic fact that the brain is a paired organ. Man's bilateral nervous system is a feature which he shares with all other vertebrate animals. Together with paired sensory receptors this may be essential for an isomorphic spatial mapping of the external world onto the brain (Young, 1962). The fact that perceived spatial location can in some circumstances be as profound a determinant of laterality effects as hemispheric specialisation of function (Goldstein and Lackner, 1974; Morais and Bertolson, 1973; 1975; Corballis, Anuza and Blake, 1978; Beaumont, 1979; Cotton, Tzeng and Hardyck, 1980; Bowers and Heilman, 1980a; Bowers, Heilman and Van Den Abell, 1981) supports the view that evolution accorded a high priority to the cortical representation of space.

The duplicate structure of the brain means that each side can register and respond to events occurring in the contralateral half of space. If the two halves of the brain were entirely autonomous then coincident events on left and right sides could lead to simultaneous but incompatible responses. One side of the body might try to go forward while the other tried to go back. In order for this indecision to be resolved some form of neural integration is required such that one side gains ascendancy over the bilateral locomotor apparatus (Kinsbourne, 1974a). To discover how the two halves of the brain interact to produce co-ordinated patterns of behaviour remains one of the most challenging problems of neuropsychology. Yet few scientists have addressed themselves to this problem (but see Denenberg, 1980). Mechanisms of inhibition must be involved for, as Jung (1962) points out, 'if inhibition did not occur in the cortex we would all be epileptics, and if it did not occur between the hemispheres we would not develop skilled voluntary movements' (p. 274).

Although normal behaviour must be the product of the integrated activities of both hemispheres, it would be strange if each side of the brain did not retain some capacity for independent action even in the presence of intact mid-line commissures. Why else are the limbs on left and right sides of the body each controlled from the hemisphere that analyses the corresponding half of visual space – and why else is there hemispheric asymmetry of function?

## THE NORMAL BRAIN: ONE OR TWO CHANNELS?

Dimond (1972) has been the foremost exponent of the view that, even in the normal intact brain, the two hemispheres act to some extent as separate channels in analysing incoming information. Part of his evidence for this view comes from experiments in which stimuli were directed to either a single hemisphere or to both hemispheres at once.

In one experiment Dimond (1971) presented a pair of letters either to the left hemisphere or to the right hemisphere (unilateral condition) or one letter of the stimulus pair went to the left hemisphere at the same time as the second letter went to the right hemisphere (bilateral condition). The apparatus used to achieve this consisted of two sheets of ground glass arranged vertically and at an angle to each other. A vertical partition was set perpendicular to each sheet of glass such that it was divided into two separate halves, thereby making four screens in all. These are numbered 1 to 4 in Figure 8. The subject sat facing a fixation light set in the angle between screens 2 and 3. The vertical partitions ensured that screen 1 was seen only by the nasal hemiretina of the left eye, screen 2 only by the temporal hemiretina of the right eye, screen 3 only by the temporal hemiretina of the left eye, and screen 4 only by the nasal hemiretina of the right eye. Thus any combination of hemiretinal areas could be stimulated at will.

Dimond (1971) found that if a pair of letters was presented to screens 1

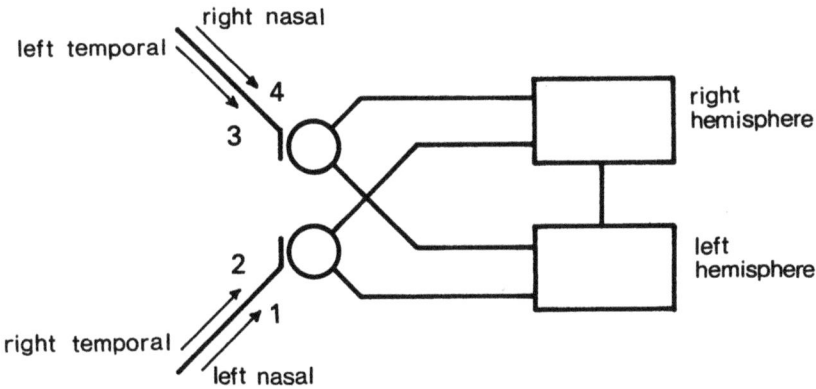

**Fig. 8** Diagram showing apparatus and visual pathways

and 4 or screens 2 and 3 this resulted in more accurate recognition than if two letters were presented to screens 1 and 2 or to screens 3 and 4. Comparable findings – i.e. better performance in the bilateral conditions – were obtained in other experiments (but see Dimond, 1969) using both recognition and reaction time as response measures (Dimond and Beaumont, 1974). The conclusion drawn from these studies was that in the bilateral condition the two halves of the brain could analyse in parallel the information presented to them. When two items were sent to a single hemisphere, however, there was a 'block to function' since each item had to be analysed in turn.

Dimond and Beaumont's conclusion has been criticised by Mollon (1976) who pointed out that the apparatus used in their experiments confounds hemifield of input with retinal eccentricity. The same applies to studies by Davis and Schmit (1971, 1973) who came to similar conclusions as Dimond and Beaumont.

In Dimond's apparatus screens 1 and 4 are further from the fixation light, and hence stimuli are further from the fovea, than screens 2 and 3. Mollon points out that there is no reason to suppose that presenting stimuli close to fixation (screens 2 and 3) on 50 per cent of bilateral trials and further from fixation on the other 50 per cent of trials is equivalent – with regard to the areas of retina stimulated – to having one stimulus close to fixation (screen 2 or 3) and the other further from fixation (screen 1 or 4) on unilateral trials. Thus according to Mollon the results obtained by Dimond and Beaumont do not justify the conclusion that superior performance in the bilateral condition is due solely to distributing the information between the two cerebral hemispheres.

Contrary to Mollon's understanding that '*both* stimuli are relatively close to the fixation point on *50 per cent* of trials' (my italics), Dimond and Beaumont do not state in the article reviewed by Mollon that in the bilateral condition stimuli were presented only to the nasal screens (1 and 4) or only to the temporal screens (2 and 3). As papers by Dimond (1970a; 1971) make clear, the combinations of screens 1 and 3 and screens 2 and 4 were also used. In terms of the distance of the stimuli from the fovea, these combinations of retinal areas are equivalent to the combination of screens 1 and 2 or screens 3 and 4. Performance is still superior in the bilateral conditions.

The main thrust of Mollon's criticism of Dimond and Beaumont's apparatus is that the degree of separation between the stimuli is not the same in the bilateral and unilateral conditions. Since interference between stimuli that are spatially contiguous can be considerable, the lower performance levels in the unilateral condition may be attributable to the closer proximity of the stimuli in this condition. Mollon is presumably thinking here of the phenomenon of lateral inhibition in the retina (Hartline and Ratliff, 1957). However, owing to the vertical partitions in Dimond's apparatus the two stimuli in the unilateral condition never stimulate the same eye. Lateral inhibition in the retina therefore can not account for the poorer level of performance, although arguably a similar phenomenon at a cortical level could do so. The influence of retinal factors is shown instead in the bilateral condition. Presenting information to left and right hemispheres simultaneously through the nasal or temporal hemiretinae leads to less accurate performance than when the same information is presented to the two hemispheres via the hemiretinae of a single eye (Dimond, 1970a; 1971). Even using a single eye to present information to two hemispheres, however, still leads to better performance than directing information to one hemisphere via different eyes. This supports Dimond's view that the two hemispheres act as separate channels in the receipt of information.

The question of whether the two hemispheres operate as two separate channels for processing information was approached in a different way by Guiard and Requin (1977). The concept of the brain as a single channel of limited processing capacity was developed largely as a result of the phenomenon known as the psychological refractory period (Welford, 1967). If two signals, each requiring a response, are presented in fairly close temporal succession then the time taken to respond to the second signal is considerably extended relative to the response time to the first signal. Guiard and Requin (1977) presented the two signals either to a single hemisphere or to both hemispheres and asked their subjects to respond either with one hand to both signals or with a separate hand for each signal. The experiment showed that the response times to the first signal (S1) were reduced by distributing the signals between the hemispheres and by requiring responses to be made by two hands rather than one alone. The number of 'false alarm' (anticipatory) responses to the second signal (S2) were also reduced. These results were seen as showing that 'the total processing capacity of the brain is appreciably increased by the interhemispheric sharing of the task'.

The reduced response time to S1 may have been a consequence of the retinal eccentricities at which the stimuli were presented in the different conditions of the experiment. In the interhemispheric condition each signal was presented 1.5 degrees from the fixation point, each signal in a different visual half-field. In the intra-hemispheric condition S1 was presented 3.5 degrees from fixation and S2 1.5 degrees from fixation in the same visual half-field. It is well established that acuity declines (e.g. Beaton and Blakemore, 1981) and reaction time increases (Poffenberger, 1912; Haun, 1978) as a stimulus is moved from the fixation point along the horizontal meridian. Thus Guiard and Requin's finding that the time taken to respond to S1 was faster in the inter-hemispheric than the intra-hemispheric condition may reflect nothing but an effect of retinal eccentricity. This does not, of course, explain the reduced frequency of anticipatory responses to S2 nor the advantage obtained when responses had to be made by each of the two hands rather than by only one. This latter result contrasts with previous findings that a response is made more quickly if it follows a response made with the same hand, than if it follows a response made by the opposite hand (Rabbitt, 1965; 1978). In Guiard and Requin's experiment subjects knew that their second response would be made with the opposite hand from the first response. This was not the case in the choice reaction time studies reported by Rabbitt. It is probable that the latter's findings reflect a central decision mechanism as to the choice of limb to be used. Guiard and Requin's result may reflect hemispheric separation of the motor programmes for response.

Beaton (1979c) has shown that whether results are obtained which support the dual-channel model depends upon the nature of the relation between input and output factors. But even where the system appears to have the characteristics of a single channel there may in fact be two in operation. Dimond (1972) has made the point that a bottleneck in the transmission of information at the point where two channels come together will lead to the entire system having the appearance of a single channel. His own work has suggested that the hemispheres analyse information in parallel. (Given the division of the visual field at the vertical meridian it would be surprising if they did not.) This does not necessarily mean that the overall processing *capacity* of the brain is thereby increased beyond what it would otherwise be.

## ARE TWO HEMISPHERES BETTER THAN ONE?

The hypothesisis that the total processing capacity of the brain might be increased by distributing information between the hemispheres was tested by Dimond and Beaumont (1971b) in an experiment in which subjects performed a manual sorting task at the same time as a visual identification task. As in their other experiments, recognition of the visual stimuli was enhanced in the bilateral condition. Dimond and Beaumont therefore concluded that 'when perceptual load is distributed between the cerebral hemispheres total output is increased.' This is a rather misleading statement in so far as there was no significant difference on the manual sorting task between unilateral and bilateral conditions of visual input. If distributing the perceptual load really did increase overall processing capacity it might have been expected that this would have shown up on the secondary task.

Certain results from animal research suggest that distributing information between the hemispheres leads to optimal levels of performance. In one experiment rats were trained on a T-maze using only the left or right cerebral hemisphere, the other being 'anaesthetised' by cortical spreading depression induced by the application of a potassium chloride solution. The animals learned to run to the left or right arm of the maze on alternate trials in order to escape from mild electric shock. Once this alternation reaction had been learned the rats were trained again using only the previously depressed hemisphere. This time they had to learn to escape from shock by jumping on to a platform. In the test phase of the experiment the two training tasks were combined. Shock was avoided by alternately jumping onto an elevated platform on the left or right hand side of the apparatus. It was found that animals who had learned the two components of the task using separate hemispheres achieved a quicker synthesis of this information than either rats who had been trained using only one hemisphere throughout, or intact rats who had not been subjected to cortical spreading depression (Burešová, Lukazewska and Bureš, 1966).

## INFORMATION PROCESSING IN SPLIT-BRAIN ANIMALS

Guiard and Requin (1978) trained split-brain monkeys to respond in a simple reaction time (RT) task to the appearance of a signal. The monkeys were first trained to respond with one hand to a stimulus appearing on the same side of a display panel. Once this was learned the monkeys were trained using the opposite hand to respond to a stimulus on that side. Finally, they were required to respond in a choice RT task in which the stimulus could appear on either side of the display panel and the response had to be made with the appropriate hand. If the hemispheres function entirely independently, without any interaction, then the animals ought to be able to perform this choice RT task as well as normal animals. This was found not to be the case.

The split-brain monkeys showed no improvement in either errors or response times over sessions. Errors, in fact, were consistently around 50 per cent, showing that the animals were practically unable to perform the task. By contrast, one normal animal and one animal with only the optic chiasm divided were each able to carry out the task successfully and quickly. When the response output requirements were changed so that the subjects had to perform a go/no-go response (i.e. respond to a signal on one side and withhold responses to signals on the opposite side) the split-brain monkeys showed a significant improvement in performance. This implies that the deficit observed originally lay in the selection of the correct response. The authors therefore argued that their results are consistent with a single channel rather than a dual channel hypothesis of split-brain function.

Gazzaniga and Young (1967) reported that two split-brain monkeys could hold in memory up to four discriminations in each hemisphere and respond accordingly, whereas normal monkeys failed after only two discriminations per hemisphere. Gazzaniga and Young suggested that commissurotomy results in an increase in the overall information processing capacity of the brain. However, Trevarthen (1974a) has pointed out that the conditions of testing were such that the normal animals would have experienced interocular rivalry – that is, each eye's percept would have competed with that of the other – which may explain their inability to sustain the task. (Animals with a split chiasm and callosum do not experience rivalry.)

In another experiment Nakamura and Gazzaniga (1977) compared total and partial split-brain and hemispherectomised monkeys with normal control animals on various matching-to-sample tasks. Using either one eye or both eyes all animals performed at equivalent levels on almost all tasks. However, a difference between the animals emerged on a nested matching-to-sample task. For example, first, a colour stimulus was presented followed by a pattern stimulus. Next two patterns were shown, one of which had to be matched to the pattern previously presented. Finally, two colours were illuminated, one of which was the same as the original colour stimulus. Thus one discrimination problem started before, and finished after, the other problem.

Using one eye alone for the above task, total split-brain animals performed less well than the other groups. Since there was no difference between the split-brain and hemispherectomised subjects in the binocular condition it was suggested that the poor performance of the split-brain animals in the monocular condition did not reflect reduced processing capacity of one hemisphere as opposed to two. Instead, Nakamura and Gazzaniga suggest, the non-seeing hemisphere of the split-brain animal actively interferes with performance by the seeing hemisphere. Allowing the non-seeing hemisphere to have access to the information presented to the seeing hemisphere, by allowing both half-brains to view the task (normal animals) or by leaving the mid-line commissures partly intact (partial splits), eliminates this interference, as does removal of the non-seeing hemisphere altogether.

Nakamura and Gazzaniga's findings suggest that learning of a difficult task by two hemispheres (binocular condition) of the split-brain monkey is superior to learning by one. An advantage in using two hemispheres rather than one in learning visual shape discriminations has also been reported for the split-brain cat by Robinson and Voneida (1970; 1973). These authors saw their results as consistent with the principal of cortical mass action proposed by Lashley (1929). This principle holds that the amount of cortex participating in a task determines the maximum level of performance on the task. However, the apparent superiority of two hemispheres over one may rather have to do with the fact that viewing was binocular in the two hemisphere condition and monocular in the unilateral condition (Lehman and Spencer, 1973).

There is evidence that during monocular viewing a chiasm-sectioned animal may attend to only one side of a stimulus, the other falling into the area of the visual field defect caused by interruption of the nasal hemiretinal fibres at the optic chiasm (Hamilton, Tieman and Winter, 1973). Testing, therefore, should utilise bilaterally symmetric stimuli. The stimuli used by Robinson and Voneida (1970; 1973) were in most cases bilaterally symmetric about a vertical axis. It is on the two tasks where they were not that monocular-binocular differences were significant. Robinson and Voneida interpreted their results in terms of task complexity, but perusal of their paper reveals equal support for an interpretation in terms of stimulus asymmetry.

A bi-temporal visual field defect in conjunction with stimulus asymmetry provides an explanation of those instances (Noble, 1966; 1968) of 'paradoxical' inter-ocular reversal of response to mirror-image stimuli in monkeys with section of the optic chiasm. It cannot, however, explain the finding of Nakamura and Gazzaniga (1977) that split-brain monkeys using one eye were inferior to animals with only partial callosal section. On the other hand, since the latter performed no better than hemispherectomised monkeys it cannot be argued that the availability of two hemispheres rather than one improves performance.

In view of this, and the difficulty in interpreting the reports of Gazzaniga and Young (1967) and Robinson and Voneida (1970; 1973) the conclusion from the animal research on visual discrimination learning must be that there are no convincing demonstrations of an overall increase in the capacity of the brain consequent upon using two hemispheres rather than one.


## INFORMATION PROCESSING IN HUMAN SPLIT-BRAIN PATIENTS

The human split-brain studies have repeatedly demonstrated the functional separation of the cerebral hemispheres, in so far as each can perceive, remember and respond to stimuli in their own half-field of vision. Furthermore, experiments using chimeric stimuli have shown that each hemisphere can *simultaneously* register its own percept (Levy, Trevarthen and Sperry, 1972). But is the operating efficiency of one hemisphere entirely unaffected by concomitant activity in the other hemisphere? Put another way, do the two halves of the split-brain behave as two separate channels in processing information?

Gazzaniga (1968) presented two split-brain subjects with either four letters all in one half-field, or with eight letters, four in each half-field. Normal control subjects did no better in the eight letter condition than in the four letter condition at picking out target letters from an array of alternatives. By contrast, the commissurotomised patients did twice as well in the bilateral compared to the unilateral condition. This suggested to Gazzaniga that the surgery had created two separately functioning channels. However, as there were twice as many opportunities for successful guessing in the bilateral condition (a forced-choice procedure was not used) this conclusion may not be justified. The control patients, it is true, did not show a similar increase in scores in the bilateral condition, but their unilateral scores were much higher anyway than those of the commissurotomised patients. Trevarthen (1974a) suggested that the low unilateral scores of the patients may indicate that they were attending to the 'wrong' side of space when the stimulus arrived.

In an experiment using multiple-choice response procedures Teng and Sperry (1973; 1974) obtained results which fail to support those of

Gazzaniga (1968). Two letters or digits were flashed to a single hemisphere in the unilateral condition, and two letters or digits to each hemisphere simultaneously in the bilateral condition. Performance was generally at a higher level during unilateral trials; on bilateral trials there was often complete neglect of the left field stimulus.

A discriminative reaction time task was utilised by Gazzaniga and Sperry (1966). Four split-brain patients were compared with normal control subjects on two simultaneous visual discriminations, one presented to each hemisphere. It was found that whereas the control subjects took considerably longer to respond to two discriminations than to one alone, the split-brain subjects responded as quickly to the two as to one. Although this might be taken as demonstrating parallel processing in the split-brain the response times of the patients were very prolonged in comparison with those of the control subjects.

When normal subjects are asked to compare visually presented letters (probe set) with a number of letters held in memory (target set) the time taken to indicate whether any probe letters match those in the memory set is a linear function both of the number of items in the memory set and the number in the probe set (Sternberg, 1966). If splitting the brain causes two channels to operate independently then there should be little or no difference in response times between a pair of letters in the probe set being distributed between the hemispheres and one letter being sent to a single hemisphere. Two letters presented to a single hemisphere, however, should take longer to be responded to than one. Kinsbourne (1974a) reports that when this experiment was carried out with two split-brain patients it was often the case that they did not respond to both letters in the bilateral condition. When they did so, response times were much longer than to a pair of letters sent to a single hemisphere. Kingsbourne does not say whether response times for two letters presented to a single hemisphere were longer than for one letter but, in any event, the results do not support the dual-channel hypothesis.

In the above experiment, neglect of the stimulus on the left side was most severe when subjects had to respond 'yes' or 'no' *at the same time* as making a response with a finger. If it was a finger on the left hand this had the effect of enhancing the efficiency of response to the left visual hemifield. With a right hand response, neglect of the LVF was total. This does not support the view that the two hemispheres can simultaneously process information without mutual interference. This was also shown in a simple experiment (Kreuter, Kinsbourne and Trevarthen, 1972) in which the ability of a patient to carry out two tasks continuously was examined as a function of task difficulty. The patient was asked to tap continuously with the index finger of one or both hands and at the same time to do various verbal tasks. Easy tasks, such as reciting the alphabet, affected only the right hand, but more difficult tasks, such as mental arithmetic, affected both hands. Thus, according to Kreuter et al., maximum effort by one hemisphere does withdraw capacity from the other, an effect which in the absence of the corpus callosum is presumably mediated by a 'capacity distributing system' located in the brain stem. (p. 460)

It seems, then, that the two hemispheres of the split-brain patient act as independent channels (see also Franco, 1977; Ellenberg and Sperry, 1980) only if the task is insufficiently taxing and attention is not required from both hemispheres at once. If the processing demands are increased, the apparent independence of the two hemispheric channels breaks down. This suggests that each hemisphere is an information processing system in its own right, invested with its own limited capacity. Both hemispheres, however, have access to some common system which allocates additional capacity to one or other hemisphere as and when the situation demands. Under normal circumstances, that is, in neurologically intact individuals, the corpus callosum presumably serves to distribute attentional resources equally between left and right sides of the brain.


## HEMISPHERE ASYMMETRY IN ATTENTION AND AROUSAL

Guiard (1980) has argued that the classical disconnection phenomena observed in split-brain animal or man may reflect not disruption of information transfer between the cerebral hemispheres, as is commonly believed, but an asymmetry of attentional processes. In questioning the relevance of the transmission of information between the hemispheres he points to the fact that the striate cortex at the two sides of the brain is not directly connected by callosal fibres. He suggests that, among other findings, the observation that split-brain patients cannot match visual stimuli across the two halves of the visual field 'casts no light on the question of whether both hemispheres were alert.'

There is direct evidence that some aspects of visual attention may be divided by commissurotomy. Holtzman, Sidtis, Volpe, and Wilson and Gazzaniga (1981) provided two split-brain patients with warning cues as to the spatial location of an impending stimulus. This shortened response times to the stimulus even if the cue was provided in the visual hemifield opposite that in which the stimulus appeared. Thus some information for spatial attention was undivided by callosal section. However, when the patients were required to indicate whether two stimuli occupied spatially corresponding positions within two matrices, response times were very much longer for between-field than within-field comparisons. This implies that the information necessary for explicit stimulus comparisons was not available bilaterally. Holtzman et al. argued that the two different aspects of spatial attention require different neural circuits. Commissurotomy interrupts one circuit but not the other.

Kinsbourne (1970; 1973; 1974a,b; 1975a) has articulated most explicitly the view that the two hemispheres maintain a balance of attention which is mediated entirely, though not exclusively, by the corpus callosum. In support of this view he discusses, amongst other evidence, the phenomenon of unilateral neglect or hemi-inattention. This refers to a tendency in some patients with unilateral cerebral lesions to ignore the side of space (or the side of the body) contralateral to the lesion. The defect may appear in a variety of forms affecting the visual, auditory or tactile modalities (for reviews see Friedland and Weinstein, 1977; Heilman and Watson, 1977; Hécaen and Albert, 1978) and may involve an emotional component. Friedland and Weinstein (1977) quote the example of

> ... an elderly lawyer who, when examined from his good side, behaved like the courtly southern gentleman that he was. When approached from his left side, however, he would make remarks like, 'When are you going to get this over with you head-shrinking son of a bitch?' (p. 3)

In its milder form unilateral neglect may be manifest as 'extinction' of a stimulus on the side opposite the lesion when the patient is stimulated bilaterally. Even in the absence of gross features of neglect, careful attention to the patient's behaviour may reveal tendencies to initiate a visual search from the side of space ipsilateral to the lesion (Chédru, Leblanc and Lhermitte, 1973).

According to Kinsbourne (1974b), the various features of the neglect syndrome reflect a dominance of the orienting or turning tendency of the intact hemisphere, which is normally balanced by an equal tendency of the opposite hemisphere. In intact organisms these tendencies balance precisely, so that the resultant orientation is to the centre. The presence of a lesion in one hemisphere reduces the degree of activation of that hemisphere, and hence the intact hemisphere dominates the patient's behaviour. An imbalance can also be induced in normal subjects by requiring them to engage in some activity preferentially mediated by one hemisphere.

Unilateral neglect has been found in some studies to be more frequent or severe following right than left hemisphere lesions (Hécaen, 1962; Gainotti, 1968; Faglioni, Scotti and Spinnler, 1969; Gainotti and Tiacci, 1971; Albert, 1973; Colombo, De Renzi and Faglioni, 1976). Although a laterality effect has not always been found (e.g. Battersby, Bender and Pollack, 1956; Costa and Vaughan, 1962), and aphasia in cases of left brain damage may mask some of the signs of neglect, there has never been any suggestion that the incidence or severity of neglect is greater following left-sided lesions.

The frequent association of unilateral neglect with lesions of the right hemisphere fits with a view that is beginning to emerge which sees the right hemisphere as having a particular role in controlling attention or arousal. Heilman and Van Den Abell (1979) presented a lateralised warning stimulus followed by a second stimulus in central vision to which subjects made a manual response. Right hand response times were reduced more by warning stimuli presented in the left visual half-field than in the right. Furthermore, a warning stimulus in the LVF reduced right hand response times more than a warning stimulus in the RVF reduced response times of the left hand. Although the nature of the warning stimulus, verbal or non-verbal, interacts with these effects (Bowers and Heilman, 1976; 1980b) the findings point to a greater influence of warning stimuli presented in the LVF.

In a study of regional cerebral blood flow using the PECT technique (see Chapter 6) greater bilateral flow increases were produced by a spatial than by a verbal task (Gur, Gur, Obrist et al., 1982). This suggests greater overall arousal by experimentally induced right hemisphere activation than by left hemisphere activation. In another investigation using PECT the subjects were presented with stimuli in one or other visual field. A warning stimulus was provided by dimming a central fixation diode prior to presentation of the lateralised stimulus. Control subjects were blind-folded and their ears were plugged. Greater right hemisphere glucose metabolism was found for experimental subjects, but not for control subjects, in the inferior parietal region but not in superior temporal or cerebellar regions of the brain. The latter two areas were chosen as control sites from which no effect of a warning stimulus was expected (Reivich, Gur, Alavi et al., 1983).

It has been found that left hemisphere EEG desynchronisation is maximal after right sided warning stimuli, but right hemisphere desynch-ronisation is equally large after left and right sided stimuli (Heilman and Van Den Abell, 1980). As EEG desynchronisation is thought to indicate cortical arousal it was concluded that the right hemisphere is particularly important for cerebral activation. It is possible to interpret a particular association between the right hemisphere and control of heart rate changes (Greenstadt, Schuman and Shapiro, 1978; Davidson, Horowitz, Schwartz and Goodman, 1981) in a similar way. On the other hand, there is evidence from lesion (Schwartz, 1967; Albert, Silverberg, Reches and Berman, 1976) and amytal (Serafetinides, Hoare and Driver, 1965, but see Rossi and Rosadini, 1967) studies that perturbation of the left hemisphere is more often followed by unconsciousness than similar insult to the right hemisphere. It may be, therefore, that the right hemisphere is better able to initiate cortical arousal and the left hemisphere to sustain it. However, studies of the vigilance performance of the two halves of the intact brain (Dimond and Beaumont, 1971a; 1973), suggest the opposite conclusion.

It is possible that there are regular fluctuations throughout the day and night in the balance of arousal between the hemispheres. Cyclical changes in cognitive performance levels throughout a day were noted by Klein and Armitage (1979). Subjects engaged in matching letter names and random dot patterns were tested at 15 minute intervals throughout an 8 hour period. It was found that as performance on one task improved there was a decline in performance on the other task. Fluctuations had an approxima-tely 90 minute cycle, the two tasks being out of phase by one half-cycle.

Cohen (1977) woke subjects during periods of rapid eye movement sleep (REM) and claimed to find a shift towards 'left hemisphere' categories of dream recall as the night progressed. This is interesting in view of several suggestions in the non-scientific literature that dreaming is a right-hemisphere function. The idea seems to have arisen in part from Humphrey and Zangwill's (1951) report that a patient with a right-sided parietal lesion apparently ceased to dream after his injury. However, questioning of a total split-brain patient wakened during REM sleep suggested that his dreams were accessible to verbal recall and hence not confined to the right hemisphere (Greenwood, Wilson and Gazzaniga, 1977). Confabulation by the left-hemisphere is, of course, a possibility but there seems no good reason to suppose that dreams are largely the product of right hemisphere activity. Although dreams may involve scenes of vivid imagery (right hemisphere?) they include people who speak. Thus although there may indeed be fluctuations of left and right hemisphere 'dominance' over dreaming the evidence suggests that we dream with the whole brain.

In contrast to the results of Cohen (1977) some data have been collected which suggest more effective right hemisphere functioning during REM than non-REM sleep. Gordon, Fromkin and Lavie (1982) woke their volunteers during both types of sleep and gave them items from a battery of tests, said to be left or right hemisphere dependent on the basis of the performance of unilaterally brain damaged subjects (Bentin and Gordon, 1979). Scores on the tests were converted into a laterality quotient, taken to reflect relative left or right hemisphere superiority. Waking subjects during REM sleep was found to shift the quotient towards comparatively better right hemisphere performance; after non-REM wakings there was a shift toward comparatively better left hemisphere performance. This suggests a fluctuation in hemispheric 'control' correlated with the appearance of REM and non-REM sleep.

## SUMMARY

The duplicate structure of the brain implies some capacity for independent action by each cerebral hemisphere, even in neurologically normal subjects. Experiments suggesting that each hemisphere can carry out its own perceptual analysis without interference from the other have been interpreted within a dual-channel theory of brain function. However, the view that parallel processing by each hemisphere thereby increases the total capacity of the brain receives little support from studies carried out with split-brain animals and human commissurotomy patients.

The corpus callosum appears to have a role in the allocation of attentional resources between the two sides of the brain. There are also indications of a cyclical pattern of arousal between the hemispheres, while certain evidence implicates the right hemisphere in cerebral activation.

# 14

# What, why and how? – some loose ends

The research reviewed in the previous chapters of this book has catalogued a vast number of left-right differences in the human brain but it is not immediately obvious what most distinguishes the functions of the left hemisphere from those of the right.

## DICHOTOMIES OF HEMISPHERE FUNCTION

According to Levy (1974a):

> The right hemisphere synthesises over space. The left hemisphere analyses over time. The right hemisphere notes visual similarities to the exclusion of conceptual similarities. The left hemisphere does the opposite. The right hemisphere perceives form, the left hemisphere detail. The right hemisphere codes sensory input in terms of images, the left hemisphere in terms of linguistic descriptions. The right hemisphere lacks a phonological analyser; the left hemisphere lacks a Gestalt synthesiser. (p. 167)

Bogen (1969b) discusses different ways in which the hemispheres have been compared with each other and himself favours the term 'propositional' to describe left hemisphere functions and 'appositional' to refer to those of the right hemisphere. Other dichotomies that have been proposed (see Chapter 4) include verbal versus non-verbal, linguistic versus spatial, serial versus parallel, analytic versus holistic, focal versus diffuse, temporal versus atemporal, and rational versus intuitive (see Bogen, 1969b; Levy, 1974a; Bradshaw and Nettleton, 1981).

There are certain problems in attempting to encapsulate some 'fundamental' hemispheric difference in terms of one or other of the above dichotomies. In the first place, virtually all investigators are agreed that cerebral asymmetry is not absolute but a matter of degree. It has not been shown, for example, that one hemisphere is totally incapable of carrying out the functions normally ascribed to its partner. Even with regard to language, the area in which left-right asymmetry is most unequivocal, it is evident that the right hemisphere possesses considerable powers of comprehension and can, under certain circumstances, demonstrate some expressive ability (see Chapters 3 and 5). While it is a truism that the term 'language' does not refer to a single function, it is also true that attempts to dichotomise left and right hemisphere functions presuppose some single organising principle. With regard to the verbal/non-verbal distinction this principle is clearly breached.

Other dichotomies are even less securely founded. A linguistic versus spatial distinction, for example, ignores not only the linguistic capacities of the right hemisphere, but the spatial capacities of the left. An individual using only his left hemisphere (following right hemispherectomy or callosal section) is not totally unable to operate within a spatially extended environment. Indeed, patients with left hemisphere lesions may be impaired in certain aspects of spatial orientation even when the right hemisphere is entirely intact (Semmes, Weinstein, Ghent and Teuber, 1955; Newcombe, 1969). This implies that the left hemisphere may preferentially mediate some aspects of spatial performance.

The most recent serious effort to find a single unifying principle describing left and right hemisphere specialisation of function is that of Bradshaw and Nettleton (1981). Their account will therefore be examined in some detail. These authors argue in favour of

> some general distinction such as the analytic/holistic one, which might partly subsume related distinctions such as focal/diffuse or serial/parallel. (p. 58)

To state their position in more detail:

> Verbal processing is largely the province of the left because of this hemisphere's possession of sequential analytic, time-dependent mechanisms. Other distinctions (e.g. focal/diffuse and serial/parallel) are special cases of an analytic/ holistic dichotomy. More fundamentally, however, the left hemisphere is characterised by its mediation of discriminations involving duration, temporal order, sequencing and rhythm, at the sensory level, and especially at the motor level. Spatial aspects characterize the right, the mapping of exteroceptive body space, and the positions of fingers, limbs, and perhaps articulators, with respect to actual and target positions. (p. 51)

The argument that the focal versus diffuse and serial versus parallel distinctions are each 'special cases' of the analytic versus holistic dichotomy is not convincing.

As formulated originally (Semmes, 1968) the focal-diffuse distinction referred to the representation of sensorimotor functions in the cortex. As such, it was a distinction made at a physiological level, not a psychological level. There is no evidence that focal organisation at the physiological level is in any way necessary or sufficient for analytic processing at the psychological level. The same holds for diffuse representation and holistic processing. There is thus no reason to suppose that the focal versus diffuse distinction is in any way a 'special case' of the analytic versus holistic processing dichotomy.

With regard to the serial-parallel distinction, Bradshaw and Nettleton have this to say:

> The term 'analytic' should perhaps properly be reserved for perceptual processing, in the context of a segmental breakdown of a visual array or an auditory grouping into separate components, features or elements. Such a breakdown may be performed either sequentially or in parallel. (p. 79)

If 'analytic' can mean either serial (sequential) or parallel processing, then how can a serial-parallel distinction be a 'special case' of the analytic-holistic dichotomy? Besides, to subsume the serial-parallel distinction within one branch of the analytic-holistic dichotomy, which is what Bradshaw and Nettleton have done, is to reject the (admittedly scant) evidence on which a *hemispheric* substrate for the serial-parallel distinction was originally suggested (Cohen, 1973).

Bradshaw and Nettleton suggest that 'the left hemisphere is characterized by its mediation of discriminations involving duration, temporal order and sequencing.' Yet they mention that a left ear advantage has been found in dichotic recognition of melodies (see Chapter 4). Surely this involves temporal analysis? Furthermore, there is a structure to language just as there is to music – both are extended in time and their 'meaning' depends upon the relations between components making up the whole. Evidence clearly implicates the left hemisphere in the appreciation of the structure between words as well as within words. Is this not 'holistic' processing? And what of those cases in which singing has been found to be preserved despite severe expressive aphasia resulting from extensive left hemisphere damage? Does this residual capacity, mediated presumably by the right hemisphere (see Chapter 5), not involve 'the mediation of discriminations involving duration, temporal order and sequencing'? (See Marshall, 1981).

It is clear from their paper that Bradshaw and Nettleton are aware of the difficulties in maintaining an entrenched position with regard to the analytic-holistic distinction or, indeed, any other. They therefore propose that 'there is a continuum of function between the hemispheres, rather than a rigid dichotomy, the differences being quantitative rather than qualitative, of degree rather than of kind.' But are not Bradshaw and Nettleton trying both 'to have their cake and eat it' in arguing for a dichotomy *and* a continuum? What constitutes a relative *degree* of analytic (or holistic) processing?

Perhaps the search for a dichotomy of function between the left and right hemispheres is bound to fail. There is, after all, no reason *why* the brain should have evolved so conveniently. Certainly, many of the phenomena of everyday life cannot be described as *either* analytic *or* holistic, temporal *or* spatial or whatever. Morgan (1981) points out, for example, that our perception of motion represents temporal and spatial events inextricably related. Perhaps, then, a single dichotomy of brain frunction is inherently improbable. Evidence of specialisation not only between the hemispheres but within each hemisphere (Newcombe and Ratcliff, 1979) suggests that different areas of left and right hemispheres may bear a different relationship to each other (Piercy, 1964; Hécaen, 1968). It may, therefore, be profoundly misleading to assume that the relationship between the hemispheres as a whole can be described in terms of any single principle.

## THE EVOLUTION OF ASYMMETRY

Evolution has imposed on the brain's symmetrical structure an asymmetry of function. It is tempting to suppose that this represents a biologically adaptive development directed towards some end. But if so, what is it? Put another way, why are functions asymmetrically organised in the brain? At the present time this question admits of no more easy solution than the question of what is lateralised.

The most common assumption as to why functions are lateralised has to do with the asymmetrical representation of language. It is suggested that unilateral hemispheric control of the bilaterally innervated vocal apparatus is required if interhemispheric conflict is to be avoided (Segalowitz and Gruber, 1977). This can hardly be the whole story since a not inconsiderable proportion of left handers presumably have bilateral speech representation (see Chapter 5) without apparently suffering from the lack of exclusive left hemisphere control. Furthermore, it is clear that the left hemisphere's specialisation extends beyond mere motor control of speech. Nonetheless, some authors have focused on non-speech motor functions of the left hemisphere and have suggested, either explicitly or implicitly, that left hemisphere control of motor sequencing is the evolutionary precursor that attracted speech and language to the left side of the brain. Such an approach may well be appropriate, given that Man is the only known animal to show a species-specific preference for the limb on one side (see Chapter 2). The question, however, still remains: what caused this bias?

Annett (1972; 1975) argues that the right sided bias in the handedness distribution of humans is an incidental by-product of a single gene $(RS+)$ that makes Man left-brained for speech (see Chapter 2). Kilshaw and Annett (1983) found that left handers tended to be faster than right handers on a peg-moving task, this finding being more evident for the non-preferred than for the preferred hand. They therefore argued that the specialisation of cerebral function associated with the typical human bias to the left hemisphere and the right hand might be due to a right hemisphere handicap rather than to a left hemisphere advantage. Family handedness data suggested (Annett, 1978) that the heterozygote $(rs+-)$ genotype is the most frequent in the population, occurring about as frequently as theoretically possible. This was taken to imply that the RS + gene is not fully dominant and that a double dose of the gene (rs + +) entails risks which have prevented the gene from becoming universal. Consequently, the rs − − genotype has persisted. Annett and Kilshaw (1983) speculate that the rs + + genotype developed language skills at the expense of the practical and visuospatial skills necessary for using tools, whereas the risks associated with the rs − − genotype lay in the realm of language development. Thus natural selection favoured the retention of both genotypes.

Gazzaniga and Le Doux (1978) outline a theory to explain right hemisphere superiority for spatial functions. Unlike other authors (e.g. Webster, 1977b) who suppose that each hemisphere gradually evolved its own functions, Gazzaniga and Le Doux regard right hemisphere spatial − and more particularly manipulo-spatial (Le Doux, Wilson and Gazzaniga, 1977b) − superiority as a development which occurred by default. Their argument goes as follows.

Language required a unilateral hemispheric substrate because

> unilateral control over speech provides a common point through which the various cognitive activities related to language can be channeled and through which their relative importance can be ranked and motor commands... programmed and executed. (p. 81)

Gazzaniga and Le Doux mention the opinion of some authorities (e.g. Geschwind, 1965) that, early in evolutionary history, language, in the form of object-naming, arose out of a capacity for cross-modal integration. Gazzaniga and Le Doux consider that the manipulation of tools also required integration of tactual, visual and spatial information. Consequently, there arose within the left hemisphere, but not within the right, competition for the same neural space.  At this point in the argument Gazzaniga and Le Doux switch from phylogenetic to ontogenetic considerations. By some strange quirk of reasoning they see language as emerging earlier in childhood development than manipulo-spatial functions. Language therefore pre-empts neural space in the left hemisphere, and manipulo-spatial functions have to take up whatever space remains available at this side of the brain. The right hemisphere, however, not being committed to language, can accommodate manipulo-spatial functions with less pressure on neural space. Hence, according to Gazzaniga and Le Doux, a right hemisphere manipulo-spatial advantage arises as a consequence of a disadvantaged left hemisphere.

Levy (1974a; 1976b; 1977d) points out that bilateral symmetry of the nervous system must have been under very strong selective pressures in several phyla for tens of millions of years. Consequently, asymmetry of function must have arisen as a result either of a relaxation of selective pressure or as a consequence of very strong pressure towards asymmetry. She favours the latter alternative, arguing that if relaxation of selective pressure were the cause then all mammals would have lateralised cognitive abilities, since selective pressures would be equally relaxed for all of them (Levy, 1976b). Yet there is little evidence for lateralisation of cognition in species other than Man (but see Chapter 6).

In developing her argument, Levy mentions the findings of Kinsbourne and others that verbal processing activates the left hemisphere, with a concomitant perceptual bias towards the right half of visual space, while non-verbal processing induces a bias towards the left half. Further, studies of lateral eye movements suggest that activation of one or other hemisphere results in eye movements in the direction of the contralateral visual field (see Chapter 4). Such biases may occur even in the absence of experimental conditions specifically designed to bring about an orientational asymmetry.

Levy (1976c) presented subjects with pictures and their left-right mirror images. A majority of right handers preferred the versions with the dominant content on the right rather than the left of the scene. Among left handers approximately 60 per cent preferred the same orientation as dextrals while the remainder preferred the reverse orientation. Levy suggested that, since the pictures were non-verbal, they induced a slight left perceptual bias in right handers which was balanced when the greater weight of the picture was on the right, such a balance being aesthetically more pleasing than an imbalance. For left handers, the direction of preference was, presumably, less pronounced as a result of their less consistent direction of cerebral asymmetry. Somewhat similar results to those of Levy were reported by Ellis and Miller (1981) who reported that whereas right handers preferred a picture to be on the left of the typed text of an advertisement, left handers showed no overall preference for one arrangement or the other.

The above findings suggest that asymmetric hemispheric activation can occur in the absence of asymmetric stimulation from the environment. Levy (1976b) believes that the resulting behavioural asymmetries carry with them the same sort of disadvantages that would follow from lack of symmetry in the sensory and motor systems themselves. Since there are strong selective pressures towards symmetry of sensory and motor systems it is doubtful that cognitive lateralisation of function resulted from relaxation of such pressures. Hence, according to Levy, it must be concluded that there are advantages associated with lateralisation, and that these outweighed the disadvantages. Such advantages must have arisen from powerful selective pressure in favour of cognitive asymmetry, despite asymmetry of sensory and motor systems being under strong negative selection.

Although cerebral asymmetry of function arose initially out of a 'fortunate mutation' (Levy, 1974a) natural selection favoured its continuance. What lateralisation achieved, according to Levy, is separation of two different patterns of neural organisation. The kind of neural circuitry best fitted to serve the needs of language and its cognitive correlates is not well adapted to the most efficient execution of visuospatial tasks. Levy's theory (see Chapter 10) therefore predicts that individuals who are not well lateralised at the cerebral level will be at some disadvantage in comparison with their well-lateralised brethren.

## HOW SHOULD LATERALITY BE MEASURED?

Differences in degree of cerebral lateralisation are inferred from differences between left and right ears or visual hemifield computed in terms of one or other of the various laterality coefficients (see Chapter 4) that have been proposed. A problem arises from the fact that the relative level of performance at left and right sides, and hence the value of the laterality coefficient, is found to vary with such factors as the signal-to-noise ratio, the exposure duration and so forth. Yet if the laterality coefficient is meant to be an index of some fixed property of the brain this should not be the case (Berlin, 1977). Even if the laterality coefficient were less susceptible to changes in experimental conditions, there remains the interpretive problem of what is meant by differences between groups or individuals in the size of any coefficient.

Differences in the size of a particular laterality coefficient are generally regarded as reflecting different degrees of cerebral lateralisation for the function in question. But if, for example, cerebral asymmetry for language is a matter of degree (Zangwill, 1960; Shankweiler and Studdert-Kennedy, 1975) what is it that varies? Possible answers to this question are the propensity of the right hemisphere to compensate for damage to the left hemisphere, or the ratio of the vocabulary of the left hemisphere to that of the right. In neither case has any performance measure of laterality been shown to relate to the brain asymmetry. The quantitative measures used in laterality research therefore provide a false sense of quantification of underlying brain processes.

Because of the above difficulty Colbourn (1978) argued that laterality should not be measured on a ratio scale, as, for example, is employed in the calculation of the 'f' coefficient recommended (see Chapter 4) by Marshall, Caplan and Holmes (1975). Colbourn believes we should not compare different individuals or groups in terms of their *degree* of laterality. Instead we should note the *direction* of laterality since this, at least in theory, bears a specific relation to underlying brain asymmetry. Different groups of subjects can then be compared in terms of the number of individuals showing a superiority of one or other side.

The laterality coefficient derived by Marshall et al. reflects the assumption that degree of laterality should not be constrained by accuracy (see Chapter 4). Richardson (1976) took issue with Marshall et al. on the grounds that we cannot predict how laterality scores should or should not vary as a function of overall accuracy. He therefore proposed a procedure for comparing degrees of laterality, based on ranking sets of scores obtained from left and right sides and applying non-parametric statistical calculations. The size of a particular lateral advantage, in other words, is not tacitly assumed to be *proportional* to the magnitude of underlying brain asymmetry. Richardson thus favours an ordinal scale for the measurement of laterality.

While it is undoubtedly true that measuring laterality on an interval or a ratio scale gives a false sense of quantification of underlying brain asymmetry, it is not necessarily wrong to use such a measure. After all, a *consistent* difference in the size of the laterality coefficient obtained by different individuals or groups presumably means something. It may indeed not mean that underlying brain asymmetry varies in proportion to the difference in size of the laterality coefficient (that is a problem of *interpretation* of experimental results). To eschew the measurement of laterality on other than an ordinal scale of measurement, as advocated by Colbourn (1978), is to rule out of court some important questions (Eling, 1981). Although we may not be able to specify precisely how degree of *brain* asymmetry correlates with degree of tachistoscopic asymmetry, or whatever, it is surely legitimate to investigate how the latter correlates with other variables. Birkett (1980), for example, carried out polynomial regression analyses, using as the independent variable a laterality coefficient derived from scores on a lateralised tachistoscopic test of random form recognition. The scores on standardised spatial tasks were the dependent variables. For left handed males and right handed females, increasing RVF superiority was associated with decreasing levels of performance on (different) tests of spatial ability. This suggests that an increase in the tendency to use left hemisphere based strategies on a non-verbal task is associated with reduced spatial ability. Such a suggestion could not even arise if laterality were not measured on at least an interval scale. Thus the level of measurement adopted by researchers is intimately tied up with the kind of theory that is developed to explain results and generate predictions. If the measurement of laterality is confined to a nominal scale the kinds of theories developed to account for the data are likely to prove conceptually arid, unable to do justice to between- and within-subject variation in performance.

Granted, then, that higher than nominal scales of measurement may properly be used in laterality research, there is no denying that interpretation of ear, visual field or other behavioural asymmetry requires more careful thought than it has sometimes received in the past. This is not only because the direction and degree of lateral asymmetries are uncomfortably labile – and hence require flexible rather than rigid models for their explanation – but also because much research on left-right differences has been conducted without regard to theoretical advances in other areas of psychology. For example, virtually all demonstrations of attenuated or zero asymmetry between left and right sides have been interpreted as showing reduced cerebral lateralisation of function. Left handers, females, dyslexics, autistics, stutterers, the socially disadvantaged, the deaf and bilingual speakers have all, at one time or another, been found to show less left-right asymmetry on some performance measure than 'normal' right handed males. In the face of such ubiquity the rather vague concept of reduced cerebral lateralisation of language forfeits any explanatory usefulness it may once have had.

It is to be hoped that the future will see attempts to achieve a more sophisticated understanding of how lateralised components of complex psychological functions interact to produce variation of performance within and between individuals. Fortunately, the signs are that it will.

# References

AARON, P. G., BAXTER, C. F., and LUCENTI, J., (1980) Developmental Dyslexia and Acquired Alexia: Two sides of the same coin? *Brain and language*, 11, 1–11.

ABIGAIL, E. R., and JOHNSON, E. G., (1976) Ear and hand dominance and their relationship with reading retardation. *Perceptual and motor skills*, 43, 1031–1036.

ABLER, W. L., (1976) Asymmetry in the skulls of fossil man: evidence of lateralized brain function? *Brain, behavior and evolution*, 13, 111–115.

ADAMS, J., (1971) Visual Perception of Direction and Number in Right and Left Visual Fields. *Cortex*, 7, 227–235.

AHERN, G. L., and SCHWARTZ, G. E., (1979) Differential lateralization for positive versus negative emotion. *Neuropsychologia*, 17, 693–698.

AJURIAGUERRA, J. DE, and TISSOT, R., (1969) The apraxias. In VINKEN, P. J., and BRUYN, G. W., (Eds.) *Handbook of clinical neurology*, Vol. 4. North Holland Publishing Company: Amsterdam, 48–66.

AKELAITIS, A. J., (1941) Studies on the corpus callosum II: Higher visual functions in each homonymous visual field following complete section of the corpus callosum. *Archives of neurology and psychiatry*, 45, 788–796.

AKELAITIS, A. J., (1942) Studies on the corpus callosum V: Homonymous defects for colour object and letter recognition (homonymous hemiamblyopia) before and after section of the corpus callosum. *Archives of neurology and psychiatry*, 48, 108–118.

AKELAITIS, A. J., (1943) Studies on the corpus callosum VII: Study of language functions (Tactile and Visual Lexia and Graphia) unilaterally following section of corpus callosum. *Journal of neuropathology and experimental neurology*, 2, 226–262.

AKELAITIS, A. J., (1944) A study of Gnosis, Praxis and Language Following Section of the Corpus Callosum and Anterior Commissure. *Journal of neurosurgery*, 1, 94–102.

AKELAITIS, A. J., (1945) Studies on the Corpus Callosum: IV Diagnostic Dyspraxia in Epileptics Following Partial and Completion Section of the Corpus Callosum. *American journal of psychiatry*, 101, 594–599.

AKELAITIS, A. J., RISTEEN, W. A., HERREN, R. Y., and VAN WAGENEN, W. P., (1942) Studies on the Corpus Callosum III: A contribution to the study of dyspraxia and apraxia following partial and complete section of the corpus callosum. *Archives of neurology and psychiatry*, 47, 971–1008.

ALBERT, M., (1973) A simple test of visual neglect. *Neurology*, 23, 655–664.

ALBERT, M. L., and OBLER, L. K., (1978) *The bilingual brain*. Academic Press: New York.

ALBERT, M. L., SILVERBERG, R., ROCHES, A., and BERMAN, H., (1976) Cerebral Dominance for Consciousness. *Archives of neurology*, 33, 453.

ALFORD, R. I., and FINNEGAN-ALFORD, K., (1981) Sex differences in asymmetry in the facial expression of emotion. *Neuropsychologia*, 19, 605–608.

ALLARD, F., and BRYDEN, M. P., (1979) The effect of concurrent activity on hemispheric asymmetries. *Cortex*, 15, 5–18.

ALLEN, M., and WELLMAN, M. M., (1980) Hand position during writing, cerebral laterality and reading: age and sex differences. *Neuropsychologia*, 18, 33–40.

ALLPORT, D. A., (1980) Attention and performance. In CLAXTON, G., (Ed.) *Cognitive psychology: new directions*. Routledge and Kegan Paul: London, 112–153.

ALPERT, M., and MARTZ, M. J., (1977) Cognitive views of schizophrenia in light of recent studies of brain asymmetry. In SHAGASS, C., GERSHON, S., and FRIEDHORF, A. J., (Eds.) *Psychopathology and brain dysfunction*. Raven Press: New York, pp. 1–13.

ALPERT, M., RUBINSTEIN, H., and KESSELMAN, M., (1976) Asymmetry of information processing in hallucinators and non-hallucinators. *Journal of nervous and mental diseases*, 162, 258–265.

ANDERSON, A. L., (1951) The Effect of Laterality Localization of Focal Brain Lesions on the Wechsler-Bellevue subtests. *Journal of clinical psychology*, 7, 149–153.

ANDERSON, S. W., (1977) Language-related asymmetries of eye-movement and evoked potentials. In HARNAD, S., DOTY, R. W., GOLDSTEIN, G., JAYNES, J., and KRAUTHAMER, G., (Eds.) *Lateralization in the nervous system*. Academic Press: New York, 403–428.

ANDREASI, J. L., DE SIMONE, J. J., FRIEND, M. A., and GROTA, P. A. (1975) Hemispheric amplitude asymmetry in the auditory evoked potential with monaural and binaural stimulation. *Psychophysiology*, 3, 169–171.

ANDREASON, N. C., and GROVE, W., (1979) The relationship between schizophrenic language, manic language, and aphasia. In GRUZELIER, J., and FLOR-HENRY, P. (Eds.) *Hemisphere asymmetries of function in psychopathology*. Elsevier/North Holland Biomedical Press: Amsterdam. pp. 373–390.

ANDREW, R. J., MENCH, J., and RAINEY, C., (1980) Right-left asymmetry of response to visual stimuli in the domestic chick. In INGLE, D. J., MANSFIELD, R. J. W., and GOODALE, M. A., (Eds.) *Advances in the analysis of visual behaviour*. MIT Press: Cambridge, Massachusetts.

ANDREWS, G., QUINN, P.T., SORBY, W. A., (1972) Stuttering: an investigation into cerebral dominance for speech, *Journal of Neurology, neurosurgery and psychiatry*, 35, 414–418.

ANDREWS, R. J., (1977) Aspects of language lateralization correlated with familial handedness. *Neuropsychologia*, 15, 769–778.

ANNETT, J., ANNETT, M., HUDSON, P. T. W., and TURNER, A., (1979) The control of movement in the preferred and non-preferred hands. *Quarterly journal of experimental psychology*, 31, 641–652.

ANNETT, J., and SHERIDAN, M. K., (1973) Effects of S-R. and R-R. compatibility on bimanual movement time. *Quarterly journal of experimental psychology*, 25, 247–252.

ANNETT, M., (1967) The Binomial Distribution of right, mixed and left handers. *Quarterly journal of experimental psychology*, 19, 327–334.

ANNETT, M., (1970a) A Classification of Hand Preference by Association Analysis. *British journal of psychology*, 61, 303–321.

ANNETT, M., (1970b) Handedness, Cerebral Dominance and the growth of intelligence. In BAKKER, D. J., and SATZ, P., (Eds.) *Specifying reading disability*. Rotterdam University Press: Rotterdam. pp. 61–79.

ANNETT, M., (1970c) The growth of manual preference and speed. *British journal of psychology*, 61, 545–558.

ANNETT, M., (1972) The distribution of manual asymmetry. *British journal of psychology*, 63, 343–358.

ANNETT, M., (1973a) Handedness in families. *Annals of human genetics*, 37, 93–105.

ANNETT, M., (1973b) Laterality of childhood hemiplegia and the growth of speech and intelligence. *Cortex*, 9, 4–33.

ANNETT, M., (1975) Hand preference and the laterality of cerebral speech. *Cortex*, 11, 305–328.

ANNETT, M., (1978) *A single gene explanation of right and left handedness and brainedness*. Lanchester Polytechnic: Coventry.

ANNETT, M., (1979) Family handedness in three generations predicted by the right shift theory. *Annals of human genetics*, 42, 479–491.

ANNETT, M., (1981) The right shift theory of handedness and developmental language problems. *Bulletin of the Orton Society*, 31, 103–121.

ANNETT, M., HUDSON, P. T. W., and TURNER, A. B., (1974) Effects of right and left unilateral ECT on naming and visual discrimination analysis in relation to handedness. *British journal of psychiatry*, 124, 260–264.

ANNETT, M., and KILSHAW, D., (1983) Right and left hand skill – II. Estimating the parameters of the distribution of L-R differences in males and females. *British journal of psychology*, 74, 269–281.

ANNETT, M., and TURNER, A., (1974) Laterality and the growth of intellectual abilities. *British journal of educational psychology*, 37–46.

ANNETT, M., and OCKWELL, A., (1980) Birth order, birth stress and handedness. *Cortex*, 16, 181–188.

ANZOLA, G. P., BERTOLINI, G., BUCHTEL, H.A., and RIZZOLATTI, G., (1977) Spatial compatibility and anatomical factors in simple and choice reaction time. *Neuropsychologia*, 15, 295–302.

APRIL, R. S., and HAN, M., (1980) Crossed aphasia in a right-handed bilingual chinese man. *Archives of neurology*, 37, 342–346.

ARGYLE, M., and COOK, M. (1976) *Gaze and mutual gaze*. Cambridge University Press: Cambridge, England.

ARNDT, S., and BERGER, D. E., (1978) Cognitive mode and asymmetry in cerebral functioning. *Cortex*, 14, 78–86.

ARRIGONI, G., and DE RENZI, E., (1964) Constructional apraxia and hemispheric locus of lesion. *Cortex*, 1, 170–197.

ASLIN, R. N., and BANKS, M. S., (1978) Early visual experience in humans: evidence for a critical period in the development of binocular vision. In PICK, H., LEIBOWITZ, H. W., SINGER, J. E., STEINSCHNEIDER, A., and STEVENSON, H., (Eds.) *Psychology: from research to practice*. Plenum: New York. pp. 227–239.

ATKINSON, J., and EGETH, H., (1973) Right hemisphere superiority in visual orientation matching. *Canadian journal of psychology*, 27, 152–158.

ATKINSON, R. C., and SHIFFRIN, R. M., (1968) Human memory: a proposed system and its control processes. In SPENCE, K. W., and SPENCE, J. T. (Eds.) *The psychology of learning and motivation: advances in research and theory*. Adacemic Press: New York.

AUSTIN, G. M., and GRANT, F. C., (1955) Physiologic Observations following Total Hemispherectomy in Man. *Journal of surgery*, 38, 239–258.

AXELROD, S., HARYADI, T., and LEIBER, L., (1977) Oral report of words and word approximations presented to the left or right visual field. *Brain and language*, 4, 550–557.

AYRES, J. J. B., (1966) Some artifactual causes of perceptual primacy. *Journal of experimental psychology*, 71, 896–901.

BADDELEY, A. D., and HITCH, G., (1974) Working Memory. In BOWER, G. H., (Ed.) *The psychology of learning and motivation*, 8, 47–90.

BADIAN, H., and WOLFF, P. H. (1977) Manual asymmetries of motor sequencing ability in boys with reading disability. *Cortex*, 13, 343–349.

BAKAN, P., (1969) Hypnotizability, laterality of eye movements and functional brain asymmetry. *Perceptual and motor skills*, 28, 927–932.

BAKAN, P., (1971) Handedness and birth order. *Nature*, 229, 195.

BAKAN, P., (1973) Left handedness and alcoholism. *Perceptual and motor skills*, 36, 514.

BAKAN, P., (1977) Left handedness and birth order revisited. *Neuropsychologia*, 15, 837–840.

BAKAN, P., DIBB, G., and REED, P., (1973) Handedness and birth stress. *Neuropsychologia*, 11, 363–366.

BAKAN, P., and SVORAD, D., (1969) Resting EEG alpha and asymmetry of reflective lateral eye movements. *Nature*, 223, 975–976.

BAKKER, D. J., (1970a) Ear-asymmetry with monaural stimulation: relations to lateral dominance and lateral awareness. *Neuropsychologia*, 8, 103–117.

BAKKER, D. J., (1970b) Temporal order perception and reading retardation. In BAKKER, D. J., and SATZ, P., (Eds.) *Specific reading disability: advances in theory and method*. Rotterdam University Press: Rotterdam, 81–96.

REFERENCES 297

BAKKER, D. J., (1973) Hemispheric specialization and stages in the learning-to-read process. *Bulletin of the Orton Society*, 23, 15–27.
BAKKER, D. J., (1979) Dyslexia in developmental neuropsychology: hemisphere specific models. In OBORNE, D. J., GRUNEBERG, M. M., and EISER, J. R., (Eds.) *Research in Psychology and Medicine*, Vol. 1, Academic Press: London, 347–353.
BAKKER, D. J., HOEFKINS, M., and VAN DER VLUGT, H., (1979) Hemispheric specialization in children as effected in the longitudinal development of ear asymmetry, *Cortex*, 15, 619–626.
BAKKER, D. J., SMINK, T., and REITSMA, P., (1973) Ear dominance and reading ability. *Cortex*, 9, 301–312.
BAKKER, D. J., TEUNISSEN, J., and BOSCH, J., (1976) Development of laterality-reading patterns. In KNIGHT, R. M., and BAKKER, D. J. (Eds.) *The neuropsychology of learning disorders*. University Park Press: Baltimore, pp. 207–220.
BARROSO, F., (1976) Hemispheric asymmetry of function in children. In RIEBER, R. W., (Ed.) *The neuropsychology and language: essays in honor of Eric Lenneberg*. Plenum Press: New York and London, 157–180.
BARRY, C., (1981) Hemispheric asymmetry in lexical access and phonological encoding. *Neuropsychologia*, 19, 473–478.
BARRY, R. J., and JAMES, A. L. (1978) Handedness in autistics, retardates and normals of a wide range. *Journal of autism and childhood schizophrenia*, 8, 315–323.
BARTHOLOMEUS, B., (1974) Effects of task requirements on ear superiority for sung speech. *Cortex*, 10, 215–223.
BARTON, M., GOODGLASS, H., and SHAI, A., (1965) Differential recognition of tachistoscopically presented English and Hebrew words in right and left visual fields. *Perceptual and motor skills*, 21, 431–437.
BASHORE, T. R., (1981) Vocal and Manual Reaction Time Estimates of Interhemispheric Transmission Time. *Psychological bulletin*, 89, 352–368.
BASSER, L. S., (1962) Hemiplegia of early onset and the faculty of speech with special reference to the effects of Hemispherectomy. *Brain*, 85, 427–460.
BASSO, A., CAPITANI, E., LAIACONA, M., and LUZZATTI, C., (1980) Factors influencing type and severity of aphasia, *Cortex*, 16, 631–636.
BATTERSBY, W. S., BENDER, M. B., and POLLACK, M., (1956) Unilateral spatial agnosia (inattention) in patients with cerebral lesions. *Brain*, 79, 68–93.
BAZHIN, E. F., WASSERMAN, L. I., and TONKONOGH, I. M., (1981) Auditory hallucinations and left temporal lobe pathology, *Neuropsychologia*, 13, 481–487.
BEAR, D., (1979) The temporal lobes: An approach to the study of organic behavioral change. In GAZZANIGA, M. S., (Ed.) *Handbook of behavioral neurobiology*, Vol. 2, Neuropsychology, Plenum: New York, 75–95.
BEATON, A. A., (1977) Unpublished Ph.D. Thesis, University of Wales.
BEATON, A. A., (1979a) Duration of the motion after-effect as a function of retinal locus and visual field. *Perceptual and motor skills*, 48, 143–146.
BEATON, A. A., (1979b) Hemisphere function and dual-task performance. *Neuropsychologia*, 17, 629–635.
BEATON, A. A., (1979c) Hemispheric emotional asymmetry in a dichotic listening task. *Acta Psychologica*, 43, 103–109.
BEATON, A., and BLAKEMORE, C., (1981) Orientation selectivity of the human visual system as a function of retinal eccentricity and visual hemifield. *Perception*, 10, 273–282.
BEATON, A. A., and MOSELEY, L. G., (1984) Anxiety and the measurement of handedness. *British journal of psychology*, 75, 275–278.
BEATON, A. A., SYKES, R. N., KING, H., and JONES, H., (1982) Unpublished report.
BEAUMONT, J. G., (1974) Handedness and hemisphere function. In DIMOND, S. J., and BEAUMONT, J. G., (Eds.) *Hemisphere function in the human brain*. Elek: London, pp. 89–120.
BEAUMONT, J. G., (1979) Lateral asymmetries of orientation to track control in extrapersonal space. *Acta Psychologica*, 43, 85–101.
BEAUMONT, J. G., and COLLEY, M., (1980) Attentional bias and visual field asymmetry. *Cortex*, 16, 391–396.

BEAUMONT, J. G., and DIMOND, S. J., (1973) Brain Disconnection and Schizophrenia. *British Journal of psychiatry*, 123, 661–662.

BEAUMONT, J. G., and DIMOND, S. J., (1975) Interhemispheric transfer of figural information in right and non-right handed subjects. *Acta Psychologica*, 39, 97–104.

BEAUMONT, J. G., MAYES, A. R., and RUGG, M. D., (1978) Asymmetry in EEG alpha coherence and power: effects of task and sex. *Electroencephalography and clinical neurophysiology*, 45, 393–401.

BEAUMONT, J. G., and McCARTHY, R., (1981) Dichotic ear asymmetry and writing posture. *Neuropsychologia*, 19, 469–472.

BEAUMONT, J. G., and RUGG, M. D., (1978) Neuropsychological laterality of function and dyslexia: a new hypothesis. *Dyslexia review*, 1, 18–21.

BELMONT, L., and BIRCH, H. G., (1965) Lateral dominance, lateral awareness and reading disability. *Child development*, 36, 57–71.

BENSON, D. F., and BARTON, M. I., (1970) Disturbances in Constructional Ability. *Cortex*, 6, 19–46.

BENSON, D. F., SEGARRA, J., and ALBERT, M. L., (1974) Visual Agnosia – Prosopagnosia. A Clinicopathological Correlation. *Archives of neurology*, 30, 4, 307–310.

BENTIN, S., (1981) On the representation of a second language in the cerebral hemispheres of right handed people. *Neuropsychologia*, 19, 599–604.

BENTIN, S., and GORDON, H. W., (1979) Assessment of cognitive asymmetries in brain damaged and normal subjects: validation of a test battery. *Journal of neurology, neurosurgery and psychiatry*, 42, 715–723.

BENTON, A. L., (1964) Contributions to aphasia before Broca. *Cortex*, 1, 314–327.

BENTON, A. L., (1965) The problem of cerebral dominance. *Canadian psychologist*, 6, 332–348.

BENTON, A. L., (1968) Different Behavioral Effects in Frontal Lobe Disease. *Neuropsychologia*, 6, 53–60.

BENTON A. L., (1969a) Disorders of spatial orientation. In VINKEN, P. J., and BRUYN, G. W., (Eds.) *Handbook of clinical neurology*. North-Holland Publishing Co.: Amsterdam, 3, 212–228.

BENTON, A. L., (1969b) Constructional Apraxia: Some Unanswered Questions. In BENTON, A. L. (Ed.) *Contributions to clinical neuropsychology*. Aldine Publishing Company: Chicago, 129–141.

BENTON, A. L., (1975) Developmental Dyslexia: Neurological aspects. In FRIEDLANDER, J., (Ed.) *Advances in neurology*, Vol. 7, Raven Press: New York, 1–47.

BENTON, A. L., and HÉCAEN, H., (1970) Stereoscopic vision in patients with unilateral cerebral disease. *Neurology*, 10, 1084–1088.

BENTON, A. L., LEVIN, H. S., and VARNEY, N. R., (1973) Tactile Perception of Direction in Normal Subjects: Implications for hemispheric cerebral dominance. *Neurology*, 23, 1248–1250.

BENTON, A. L., MEYERS, R., and POLDER, G. J., (1962) Some aspects of handedness. *Psychiatrica et neurologica (Basel)*, 144, 321–337.

BENTON, A. L., VARNEY, N. R., and HAMSHER, K. De S., (1978) Lateral differences in tactile directional perception, *Neuropsychologia*, 16, 109–114.

BERENBAUM, S. A., and HARSHMAN, R. A., (1980) On testing group differences in cognition resulting from differences in lateral specialisation: reply to Fennell et al. *Brain and language*, 11, 209–220.

BERENT, S., (1977) Functional asymmetry of the human brain in the recognition of faces. *Neuropsychologia*, 15, 829–831.

BERENT, S., COHEN, B. D., and SILVERMAN, A. J., (1975) Changes in verbal and nonverbal learning after a single left or right unilateral electroconvulsive treatment. *Biological psychiatry*, 10, 95–100.

BERLIN, C., (1977) Hemispheric asymmetry in auditory tasks. In HARNAD, S., DOTY, R. W., GOLDSTEIN, L., JAYNES, J., and KRAUTHAMER, G., (Eds.) *Lateralization in the nervous system*. New York: Academic Press. Chap. 17, 285–302.

BERLIN, C., LOWE-BELL, L. S., CULLEN, J. K., THOMPSON, C. L., and LOOVIS, C.

F., (1973) Dichotic speech perception: an interpretation of right ear advantage and temporal offset effects. *Journal of the acoustical society of America*, 53, 699–709.

BERLIN, C., and MCNEIL, M. R., (1976) Dichotic listening. In LASS, N., (Ed.) *Contemporary issues in experimental phonetics*. Thomas, C. C., Springfield, Chap. 10, 327–385.

BERLUCCHI, G., (1972) Anatomical and physiological aspects of visual functions of corpus callosum. *Brain research*, 37, 371–392.

BERTOLINI, G., ANZOLA, G. P., BUCHTEL, H. A., and RIZZOLATTI, G., (1978) Hemispheric differences in the discrimination of the velocity and duration of a simple visual stimulus. *Neuropsychologia*, 16, 213–220.

BESNER, D., and COLTHEART, M., (1979) Ideographic and alphabetic processing in skilled reading of English. *Neuropsychologia*, 17, 467–472.

BESNER, D., and GRIMSELL, D., and DAVIS, R., (1979) The Mind's Eye and the comparative judgement of number. *Neuropsychologia*, 13, 373–380.

BEVER, T. G., (1975) Cerebral asymmetries in humans are due to the differentiation of two incompatible processes: holistic and analytic. *Annals of the New York Academy of Sciences*, 263, 251–262.

BEVER, T. G., and CHIARELLO, R. J., (1974) Cerebral Dominance in Musicians and Non-Musicians, *Science*, 185, 537–539.

BEVER, T. G., HURTIG, R. R., and HANDEL, A. B., (1976) Analytic Processing Elicits Right Ear Superiority in Monaurally Presented Speech. *Neuropsychologia*, 14, 175–181.

BEVERIDGE, R., and HICKS, R. A., (1976) Lateral eye movement, handedness and sex. *Perceptual and motor skills*, 42, 466.

BIGELOW, L. B., NASRALLAH, H. A., and RAUSCHER, F. P., (1983) Corpus Callosum thickening in chronic schizophrenia. *British journal of psychiatry*, 142, 284–286.

BERLUCCHI, G., (1981) Recent advances in the analysis of the neural substrates of interhemispheric communication. In POMPEIANO, D., and MARSAN, C. A., (Eds.) *Brain mechanisms of perceptual awareness and purposeful behaviour*. International Brain Research Organization Monograph Series. Vol. 8. Raven Press, 133–152.

BERLUCCHI, G., BRIZZOLARA, D., MARZI, C. A., RIZZOLATTI, G., and UMILTÀ, C., (1974) Can lateral asymmetries in attention explain interfield differences in visual perception? *Cortex*, 10, 177–185.

BERLUCCHI, G., CREA, F., DI STEFANO, M., and TASSINARI, G., (1977) Influence of spatial stimulus – response compatibility on reaction time of ipsilateral and contralateral hand to lateralized light stimuli. *Journal of experimental psychology: human perception and performance*, 3, 505–517.

BERLUCCHI, G., GAZZANIGA, M. S., and RIZZOLATTI, G., (1967) Microelectrode analysis of transfer of visual information by the corpus callosum. *Archives Italiennes de biologie*, 105, 583–596.

BERMAN, A., (1971) The problem of assessing cerebral dominance and its relationship to intelligence. *Cortex*, 2, 372–386.

BERRINI, R., DELLA SALLA, S., SPINNLER, H., STERZI, R., and VALLAR, G., (1982) In eliciting hemisphere asymmetries which is more important: the stimulus input side or the recognition side? A tachistoscopic study on normals. *Neuropsychologia*, 20, 91–94.

BIRKETT, P., (1977) Measures of laterality and theories of hemispheric processes. *Neuropsychologia*, 15, 693–696.

BIRKETT, P., (1978) Hemispheric differences in the recognition of nonsense shapes: cerebral dominance or strategy effects, *Cortex*, 14, 245–249.

BIRKETT, P., (1980) Predicting spatial ability from hemispheric 'non-verbal' lateralisation: sex, handedness and task differences implicate encoding strategy effects. *Acta psychologica*, 46, 1–14.

BIRKETT, P., (1981) Familial handedness and sex differences in strength of hand preference. *Cortex*, 17, 141–146.

BISHOP, D. V. M., (1980a) Handedness, clumsiness and cognitive ability. *Developmental medicine and child neurology*, 22, 569–579.

BISHOP, D. V. M., (1980b) Measuring familial sinistrality. *Cortex*, 16, 311–313.

BISHOP, D. V. M., and BUTTERWORTH, G. E., (1980) Verbal-Performance discrepancies: relationship to birth risk and specific reading retardation. *Cortex*, 16, 375–389.

BISHOP, D. V. M., JANCEY, C., and STEEL, A. (1979) Orthoptic status and reading disability. *Cortex*, 15, 659–666.

BISHOP, E. R., MOBLEY, M. C. and FARR, W. F., (1978) Lateralization of conversion symptoms. *Comprehensive psychiatry*, 19, 393–396.

BISHOP, P. O., and HENRY, G. H., (1971) Spatial vision. *Annual review of psychology*, 22, 119–160.

BLACK, F. W., (1975) Unilateral brain lesions and MMPI performance: a preliminary study. *Perceptual and motor skills*, 40, 87–93.

BLAKE, R., OVERTON, R., and LEMA-STERN, S., (1981) Inter-ocular transfer of visual after-effects. *Journal of experimental psychology*, 7, 367–381.

BLAKEMORE, C., (1969) Binocular depth discrimination and the naso-temporal division. *Journal of physiology*, 205, 471–497.

BLAKEMORE, C., (1978) Maturation and modification in the developing visual system. In HELD, R., LEIBOWITZ, H. W., and TEUBER, H. L., (Eds.) *Handbook of sensory physiology*. Vol. VIII. Perception. Springer-Verlag: New York, 377–436.

BLUMSTEIN, S. E., (1974) The use and theoretical implications of the dichotic technique for investigating distinctive features. *Brain and language*, 1, 337–350.

BLUMSTEIN, S. E., and CASPER, W. E., (1974) Hemispheric Processing of Intonation Contours. *Cortex*, 10, 146–158.

BLUMSTEIN, S. E., GOODGLASS, H., and TARTTER, V., (1975) The reliability of ear advantage in dichotic listening. *Brain and language*, 2, 226–236.

BOCCA, E., CALEARO, C., CASSINARI, V., and MIGLIAVACCA, F., (1955) Testing 'Cortical' hearing in temporal lobe tumours. *Acta otolaryngologica*, 45, 289–304.

BODER, E., (1973) Developmental dyslexia: a diagnostic approach based on 3 atypical reading-spelling patterns. *Developmental medicine and child neurology*, 15, 663–687.

BOGEN, J. E., (1969a) The other side of the Brain I: Dysgraphia and Dyscopia following Cerebral Commissurotomy. *Bulletin of the Los Angeles neurological societies*, 34, 73–105.

BOGEN, J. E., (1969b) The other side of the Brain II: an appositional mind. *Bulletin of the Los Angeles neurological societies*, 34, 135–162.

BOGEN, J. E., (1977) Linguistic Performance in the short-term following cerebral commissurotomy. In *Studies in neurolinguistics* (Vol. 3) Eds. WHITAKER, H., and WHITAKER, H. A. Academic Press, New York, Chap. 6, 193–224.

BOGEN, J. E., (1979) The callosal syndrome. In HEILMAN, K., and VALENSTEIN, E., (Eds.) *Clinical neuropsychology*. Chap. 11, 308. Oxford University Press: New York/Oxford.

BOGEN, J. E., DeZURE, R., TENHOUTEN, W. D., and MARSH, J. F., (1972) The other side of the BRAIN IV: The A/P Ratio. *Bulletin of the Los Angeles neurological societies*, 37, 49–61.

BOGEN, J. E., FISHER, E. D., and VOGEL, P. J., (1965) Cerebral Commissurotomy: a second case report. *Journal of the American medical association*, 194, 1328–1329.

BOGEN, J. E., and GAZZANIGA, M. S., (1965) Cerebral Commissurotomy in Man: Minor Hemisphere Dominance for Certain Visuospatial Functions. *Journal of neurosurgery*, 23, 394–399.

BOGEN, J. E., and VOGEL, P. J., (1962) Cerebral Commissurotomy in Man: preliminary report. *Bulletin of Los Angeles neurological society*, 27, 169–172.

BOGEN, J. E., and VOGEL, P. J., (1974) Neurologic status in the long term following cerebral commissurotomy. In MICHEL, F., and SCHOTT, B., (Eds.) *Les syndromes de disconnexion calleuse chez l'home. Actes du colloque international de Lyon, Hôpital Neurologique.*

BOKLAGE, C. E., (1977) Schizophrenia, brain asymmetry development and twinning: cellular relationship with etiological and possible prognostic implications. *Biological psychiatry*, 12, 19–35.

BOKLAGE, C. E., ELSTON, R. C., and POTTER, R. H., (1979) Cellular origins of functional asymmetries: evidence from schizophrenia, handedness, fetal membranes and teeth in twins. In GRUZELIER, J., and FLOR-HENRY, P., (Eds.) *Hemisphere asymmetries of function in psychopathology*. Elsevier/North Holland Biomedical Press: Amsterdam, 79–104.

BOLES, D. B., (1979) Laterally biased attention with concurrent verbal load: multiple failures to replicate. *Neuropsychologia*, 17, 353–361.

BOROD, J. C., and CARON, H. S., (1980) Facedness and emotion related to lateral dominance, sex and expression type. *Neuropsychologia*, 18, 237–241.

BOROWY, T., and GOEBEL, R., (1976) Cerebral Laterality of Speech: the effects of age, sex, race and socio-economic class. *Neuropsychologia*, 14, 363–370.

BOTKIN, A. L., SCHMALTZ, L. W., and LAMB, D. H., (1977) "Overloading" the left hemisphere in right-handed subjects with verbal and motor tasks. *Neuropsychologia*, 15, 591–596.

BOUCHER, J., (1977) Hand preference in autistic children and their parents. *Journal of autism and childhood schizophrenia*, 7, 177–198.

BOUMA, H., and LEGEIN, Ch. P., (1977) Foveal and parafoveal recognition of letters and words by dyslexics and average readers. *Neuropsychologia*, 15, 69–80.

BOUMA, H., and LEGEIN, Ch. P., (1980) Dyslexia: a specific recoding deficit? An analysis of response latencies for letters and words in dyslectics and in average readers. *Neuropsychologia*, 18, 285–298.

BOWER, G. H., (1970) Organizational factors in memory. *Cognitive psychology*, 1, 18–46.

BOWERS, D., and HEILMAN, K. M., (1976) Material Specific Hemispherical Arousal. *Neuropsychologia*, 14, 123–127.

BOWERS, D., and HEILMAN, K. M., (1980a) Pseudo-neglect: effects of hemispace on a tactile line bisection task. *Neuropsychologia*, 18, 491–498.

BOWERS, D., and HEILMAN, K. M., (1980b) Material specific hemispheric activation. *Neuropsychologia*, 18, 309–319.

BOWERS, D., HEILMAN, K. M., SATZ, P., and ALTMAN, A., (1978) Simultaneous performance on verbal, non-verbal and motor tasks by right-handed adults. *Cortex*, 14, 540–556.

BOWERS, D., HEILMAN, K. M., and VAN DEN ABELL, T., (1981) Hemispace – VHF compatibility. *Neuropsychologia*, 19, 757–765.

BRACKENRIDGE, C. J., (1981) Secular variation in handedness over ninety years. *Neuropsychologia*, 19, 459–462.

BRADSHAW, J. L., (1980) Right hemisphere language: familial and non familial sinistrals, cognitive deficits and writing hand position in sinistrals and concrete, abstract, imageable – non-imageable dimensions in word recognition. A review of interrelated issues. *Brain and language*, 10, 172–188.

BRADSHAW, J. L., and GATES, E. A., (1978) Visual field differences in verbal tasks: effects of task familiarity and sex of subject. *Brain and language*, 5, 166–187.

BRADSHAW, J. L., GATES, A., and NETTLETON, N. C., (1977) Bihemispheric involvement in lexical decisions: handedness and a possible sex difference. *Neuropsychologia*, 75, 277–285.

BRADSHAW, G. J., HICKS, R. E., and ROSE, B., (1979) Lexical discrimination and letter-string identification in the two visual fields. *Brain and language*, 8, 10–18.

BRADSHAW, J. L., and NETTLETON, N. C., (1981) The nature of hemispheric specialization in man. *The behavioral and brain sciences*, 4, 51–91.

BRADSHAW, J. L., NETTLETON, N. C., and TAYLOR, M. J., (1981a) The use of laterally presented words in research into cerebral asymmetry: is directional scanning likely to be a source of artifact? *Brain and language*, 14, 1–14.

BRADSHAW, J. L., NETTLETON, N. C., and TAYLOR, M. J., (1981b) Right hemisphere language and cognitive deficit in sinistrals. *Neuropsychologia*, 19, 113–132.

BRADSHAW, J. L., and TAYLOR, M. J., (1979) A word-naming deficit in non-familial sinistrals? Laterality effects of vocal responses to tachistoscopically presented letter strings. *Neuropsychologia*, 17, 21–32.

BRADY, J. P., and BERSON, J., (1975) Stuttering, dichotic listening and cerebral dominance. *Archives of general psychiatry*, 32, 1449–1452.

BRAMWELL, B., (1899) On 'crossed' aphasia and the factors which go to determine whether the 'leading' or 'driving' speech centres shall be located in the left or in the right hemisphere of the brain with notes of a case of 'crossed' aphasia (aphasia with right sided hemiplegia) in a left handed man. *Lancet*, 1, 1473–1479.

BRANCH, C., MILNER, B., and RASMUSSEN, T., (1964) Intracarotid sodium amytal for the lateralization of cerebral speech dominance. Observations in 123 patients. *Journal of neurosurgery*, 21, 399–405.

BREMER, F., (1966) Etude Electrophysiologique D'un Transfert Interhémisphérique Callosal. *Archives Italienne de Biologie*, 104, 1–29.

BRIGGS, G. G., (1975) A comparison of attentional and control shift models of the performance of concurrent tasks. *Acta psychologia*, 39, 183–191.

BRIGGS, G. G., and NEBES, R. D., (1975) Patterns of hand preference in a student population. *Cortex*, 11, 230–238.

BRIGGS, G. G., and NEBES, R. D., (1976) The effects of handedness, family history and sex on the performance of a dichotic listening task. *Neuropsychologia*, 14, 129–133.

BRIGGS, G. G., NEBES, R. D., and KINSBOURNE, M., (1976) Intellectual differences in relation to personal and family handedness. *Quarterly journal of experimental psychology*, 28, 591–601.

BRESSON, F., MAURY, L., LE-BONNIEC, G. P., and DE SCHONEN, S., (1977) Organization and lateralization of reaching in infants. An instance of asymmetric functions in hands collaboration. *Neuropsychologia*, 15, 311–320.

BRINKMAN, J., and KUYPERS, H. G. J. M., (1972) Split-brain monkeys: cerebral control of ipsilateral and contralateral arm, hand and finger movements. *Science*, 176, 536–539.

BRINKMAN, J., AND KUYPERS, H. G. J. M., (1973) Cerebral control of contralateral and ipsilateral arm, hand and finger movements in the split-brain rhesus monkey. *Brain*, 96, 653–674.

BROADBENT, D. E., (1954) The role of auditory localization in attention and memory. *Journal of experimental psychology*, 47, 191–196.

BROADBENT, D. E., (1958) *Perception and communication*, Pergamon: London.

BROADBENT, D. E., (1971) *Decision and stress*. Academic Press: London.

BROCA, P., (1861) Remarques sur le siège de la faculté du langage articulé, suivies d'une observation d'aphémie. *Bulletin de la societe anatomique de Paris*, 6, 343–357.

BROCA, P., (1865) Sur la faculté du langage articulé. *Bulletin de la societe d'anthropologie*, 6, 493–494.

BROMAN, M., (1978) Reaction-time differences between the left and right hemispheres for face and letter discrimination in children and adults. *Cortex*, 14, 578–591.

BROOKSHIRE, R. H., (1975) Recognition of auditory sequences by aphasic right hemisphere damaged and non-brain damaged subjects. *Journal of communication disorders*, 8, 51–59.

BROWN, J. W., (1975) On the neural organization of language: thalamic and cortical relationships. *Brain and language*, 2, 18–30.

BROWN, J., (1979) Language representation in the brain. Chap. 6 in *Neurobiology of social communication in primates*. (Eds.) STEKLIS, M. D., and RALEIGH, M. J., Academic Press: New York, 133–195.

BROWN, J. R., and SIMONSON, J., (1957) A clinical study of 100 aphasic patients. (1) Observations on lateralization and localization of lesions. *Neurology*, 7, 777–783.

BROWN, J. W., HÉCAEN, H., (1976) Lateralization and language representation. *Neurology*, 26, 183–189.

BROWN, J. W., and WILSON, F. R., (1973) Crossed aphasia in a dextral. *Neurology*, 23(9), 907–911.

BRUST, J. C. M., PLANK, C., BURKE, A., GUOBADIA, M. M. I., and HEALTON, E. B., (1982) Language disorder in a right-hander after occlusion of the right anterior cerebral artery. *Neurology*, 32, 492–497.

BRUYER, R., (1981) Asymmetry of facial expression in brain damaged subjects. *Neuropsychologia*, 19, 615–624.

BRYDEN, M. P., (1965) Tachistoscopic recognition, handedness and cerebral dominance. *Neuropsychologia*, 3, 1–8.

BRYDEN, M. P., (1966) Left-Right Differences in Tachistoscopic Recognition: Directional Scanning or Cerebral Dominance? *Perceptual and motor skills*, 23, 1127–1134.

BRYDEN, M. P., (1967) An evaluation of some models of laterality effects in dichotic listening. *Acta oto-laryngologica*, 63, 595–604.

BRYDEN, M. P., (1968) Symmetry of letters as a factor in tachistoscopic recognition. *American journal of psychology*, 81, 513–524.

BRYDEN, M. P., (1969) Binaural Competition and Division of Attention as Determinants of the Laterality Affect in Dichotic Listening. *Canadian journal of psychology*, 23, 101–113.

BRYDEN, M. P., (1970) Laterality effects in dichotic listening: relations with handedness and reading ability in children. *Neuropsychologia*, 8, 443–450.

BRYDEN, M. P., (1973) Perceptual asymmetry in vision: relation to handedness, eyedness and speech lateralization. *Cortex*, 9, 419–435.

BRYDEN, M. P., (1975) Speech lateralization in Families: A preliminary study using dichotic listening. *Brain and language*, 2, 201–211.

BRYDEN, M. P., (1977) Measuring handedness with questionnaires. *Neuropsychologia*, 15, 617–624.

BRYDEN, M. P., (1978) Strategy effects in the assessment of hemispheric asymmetry. In G. Underwood (Ed.) *Strategies of information processing.* Academic Press: New York, 117–149.

BRYDEN, M. P., (1979) Evidence for sex-related differences in cerebral organization. In WITTIG, M. A., and PETERSEN, A. C., (Eds.) *Sex-related differences in cognitive functioning.* Academic Press: New York. Chap. 5, 121–143.

BRYDEN, M. P., and ALLARD, F., (1976) Visual hemifield differences depend on typeface. *Brain and language*, 3, 191–200.

BRYDEN, M. P., and ALLARD, F., (1978) Dichotic listening and the development of linguistic processes. In KINSBOURNE, M., (Ed.) *Asymmetrical function of the brain.* Cambridge University Press: Cambridge, 392–404.

BRYDEN, M. P., and ALLARD, F., (1981) Do auditory perceptual asymmetries develop? *Cortex*, 1, 313–318.

BRYDEN, M. P., LEY, R. G., and SUGARMAN, J. H., (1982) A left ear advantage for identifying the emotional quality of tonal sequences. *Neuropsychologia*, 20, 83–88.

BRYDEN, M. P., and RAINEY, C. A., (1963) Left-Right Differences in Tachistoscopic Recognition. *Journal of experimental psychology*, 66, 568–571.

BRYDEN, M. P., and SPROTT, D. A., (1981) Statistical determination of degrees of laterality. *Neuropsychologia*, 19, 571–581.

BUCHSBAUM, M., CARPENTER, W. T., FEDIO, P., GOODWIN, F. K., MURPHY, D. L., and POST, R. M., (1979) Interhemispheric differences in evoked potential enhancement by selective attention to hemiretinally presented stimuli in schizophrenic, affective and post-temporal lobectomy patients. In GRUZELIER, J., and FLOR-HENRY, P., (Eds.) *Hemisphere asymmetries of function in psychopathology.* Elsevier/North Holland Biomedical Press: Amsterdam, 317–328.

BUCHSBAUM, M., and FEDIO, P., (1970) Hemispheric differences in evoked potentials to verbal and non-verbal stimuli in the left and right visual fields. *Physiology and behavior*, 5, 205–207.

BUCHSBAUM, M., INGVAR, D., KESSLER, R., et al., (1982) Cerebral glucography with positron tomography. *Archives of general psychiatry*, 39, 251–259.

BUCK, R., and DUFFY, R. J., (1980) Nonverbal communication of affect in brain-damaged patients. *Cortex*, 16, 351–362.

BUFFERY, A. W. H., (1974) Asymmetric Lateralisation of Cerebral Functions and the effects of unilateral brain surgery in epileptic patients. In S. J. DIMOND and J. G. BEAUMONT (Eds.) *Hemisphere function in the guman brain.* Elek: London, 204–234.

BUFFERY, A. W. H., and GRAY, J. A., (1972) Sex Differences in the Development of Spatial and Linguistic skils. In OUNSTED, C., and TAYLOR, D. C., (Eds.) *Gender differences: their ontogeny and significance.* Churchill Livingstone, Edinburgh, 123–157.

BUREŠ, J., BUREŠOVÀ, O., and KRIVÁNEK, J., (1981) An asymmetric view of brain laterality. *The behavioral and brain sciences*, 4, 22–23.

BUREŠOVÀ, O., LUKASZEWSKA, I., and BUREŠ, J., (1966) Interhemispheric synthesis of goal alternation and jumping escape reactions. *Journal of comparative and physiological psychology*, 65, 90–94.

BURKLAND, C. W., and SMITH, A., (1977) Language and the cerebral hemispheres. *Neurology*, 27, 627–633.

BUTLER, C. R., (1968) A memory record for visual discrimination habits produced in both cerebral hemispheres of monkey when only one hemisphere has received direct visual information. *Brain research*, 10, 152–167.

BUTLER, D. C., and MILLER, L. K., (1979) Role of order of approximation to English and letter array length in the development of visual laterality. *Developmental psychology*, 15, 522–529.

BUTLER, S. R., and GLASS, A., (1974) Asymmetries in the CNV over left and right hemispheres while subjects await numeric information. *Biological psychology*, 2, 1–16.

BUTLER, S. R., and NORSELL, U., (1968) Vocalization possibly initiated by the minor hemisphere. *Nature*, 220, 793.

BYRNE, B., (1974) Handedness and Musical Ability. *British journal of psychology*, 65, 279–281.

BYRNE, B., and SINCLAIR, J., (1979) Memory for tonal sequence and timbre: a correlation with familial handedness. *Neuropsychologia*, 17, 539–542.

CACIOPPO, J. T., AND PETTY, R. E., (1981) Lateral asymmetry in the expression of cognition and emotion. *Journal of experimental psychology*, 7, 333–341.

CALLAWAY, E., (1975) *Brain electrical potentials and individual psychological differences*. Grune and Stratton: New York.

CAMERON, R. F., CURRIER, R. D., and HAERER, A. F., (1971) Aphasia and literacy. *British journal of disorders of communication*, 6, 161–163.

CAMPBELL, A. L., BOGEN, J. E., and SMITH, A., (1981) Disorganization and re-organization of cognitive and sensori-motor functions in cerebral commissurotomy: compensatory roles of the forebrain commissures and cerebral hemispheres in Man. *Brain*, 104, 493–512.

CAMPBELL, R., (1978) Asymmetries in interpreting and expressing a posed facial expression. *Cortex*, 14, 327–342.

CAPLAN, P., and KINSBOURNE, M., (1976) Baby drops the rattle: asymmetry of duration of grasp by infants. *Child development*, 47, 532–534.

CAPLAN, B., and KINSBOURNE, M., (1981) Cerebral lateralization, preferred cognitive mode, and reading ability in normal children. *Brain and language*, 14, 349–370.

CARAMAZZA, A., GORDON, J., ZURIF, E. B., and DE LUCA, D., (1976) Right hemispheric damage and verbal problem solving behaviour. *Brain and language*, 3, 41–46.

CARLSON, J., NETLEY, C., HENDRICK, E. B., and PRICHARD, J. S. (1968) A re-examination of intellectual disabilities in hemispherectomized patients. *Transactions of the American neurological association*, 93, 198–201.

CARMON, A., and BECHTOLDT, H. P., (1969) Dominance of the right cerebral hemisphere for stereopsis. *Neuropsychologia*, 7, 29–39.

CARMON, A., and BENTON, A. L., (1969) Tactile perception of direction and number in patients with unilateral cerebral disease. *Neurology*, 19, 525–532.

CARMON, A., and GOMBOS, G. M., (1970) A physiological vascular correlate of hand preference: possible implications with respect to hemispheric cerebral dominance. *Neuropsychologia*, 8, 119–128.

CARMON, A., GORDON, H. W., BENTAL, E., and HARNESS, B. Z., (1977) Retraining in literal alexia: substitution of a right hemisphere perceptual strategy for impaired hemispheric processing. *Bulletin of the Los Angeles neurological societies*, 42, 41–50.

CARMON, A., HARISHANU, Y., LOWINGER, E., and LAVY, S., (1972) Asymmetries in hemispheric blood volume and cerebral dominance. *Behavioral biology*, 7, 853–859.

CARMON, A., and NACHSON, I., (1971) Effect of unilateral brain damage on perception of temporal order. *Cortex*, 7, 410–418.

CARMON, A., and NACHSON, I., (1973a) Hemified differences in binocular fusion. *Perceptual and motor skills*, 36, 175–184.

CARMON, A., and NACHSON, I., (1973b) Ear asymmetry in perception of emotional non-verb stimuli. *Acta psychologica*, 37, 351–357.

CARMON, A., NACHSON, I., and STARINSKY, R., (1976) Developmental aspects of visual hemifield differences in perception of verbal material. *Brain and language*, 3, 463–469.

CARMON, I., LAVY, S., GORDON H., and PORTNOY, Z., (1975) Hemispheric differences in rcBF during verbal and non-verbal tasks. In INGVAR, D. H. and LASSEN, N. (eds.) *Brain work: Alfred Bewgon Symposium VIII.* Copenhagen: Munksgaard: 414–423.

CARR, M. S., JACOBSON, T., and BOLLER, F., (1981) Crossed aphasia: analysis of four cases. *Brain and language*, 14, 190–202.

CARR, S. A., (1980) Interhemispheric transfer of stereognostic information in chronic schizophrenics. *British journal of psychiatry*, 136, 53–58.

CARTER, G. L., and KINSBOURNE, M., (1979) The ontogeny of right cerebral lateralization of spatial mental set. *Developmental psychology*, 15, 241–245.

CARTER-SALTZMAN, L., (1979) Patterns of cognitive functioning in relation to handedness and sex-related differences. In WITTIG, M. A., and PETERSON, A. C., (Eds.) *Sex-related differences in cognitive functioning*. Academic Press: New York, 97–118.

CARTER-SALTZMAN, L., (1980) Biological and sociocultural effects on handedness: comparison between biological and adoptive families. *Science*, 209, 1263–1265.

CASTRO-CALDAS, A. N., and SILVEIRA BOTELHO, M. A., (1980) Dichotic listening in the recovery of aphasia after stroke. *Brain and language*, 10, 145–151.

CAUDREY, D. J., and KIRK, K., (1979) The perception of speech in schizophrenia and affective disorders. In GRUZELIER, J., and FLOR-HENRY, P., (Eds.) *Hemisphere asymmetries of function in psychopathology*. Elsevier/North Holland Biomedical Press: Amsterdam, 581–601.

CELESIA, G. G., (1976) Organization of auditory cortical areas in man. *Brain*, 99, 403–414.

CHAN, K. S. F., HSU, F. K., CHAN, S. T., and CHAN, Y. B., (1960) Scrotal asymmetry and handedness. *Journal of anatomy*, 94, 543–548.

CHANEY, P. B., WEBSTER, J. C., (1966) Information in certain multidimensional sounds. *Journal of the acoustical society of America*, 40, 447–455.

CHAURASIA, B. D., and GOSWAMI, H. K., (1975) Functional asymmetry in the face. *Acta anatomica*, 91, 154–160.

ČERNÁCEK, J., and PODIVINSKÝ, F., (1971) Ontogenesis of handedness and somatosensory cortical responses, *Neuropsychologia*, 9, 219–232.

CHAMBERLAIN, H. D., (1928) The inheritance of left handedness. *Journal of heredity*, 19, 557–559.

CHÉDRU, F., LEBLANC, M., and L'HERMITTE, F., (1973) Visual searching in normal and brain damaged subjects (contribution to the study of unilateral inattention). *Cortex*, 9, 94–111.

CHESHER, E. C., (1936) Some observations concerning the relation of handedness to the language mechanism. *Bulletin of the neurological institute of New York*, 4, 556–562.

CHI, Je G., DOOLING, E. C., and GILLIES, F. H., (1977) Left-Right asymmetries of the Temporal Speech areas of the human fetus. *Archives of neurology*, 34, 346–348.

CHIARELLO, C., (1980) A house divided? Cognitive functioning with callosal agenesis. *Brain and language*, 11, 128–158.

CHURCHILL, J. A., IGNA, E., and SENF, R., (1962) The association of position at birth and handedness. *Paediatrics*, 29, 307–309.

CICONE, M., WAPNER, W., and GARDNER, H., (1980) Sensitivity to emotional expressions and situations in organic patients. *Cortex*, 16, 145–158.

CIOFFI, J., and KANDEL, G., (1979) Laterality of stereognostic accuracy of children for words, shapes and bigrams: a sex difference for bigrams. *Science*, 204, 1432–1434.

CLARK, M. M., (1970) *Reading difficulties in schools*. Penguin: Harmondsworth.

CLYMA, E. A., (1975) Unilateral Electroconvulsive Therapy: How to determine which hemisphere is dominant. *British journal of psychiatry*, 126, 372–379.

COHEN, A. I., (1966) Hand preference and developmental status of infants. *Journal of genetic psychology*, 108, 337–345.

COHEN, B. D., NOBLIN, C. D., and SILVERMAN, A. J., (1968) Functional asymmetry of the human brain. *Science*, 162, 475–477.

COHEN, D., (1977) Changes in REM dream content during the night: implications for a hypothesis about changes in cerebral dominance across REM periods. *Perceptual and motor skills*, 44, 1267–1277.

COHEN, G., (1972) Hemispheric differences in a letter classification task. *Perception and psychophysics*, 11, 139–142.

COHEN, G., (1973) Hemispheric differences in serial vs. parallel processing. *Journal of experimental psychology*, 97, 349–356.

COHEN, G., (1975a) Hemisphere differences in the effectings of cuing in visual recognition tasks. *Journal of experimental psychology*, 1, 366–373.

COHEN, G., (1975b) Hemispheric differences in the utilization of advance information in attention and performance. RABBITT, P. M. A., and DORNIC, S., (Eds.) Academic Press: London, Chap. 3, 20–32.

COHEN, G., (1976) Components of the laterality effect in letter recognition: asymmetries in iconic storage. *Quarterly journal of experimental psychology*, 28, 105–114.

COHEN, G., (1979) Comment on 'Information Processing in the Cerebral Hemispheres: selective activation and capacity limitations' by Hellige, Cox and Litvac. *Journal of experimental psychology* (Gen.), 108, 309–315.

COHEN, G., (1982) Theoretical interpretations of lateral asymmetries. In BEAUMONT, J. G., (Ed.) *Divided visual field studies of cerebral organisation*. Academic Press: London, 87–111.

COHEN, G., and FREEMAN, R., (1978) Individual differences in reading strategies in relation to cerebral asymmetry. In REQUIN, J., (Ed.) *Attention and performance VII*. Laurence Erlbaum Associates: New Jersey, Chap. 23, 411–426.

COHEN, H. D., ROSEN, R. C., and GOLDSTEIN, L., (1976) Electroencephalographic laterality changes during human sexual orgasm. *Archives of sexual behaviour*, 5, 189–199.

COHN, R., (1971) Differential cerebral processing of noise and verbal stimuli. *Science*, 172, 599–601.

COLBOURN, C. J., (1978) Can laterality be measured? *Neuropsychologia*, 16, 253–290.

COLBOURN, C., (1982) Divided visual field studies of psychiatric patients. In BEAUMONT, J. G., (Ed.) *Divided visual field studies of cerebral organisation*. Academic Press: London, 233–247.

COLBOURN, C. J., and LISHMAN, W. A., (1979) Lateralization of function and psychotic illness: a left hemisphere deficit? In GRUZELIER, J., and FLOR-HENRY, P., (Eds.) *Hemisphere asymmetries of function in psychopathology*. Elsevier/North Holland Biomedical Press: Amsterdam, 539–559.

COLBY, K. M., and PARKINSON, C., (1977) Handedness in autistic children. *Journal of autism and childhood schizophrenia*, 7, 3–9.

COLE, M., and PEREZ-CRUET, J., (1964) Prosopagnosia. *Neurology*, 2, 237–246.

COLLINS, R. L., (1970) The sound of one paw clapping: an inquiry into the origin of left-handedness. In LINDZEY, G., and THIESSEN, D., (Eds.) *Contributions to behavior-genetic analysis: The mouse as a prototype*. Appleton-Century-Crofts: New York, 115–136.

COLLINS, R. L., (1975) When left-handed mice live in right handed worlds. *Science*, 187, 181–184.

COLLINS, R. L., (1977) Toward an admissible genetic model for the inheritance of the degree and direction of asymmetry. In HARNAD, S., DOTY, R. W., GOLDSTEIN, L., JAYNES, J., and KRAUTHAMER, G., (Eds.) *Lateralization in the nervous system*. Academic Press: New York, 137–150.

COLOMBO, A., DE RENZI, E., and FAGLIONI, P., (1976) The occurrence of visual neglect in patients with unilateral cerebral disease. *Cortex*, 12, 221–231.

COLTHEART, M., (1972) Visual information processing: DODWELL, P. C., (Ed.) *New horizons in psychology*, 2. Penguin: London, 62–85.

COLTHEART, M., (1980a) Deep dyslexia: a review of the syndrome. In COLTHEART, M., PATTERSON, K., and MARSHALL, J. C., (Eds.) *Deep dyslexia*. Routledge and Kegan Paul: London, 22–47.

COLTHEART, M., (1980b) Deep dyslexia: a right hemisphere hypothesis. In COLTHEART, M., PATTERSON, K., and MARSHALL, J. C., (Eds.) *Deep dyslexia*. Routledge and Kegan Paul, 326–380.

COLTHEART, M., and ARTHUR, B., (1971) Tachistoscopic hemifield effects with hemifield report. *American journal of psychology*, 84, 355–364.

COLTHEART, M., LEA, C. D., and THOMPSON, K., (1974) In defence of iconic memory. *Quarterly journal of experimental psychology*, 26, 633–641.

COLTHEART, M., PATTERSON, K., and MARSHALL, J. C., (1980) (Eds.) *Deep dyslexia*, Routledge and Kegan Paul, London.

COMBS, A. L., HOBLICK, P. J., CZARNECKI, J., and KAMLER, P , (1977) Relationships of lateral eye movement to cognitive mode, hemispheric interaction and choice of college major. *Perceptual and motor skills*, 45, 983–990.

CONNOLLY, J. F., (1982) The corpus callosum and brain function in schizophrenia. *British journal of psychiatry*, 140, 429–430.

CONNOLLY, J. F., GRUZELIER, J. H., KLEINMAN, K. M., and HIRSCH, S., (1979) Lateralized abnormalities in hemisphere specific tachistoscopic tasks in psychiatric patients and controls. In GRUZELIER, J., and FLOR-HENRY, P., (Eds.) *Hemisphere asymmetries of function in psychopathology*. Elsevier/North Holland Biomedical Press: Amsterdam, 491–509.

CORBALLIS, M. C., ANUZA, T., and BLAKE, L., (1978) Tachistoscopic perception under head tilt. *Perception and psychophysics*, 24, 274–284.

CORBALLIS, M. C., and BEALE, I. L., (1970) Bilateral symmetry and behaviour. *Psychological review*, 77, 451–464.

CORBALLIS, M. C., and BEALE, I. L., (1976) *The psychology of left and right*, Halsted: New York.

CORBALLIS, M. C., and MORGAN, M. J., (1978) On the biological basis of human laterality: 1. Evidence for a maturational left-right gradient. *The behavioral and brain sciences*, 2, 261–336.

COREN, S., and KAPLAN, C. P., (1973) Patterns of ocular dominance. *American journal of optometry and archives of the American academy of optometry*, 50, 283–292.

COREN, S., and PORAC, C., (1977) Fifty centuries of right-handedness: the historical record. *Science*, 198, 631–632.

COREN, S., and PORAC, C., (1978) The validity and reliability of self-report items for the measurement of lateral preference. *British journal of psychology*, 69, 207–211.

COREN, S., and PORAC, C., (1979) Normative data on hand position during writing. *Cortex*, 15, 679–682.

CORKIN, S., (1965) Tactually guided maze language in man: effects of unilateral cortical excisions and bilateral hippocampal lesions. *Neuropsychologia*, 3, 339–351.

COSTA, L. D., (1962) Visual reaction time of patients with cerebral disease as a function of length and constancy of preparatory interval. *Perceptual and motor skills*, 14, 391–397.

COSTA, L. D., and VAUGHAN, H. G., (1962) Performance of patients with lateralized cerebral lesions I: Verbal and Perceptual Tests. *Journal of nervous and mental diseases*, 134, 162–168.

COTTON, B., TZENG, O. J. L., and HARDYCK, C., (1980) Role of cerebral hemispheric processing in the visual half-field stimulus – response compatability effect. *Journal of experimental psychology*. Human Perception and Performance, 6, 13–23.

CRAIG, J. D., (1979a) Asymmetries in Processing Auditory Non-Verbal Stimuli? *Psychological bulletin*, 86, 1339–1349.

CRAIG, J. D., (1979b) The effect of musical training and cerebral asymmetries on perception of an auditory illusion. *Cortex*, 15, 671–678.

CRAIG, J. D., (1980) A dichotic rhythm task: advantage for the left-handed. *Cortex*, 16, 613–620.

CREMER, M., and ASHTON, R., (1981) Motor performance and concurrent cognitive tasks *Journal of motor behavior*, 13, 187–196.

CRITCHLEY, M., (1962) Speech and speech loss in relation to the duality of the brain. In MOUNTCASTLE, V. B. (Ed,) *Interhemispheric relations and cerebral dominance*. Johns Hopkins University Press: Baltimore, 208–213.

CRITCHLEY, M., (1970) *The dyslexic child*. Heinemann: London.

CROCKETT, H. G., and ESTRIDGE, N. M., (1951) Cerebral hemispherectomy: a clinical, surgical and pathologic study of four cases. *Bulletin of the Los Angeles neurological societies*, 16, 71–87.

CRONIN, D., BODLEY, P., POTTS, L., MATHER, M. D., GARDNER, R. K., and TOBIN, J. C., (1970) Unilateral and bilateral ECT: a study of memory disturbance and relief from depression. *Journal of neurology, neurosurgery and psychiatry*, 33, 705–713.

CROUCH, W. W., (1976) Dominant direction of conjugate lateral eye movements and responsiveness to facial cues. *Perceptual and motor skills*, 42, 167–174.

CROVITZ, H., and ZENER, K., (1962) A group test for assessing hand-and-eye dominance. *American journal of psychology*, 75, 271–276.

CURCIO, F., MACKAVEY, W., and ROSEN, J., (1974) Role of visual acuity in tachistoscopic recognition of three letter words. *Perceptual and motor skills*, 38, 755–761.

CURRY, F. K. W., (1967) A comparison of left-handed and right-handed subjects on verbal and non-verbal dichotic listening tasks. *Cortex*, 3, 343–352.

CURRY, F. K. W., (1968) A comparison of the performances of a right hemispherectomized subject and 25 normals on four dichotic listening tasks. *Cortex*, 4, 144–153.

CURRY, F. K. W., and GREGORY, H., (1969) The performance of stutterers on dichotic listening tasks thought to reflect cerebral dominance. *Journal of speech and hearing research*, 12, 73–82.

CURRY, F. K. W., and RUTHERFORD, D. R., (1967) Recognition and recall of dichotically presented verbal stimuli by right and left-handed persons. *Neuropsychologia*, 5, 119–126.

CURTISS, S., (1977) *A psycholinguistic study of a modern day 'Wild Child'*. Academic Press: New York.

DABBS, J. M., (1980) Left-right differences in cerebral blood flow and cognition. *Psychophysiology*, 17, 548–551.

DALBY, J. T., (1980) Hemispheric timesharing: verbal and spatial loading with concurrent unimanual activity. *Cortex*, 16, 567–573.

DALBY, J. T., and GIBSON, D., (1981) Functional cerebral lateralization in subtypes of disabled readers. *Brain and language*, 14, 34–48.

DAMASIO, A. R., ALMEIDA LIMA, P., and DAMASIO, H., (1975) Nervous function after right hemispherctomy. *Neurology*, 25, 89–93.

DAMASIO, A. R., CASTRO-CALDAS, A., GROSSO, J. T., and FERRO, M., (1976) Brain specialisation for language does not depend on literacy. *Archives of neurology*, 33, 300–302.

DAMASIO, A. R., and DAMASIO, H., (1977) Musical Faculty and Cerebral Dominance. In *Music and the brain*, (Eds.) CRITCHLEY, M., and HENSON, R. A. The Camelot Press, Southampton, Chap. 9, 141–155.

DAMASIO, H., MAURER, R. G., DAMASIO, A. R., and CHUI, H. C., (1980) Computerized tomographic scan findings in patients with autistic behavior. *Archives of neurology*, 37, 504–510.

D'ANGELO, E. J., (1981) Reversed cerebral asymmetries as a potential risk factor in autism: a reconsideration. *Perceptual and motor skills*, 53, 101–102.

DANTA, G., HILTON, R. C., and O'BOYLE, D. J., (1978) Hemisphere function and binocular depth perception, *Brain*, 101, 569–589.

DART, R. A., (1949) The predatory implement technique of australopithecus. *American journal of physical anthropology*, 7, 1–38.

DARWIN, C., (1872) The expressions of the emotions in man and animals. Murray: London.

DARWIN, C. J., (1974) Ear differences and hemispheric specialisation. In SCHMITT, F. O., and WORDEN, F. G., (Eds.) *The neurosciences: third study program*. Cambridge, Massachusetts: M.I.T. Press, Chap. 5, 57–63.

DAVIDOFF, J. B., (1975) Hemispheric differences in the perception of lightness. *Neuropsychologia*, 13, 121–124.

DAVIDOFF, J. B., (1976) Hemispheric sensitivity differences in the perception of colour. *Quarterly journal of experimental psychology*, 28, 387–394.

DAVIDOFF, J. B., (1977) Hemispheric differences in dot detection. *Cortex*, 13, 434–444.

DAVIDOFF, J. B., BEATON, A. A., DONE, D. J., and BOOTH, H., (1982) Information extraction from brief verbal displays. Half-field and serial position effects for children, normal and illiterate adults. *British journal of psychology*, 73, 29–39.

DAVIDOFF, J., DONE, J., and SCULLY, J., (1981) What does the lateral ear advantage relate to? *Brain and language*, 12, 332–346.

DAVIDSON, R. J., HOROWITZ, M. E., SCHWARTZ, G. E., and GOODMAN, D. M., (1981) Lateral differences in the latency between finger tapping and heart beat. *Psychophysiology*, 18, 36–41.

DAVIDSON, R. J., and SCHWARTZ, G. E., (1976) Patterns of cerebral lateralization during cardiac biofeedback versus the self-regulation of emotions: sex differences. *Psychophysiology*, 13, 62–68.

DAVIDSON, R., SCHWARTZ, G., PUGASH, E., and BROMFIELD, E., (1976) Sex differences in patterns of EEG asymmetry. *Biological psychology*, 4, 119–138.

DAVIS, A. E., and WADA, J. A., (1977) Hemispheric asymmetries in humans infants spectral analysis of flash and click evoked potentials. *Brain and language*, 4, 23–31.

DAVIS, R., and SCHMIT, V., (1971) Timing the transfer of information between the hemispheres in man. *Acta psychologica*, 35, 335–346.

DAVIS, R., and SCHMIT, V., (1973) Visual and verbal coding in the interhemispheric transfer of information. *Acta Psychologica*, 37, 229–240.

DAWSON, G., WARRENBURG, S., and FULLER, P., (1982) Cerebral lateralization in individuals diagnosed as autistic in early childhood. *Brain and language*, 15, 353–368.

DAWSON, J. L. M., (1973) Temne-Arunta hand/eye dominance and susceptibility to geometric illusions. *Perceptual and motor skills*, 37, 659–667.

DAX, M., (1836) Lésions de la moitié gauche de l'encephale coincident avec l'oubli des signes de la pensée. In G. DAX (1878). *L'aphasie*. Montpelier.

DAY, J., (1977) Right-hemispheric language processing in normal right handers. *Journal of experimental psychology* (Human Perception and Performance), 3, 518–528.

DAY, J., (1979) Visual half-field word recognition as a function of syntactic class and imageability. *Neuropsychologia*, 17, 515–519.

DAY, M. E., (1964) An eye-movement phenomenon relating to attention, thought and anxiety. *Perceptual and motor skills*, 19, 443–446.

DAY, P. S., and ULATOWSKA, A. K., (1980) Perceptual, cognitive and linguistic development after early hemispherectomy: two case studies. *Brain and language*, 7, 17–33.

DEE, H. L., (1971) Auditory asymmetry and manual preference. *Cortex*, 7, 236–245.

DEE, H. L., and FONTENOT, D. J., (1973) Cerebral dominance and lateral differences in perception and memory. *Neuropsychologia*, 11, 167–173.

DEE, H. L., and HANNAY, A. J., (1973) Asymmetry in perception: attention versus other determinants. *Acta psychologica*, 37, 241–247.

DEGLIN, V. L., and NIKOLAENKO, N. N., (1975) Role of the dominant hemisphere in the regulation of emotional states. *Human physiology*, 1, 394–402.

DEJERINE, J., (1892) Contribution a l'étude anatamo-pathologique et clinique des différentes variétés de cécité verballe. *Memoires de la société de biologie*, 4, 61–90.

D'ELIA, G., LORENTZSON, S., RAOTMA, H., and WIDEPALM, K., (1976) Comparison of unilateral dominant and non-dominant ECT on verbal and non-verbal memory. *Acta psychiatrica Scandinavica*, 53, 85–94.

DENCKLA, M. B., (1972) Color-naming defects in dyslexic boys. *Cortex*, VIII, 164–176.

DENCKLA, M. B., (1974) Development of motor co-ordination in normal children. *Developmental medicine and child neurology*, 16, 729–741.

DENCKLA, M. B., and RUDEL, R. G., (1976) Rapid 'automatized' naming: dyslexia differentiated from other disabilities. *Neuropsychologia*, 14, 471–480.

DENENBERG, V. H., (1980) General systems theory, brain organization and early experiences. *American journal of physiology*, 238, 3–13.

DENENBERG, V. H., (1981) Hemispheric laterality in animals and the effects of early experience. *The behavioral and brain sciences*, 4, 1–49.

DENENBERG, V. H., GARBANTI, J., SHERMAN, G. YUTZEY, D., and KAPLAN, R., (1978) Infantile stimulation induces brain lateralization in rats. *Science*, 201, 1150–1152.

DENNIS, M., (1976) Impaired sensory and motor differentiation with corpus callosum agenesis: a lack of callosal inhibition during ontogeny? *Neuropsychologia*, 14, 455–469.

DENNIS, M., (1981) Language in a congenitally acallosal brain. *Brain and language*, 12, 33–53.

DENNIS, M., and WHITAKER, H. A., (1977) Hemispheric equipotentiality and language acquisition. In SEGALOWITZ, S. J., and GRUBER, F. A., (Eds.) *Language development and neurological theory*. Academic Press: New York, Chap. 8, 93–106.

DENNIS, W., (1985) Early graphic evidence of dextrality in man. *Perceptual and motor skills*, 8, 147–149.

DENNY-BROWN, D., MEYER, J. S., and HORENSTEIN, S., (1952) The significance of perceptual rivalry resulting from parietal lesion. *Brain*, 75, 433–471.

DE RENZI, E., and FAGLIONI, P., (1965) The comparative efficiency of intelligence and vigilance tests in detecting hemispheric cerebral damage. *Cortex*, 1, 410–433.

DE RENZI, E., FAGLIONI, P., and FERRARI, P., (1980) The influence of sex and age on the incidence and type of aphasia. *Cortex*, 16, 627–630.

DE RENZI, E., FAGLIONI, P., and SCOTTI, G., (1968) Tactile spatial impairment and unilateral cerebral damage. *Journal of nervous and mental disease*, 146, 468–475.

DE RENZI, E., FAGLIONI, P., and SCOTTI, G., (1970) Hemispheric contribution to exploration of space through the visual and tactile modality. *Cortex*, 6, 191–203.

DE RENZI, E., FAGLIONI, P., and SCOTTI, G., (1971) Judgement of spatial orientation in patients with focal brain damage. *Journal of neurology, neurosurgery and psychiatry*, 34, 489–495.

DE RENZI, E., FAGLIONI, P., and SORGATO, P., (1982) Modality – specific and supra modal mechanisms of apraxia, 105, 301–312.

DE RENZI, E., SCOTTI, G., (1969) The influence of spatial disorders in impairing tactual discrimination of shapes. *Cortex*, 5, 53–62.

DEUTSCH, D., (1974) An auditory illusion. *Nature*, 251, 307–309.

DEUTSCH, D., (1978) Pitch memory: an advantage for the left handed. *Science*, 199, 559–560.

DEWSON, J. H., (1977) Preliminary evidence of hemispheric asymmetry of auditory functions in monkeys. In HARNAD, S. A., DOTY, R. W., GOLDSTEIN, L., JAYNES, and KRAUTHAMER, J., (1977) *Lateralization in the nervous system*. Academic Press: New York, Chap. 4., 63–71.

DI CHIRO, G., (1962) Angiographic patterns of cerebral convexity and superficial dural sinuses. *American journal of roentology*, 87, 308–321.

DICK, A. O., and MEWHORT, J. K., (1967) Order of report and processing in tachistoscopic recognition. *Perception and psychophysics*, 2, 573–576.

DIMOND, S. J., (1969) Hemisphere function and immediate memory. *Psychonomic science*, 16, 111–112.

DIMOND, S. J., (1970a) Hemispheric refractoriness and the control of reaction time. *Quarterly journal of experimental psychology*, 24, 610–617.

DIMOND, S. J., (1970b) Reaction times and response competition time in right and left handers. *Quarterly journal of experimental psychology*, 22, 513–520.

DIMOND, S. J., (1971) Hemisphere function and word registration. *Journal of experimental psychology*, 87, 183–185.

DIMOND, S. J., (1972) *The double brain*, London: Churchill Livingstone.

DIMOND, S. J., (1976) Depletion of attentional capacity after total commissurotomy in man. *Brain*, 99, 347–356.

DIMOND, S. J., (1978) Symmetry and asymmetry in the vertebrate brain. In *Brain behaviour and evolution*. (Eds.) PLOTKIN, H., and OAKLEY, D. Methuen: London, 189–218.

DIMOND, S. J., (1979a) Tactual and auditory vigilance in split-brain man. *Journal of neurology, neurosurgery and psychiatry*, 42, 70–74.

DIMOND, S. J., (1979b) Performance by split-brain humans on lateralized vigilance tasks. *Cortex*, 15, 43–50.

DIMOND, S. J., and BEAUMONT, J. G., (1971a) Hemisphere function and vigilance. *Quarterly journal of experimental psychology*, 23, 443–448.

DIMOND, S. J., and BEAUMONT, J. G., (1971b) The use of two cerebral hemispheres to increase brain capacity. *Nature*, 232, 270–271.

DIMOND, S. J., and BEAUMONT, J. G., (1973) Difference in the vigilance performance of the right and left hemispheres. *Cortex*, 9, 259–265.

DIMOND, S. J., and BEAUMONT, J. G., (1974) Experimental studies of hemispheric function. In DIMOND, S. J., and BEAUMONT, J. G., (Eds.) *Hemisphere function in the human brain*. Elek: London, 481–88.

DIMOND, S. J., BUREŠ, J., FARRINGTON, L. J., and E. Y. M. BROUWERS, (1975) The use of contact lenses for the lateralisation of visual input in man. *Acta psychologica*, 35, 341–349.

DIMOND, S. J., and FARRINGTON, L., (1977) Emotional response to films shown to the right or left hemisphere of the brain measured by heart rate. *Acta psychologica*, 41, 255–260.

DIMOND, S. J., FARRINGTON, L., and JOHNSON, P., (1976) Differing emotional response from right and left hemispheres. *Nature*, 261, 690–692.

DIMOND, S. J., SCAMMELL, R. E., BROUWERS, E. Y. M., and WEEKS, R., (1977) Functions of the centre section (trunk) of the corpus callosum in man. *Brain*, 100, 543–562.

DIMOND, S. J., SCAMMELL, R. E., PRYCE, I. G., HUWS, D., and GRAY, C., (1979) Callosal transfer and left-hand anomia in schizophrenia. *Biological psychiatry*, 14, 735–739.

DIXON, N., and JEEVES, M. A., (1970) The interhemispheric transfer of movement after-effects: a comparison between acallosal and normal subjects. *Psychonomic science*, 20, 201–203.

DODDS, A. G., (1978) Hemispheric differences in tactuo-spatial processing. *Neuropsychologia*, 16, 247–254.

DOKTOR, R, and BLOOM, D. M., (1977) Selective lateralization of cognitive style related to occupation as determined by EEG alpha asymmetry. *Psychophysiology*, 14, 385–387.

DONCHIN, E., KUTAS, M., and MCCARTHY, G., (1977) Electro-cortical indices of hemispheric utilization. In HARNAD, S., et al., (Eds.) *Lateralization in the nervous system.* Academic Press: New York, Chap. 19, 339–384.

DONE, D. J., and MILES, T. R., (1978) Learning, memory and dyslexia. In GRUNEBERG, M. M., MORRIS, P. E., and SYKES, R. N., (Eds.) *Practical aspects of memory.* Academic Press: London, 553–560.

DORMAN, M. F., and GEFFNER, D. S., (1974) Hemispheric specialization for space perception in 6-yr-old black and white children from low and middle socio-economic classes. *Cortex*, 1974, 10, 171–176.

DORMAN, M F., and PORTER, R. J., (1975) Hemispheric lateralization for speech perception in stutterers. *Cortex*, 1975, 11, 181–185.

DUNLOP, P., (1976) The changing role of orthoptics in dyslexia. *British orthoptic journal*, 33, 22–28.

DUNLOP, D. B., and DUNLOP, P., (1974) New concepts of visual laterality in relation to dyslexia. *Australian journal of ophthalmology*, 2, 101–112.

DUNLOP, D. B., DUNLOP, P., and FENELON, B., (1973) Vision laterality analysis in children with reading disability: the results of new techniques of examination. *Cortex*, 9, 227–236.

DURNFORD, M., and KIMURA, D., (1971) Right hemisphere specialization for depth perception reflected in visual field differences, *Nature*, 231, 394–395.

DUSEK, C., and HICKS, R. A., (1980) Multiple birth risk factors and handedness in elementary school children. *Cortex*, 16, 471–478.

DVIRSKII, A. E., (1976) Functional asymmetry of the cerebral hemispheres in clinical types of schizophrenia. *Neuroscience and behavioral physiology*, 7, 236–239

DWYER, J. H., and RINN, W. E., (1981) The role of the right hemisphere in contextual inference. *Neuropsychologia*, 19, 479–482.

EASON, R. G., GROVES, P., WHITE, C. T., and ODEN, D., (1967) Evoked cortical potentials: relation to visual field and handedness. *Science*, 156, 1643–1646.

EATON, E. M., (1979) Hemisphere related visual information processing in acute schizophrenia before and after neuroleptic treatment. In GRUZELIER, J. and FLOR-HENRY, P., (Eds.) *Hemisphere asymmetries of function in psychopathology.* Elsevier/North Holland Biomedical Press: Amsterdam, 511–526.

ECCLES, J., (1973) *The understanding of the brain.* McGraw Hill: New York.

EHRLICHMAN, H., and WEINBERGER, A., (1978) Lateral eye movements and hemispheric asymmetry: a critical review. *Psychological bulletin*, 85, 1080–1101.

EIDELBERG, D., and GALABURDA, A. M., (1982) Symmetry and asymmetry in the human posterior thalamus. 1. Cytoarchitectonic analysis in normal persons. *Archives of neurology*, 325–332.

EIMAS, P. D., SIQUELAND, E. R., JUSCZYK, P., and VIGORITO, J., (1971) Speech perception in infants. *Science*, 171, 304–306.

EISONSON, J., (1962) Language and intellectual modifications associated with right cerebral damage. *Language and speech*, 5, 49–53.

EKMAN, P., (1980) Asymmetry in facial expression. *Science*, 209, 833–834.

EKMAN, P., HAGER, J. C., and FRIESEN, W., (1981) The symmetry of emotional and deliberate facial actions. *Psychophysiology*, 18, 101–106.

EKMAN, P., and OSTER, H., (1979) Facial expressions of emotion. *Annual review of psychology*, 30, 527–554.

ELING, P., (1981) On the theory and measurement of laterality. *Neuropsychologia*, 19, 321–324.

ELING, P., MARSHALL, J. C., and VAN GALEN, G., (1981) The development of language lateralization as measured by dichotic listening. *Neuropsychologia*, 19, 767–774.

ELLENBERG, L., and SPERRY, R. W., (1979) Capacity for holding sustained attention following commissurotomy. *Cortex*, 15, 421–438.

ELLENBERG, L., and SPERRY, R. W., (1980) Lateralized division of attention in the commissurotomized and intact brain. *Neuropsychologia*, 18, 411–418.

ELLIS, A. W., (1979) Developmental and acquired dyslexia: some observations on JORM (1979). *Cognition*, 7, 413–420.

ELLIS, A. W., and MILLER, D., (1981) Left and wrong in adverts: neuropsychological correlates of aesthetic preference. *British journal of psychology*, 72, 225–230.

ELLIS, H. D., (1975) Recognising faces. *British journal of psychology*, 66, 409–426.

ELLIS, H. D., and SHEPHERD, J. W., (1974) Recognition of abstract and concrete words presented in left and right visual fields. *Journal of experimental psychology*, 103, 1035–1036.

ELLIS, H. D., and SHEPHERD, J. W., (1975) Recognition of upright and inverted faces in left and right visual fields. *Cortex*, 11, 3–7.

ELLIS, H. D., and YOUNG, A. W., (1977) Age of acquisition and recognition of nouns present in the left and right visual fields: a failed hypothesis. *Neuropsychologia*, 15, 825–827.

ELLIS, N. C., and HENNELLY, R. A., (1980) A bilingual word-length effect: implications for intelligence testing and the relative ease of calculation in Welsh and English. *British journal of psychology*, 71, 43–53.

ELLIS, N. C., and MILES, T. R., (1978) Visual information processing in dyslexic children. In GRUNEBERG, M. M., MORRISS, P. E., and SYKES, R. N., (Eds.) *Practical aspects of memory*. Academic Press: London, 561–577.

ELMAN, J. L., TAKAHASHI, K., and TOHSAKU, Y. H., (1981) Lateral asymmetries for the identification of concrete and abstract Kanji. *Neuropsychologia*, 19, 407–412.

EME, R., STONE, S., and IZRAL, R., (1978) Spatial deficit in familial left-handed children. *Perceptual and motor skills*, 47, 919–922.

ENDO, M., SHIMOZU, A., and HORI, T., (1978) Functional asymmetry of visual fields for Japanese words in KANA (syllable based) writing and random-shape recognition in Japanese subjects. *Neuropsychologia*, 16, 291–298.

ENDO, M., SHIMIZU, A., and NAKAMURA, I., (1981) The influence of Hangul learning upon laterality differences in Hangul word recognition by native Japanese subjects. *Brain and language*, 14, 114–119.

ENTUS, A. K., (1977) Hemispheric asymmetry in processing of dichotically presented speech and non-speech stimuli by infants. In SEGALOWITZ, S. J., and GRUBER, F. A., (Eds.) *Language development and neurological theory*. Academic Press: New York and London. Chap. 6, 63–73.

ERTL, J., and SCHAFER, E. W. P., (1967) Cortical activity preceding speech. *Life sciences*, 6, 473–479.

ESPIR, M. L. E., and ROSE, F. C., (1976) *The basic neurology of speech*. Blackwell: Oxford.

ETTLINGER, G., and BLAKEMORE, C. B., (1969) The behavioral effects of commissural section. In BENTON, A. L., (Ed.) *Contributions to clinical neuropsychology*. Aldine: Chicago, 30–72.

ETTLINGER, G., BLAKEMORE, C. B., MILNER, A. D., and WILSON, J., (1972) Agenesis of the corpus callosum: a behavioural investigation. *Brain*, 95, 327–346.

ETTLINGER, G., BLAKEMORE, C. B., MILNER, A. D., and WILSON, J., (1974) Agenesis of the corpus callosum: a further behavioural investigation. *Brain*, 97, 225–234.

ETTLINGER, G., JACKSON, C. V., and ZANGWILL, D. L., (1956) Cerebral dominance in sinistrals. *Brain*, 79, 569–588.

FAGAN-DUBIN, L., (1974) Lateral dominance and development of cerebral specialization. *Cortex*, 10, 69–74.

FAGLIONI, P., SCOTTI, G., and SPINNLER, H., (1968) Impaired recognition of written letters following unilateral hemispheric damage. *Cortex*, 5, 120–133.

FAGLIONI, P., SCOTTI, G., and SPINNLER, H., (1971) The performance of brain damaged patients in spatial localization of visual and tactile stimuli. *Brain*, 94, 443–454.

FAIRWEATHER, H., (1976) Sex differences in cognition. *Cognition*, 4, 231–280.

FALEK, A., (1959) Handedness: A family study. *American journal of human genetics*, 11, 52–62.

FALZI, G., PERRONE, P., and VIGNOLO, L. A., (1982) Right-left asymmetry in anterior speech regions. *Archives of neurology*, 39, 239–240.

FARKAS, T., WOLF., A. P., FOWLER, J., et al., (1981) Regional brain glucose metabolism in Schizophrenia. *Journal of cerebral blood flow and metabolism* 1 (Supplement): S496.

FECHNER, G. T., (1860) *Elemente der Psychophysik* Vol. 2, Leipzig: Breitkopf and Hartel.

FENNELL, E. B., BOWERS, D., and SATZ, P., (1977) Within modal and cross-modal reliabilities of two laterality tests. *Brain and language*, 4, 63–69.

FENNELL, E. B., SATZ, P., VAN DEN ABELL, T., BOWERS, D., and THOMAS, R., (1978) Visuospatial competency, handedness and cerebral dominance. *Brain and language*, 5, 206–214.

FENNELL, E. B., SATZ, P., and WISE, R., (1967) Laterality differences in the perception of pressure. *Journal of neurology, neurosurgery and psychiatry*, 30, 337–340.

FERRIS, G. S., and DORSEN, M. M., (1975) Agenesis of the corpus callosum: 1. Neuropsychological studies. *Cortex*, 11, 95–122.

FILBEY, R. A., and GAZZANIGA, M. S., (1969) Splitting the normal brain with reaction time. *Psychonomic science*, 17, 335–336.

FINCH, G., (1941) Chimpanzee handedness, *Science*, 94, 117–118.

FINLAYSON, M. A. J., (1976) A behavioural manifestation of the development of interhemispheric transfer of language in children. *Cortex*, 12, 290–295.

FISCHER, F. W., LIBERMAN, I. Y., and SHANKWEILER, D., (1978) Reading reversals and developmental dyslexia: a further study. *Cortex*, 14, 496–510.

FITTS, P. M., and SEEGER, C. M., (1953) S-R compatibility: spatial characteristics of stimulus and response codes. *Journal of experimental psychology*, 46, 199–210.

FLANERY, R. C., and BALLING, J. D., (1979) Developmental changes in hemispheric specialization for tactile spatial ability. *Developmental psychology*, 15, 364–372.

FLEMINGER, J. L., and BUNCE, L., (1975) Investigation of cerebral dominance in left handers and right handers using unilateral electro-convulsive therapy. *Journal of neurology, neurosurgery and psychiatry*, 38, 541–545.

FLEMINGER, J. L., DALTON, R., and STANDAGE, K. F., (1977a) Handedness in Psychiatric Patients. *British journal of psychiatry*, 131, 448–452.

FLEMINGER, J. L., DALTON, R., and STANDAGE, K. F., (1977b) Age as a factor in the handedness of adults. *Neuropsychologia*, 15, 471–473.

FLEMINGER, J. L., HORNE, D. J. de L., and NOTT, P. N., (1970) Unilateral electro-convulsive therapy and cerebral dominance: effect of left and right-sided electrode placement on verbal memory. *Journal of neurology, neurosurgery and psychiatry*, 33, 408–411.

FLEMINGER, J. L., McCLURE, G. M., and DALTON, R., (1980) Lateral response to suggestion in relation to handedness and the side of psychogenic symptoms. *British journal of psychiatry*, 136, 562–566.

FLICK, G., (1966) Sinistrality revisited: a perceptual motor approach. *Child development*, 37, 613–622.

FLOR-HENRY, P., (1969) Psychosis and temporal lobe epilepsy: a controlled investigation. *Epilepsia*, 10, 363–395.

FLOR-HENRY, P., (1974) Psychosis, neurosis and epilepsy: developmental and gender-related effects and their aetiological contribution. *British journal of psychiatry*, 124, 144–150.

FLOR-HENRY, P., (1976) Lateralized temporal-limbic dysfunction and psychopathology. *Annals of the New York Academy of Sciences*, 280, 777–795.

FLOR-HENRY, P., (1979a) On certain aspects of the localization of the cerebral systems regulating and determining emotion. *Biological psychiatry*, 14, 677–698.

FLOR-HENRY, P., (1979b) Laterality, shifts of cerebral dominance, sinistrality and psychosis. In GRUZELIER, J. and FLOR-HENRY, P., (Eds.) *Hemisphere Asymmetries of Function in Psychopathology*. Elsevier/North Holland Biomedical Press: Amsterdam, 3–19.

FONTENOT, D. J., (1973) Visual field differences in the recognition of verbal and non-verbal stimuli in man. *Journal of comparative and physiological psychology*, 85, 564–569.

FONTENOT, D. J., and BENTON, A. L., (1971) Tactile perception of direction in relation to hemispheric locus of lesion. *Neuropsychologia*, 9, 83–88.

FONTENOT, D. J., and BENTON, A. L., (1972) Perception of direction in right and left visual fields. *Neuropsychologia*, 10, 447–452.

FORGAYS, D. G., (1953) The development of a differential word recognition. *Journal of experimental psychology*, 45, 165–168.

FRENCH, L. A., JOHNSON, D. R., BROWN, I. A., and VAN BERGEN, F. B., (1955) Cerebral hemispherectomy for control of intractable convulsive seizures. *Journal of neurosurgery*, 12, 154–164.

FRANCO, L., (1977) Hemispheric interaction in the processing of concurrent tasks in commissurotomy subjects. *Neuropsychologia*, 15, 707–710.

FREIDES, D., (1977) Do dichotic listening procedures measure lateralization of information processing or retrieval strategy? *Perception and psychophysics*, 21, 259–263.

FRIED, I., (1979) Cerebral dominance sub-types of developmental dyslexia. *Bulletin of the Orton society*, 29, 101–112.

FRIED, I., TANGUAY, P. E., BODER, E., DOUBLEDAY, C., and GREENSITE, M., (1981) Developmental dyslexia: electrophysiological evidence of clinical sub-groups. *Brain and language*, 12, 14–22.

FRIEDLAND, R. P., and WEINSTEIN, E. A., (1977) Hemi-inattention and hemisphere specialization: introduction and historical review. In WEINSTEIN, E. A. and FRIEDLAND, R. P., (Eds.) *Hemi-inattention and hemisphere specialization.* New York: Raven Press, 1–31.

FRIEDLANDER, W. J., (1971) Some aspects of eyedness. *Cortex*, 7, 357–371.

FRUMKIN, L. R., RIPLEY, H. S., and COX, G. B., (1978) Changes in cerebral hemispheric lateralization with hypnosis. *Biological psychiatry*, 13, 741–750.

GAEDE, S. E., PARSONS, O. A., and BERTERA, J. H., (1978) Hemispheric differences in music perception: aptitude vs. experience. *Neuropsychologia*, 16, 369–374.

GAINOTTI, G., (1968) Les manifestations de negligence et d'inattention pour l'hémispace. *Cortex*, 4, 64–91.

GAINOTTI, G., (1969) So-called catastrophic reactions and indifferent behavior during cerebral trauma: *Neuropsychologia*, 7, 195–204.

GAINOTTI, G., (1972) Emotional behaviour and hemispheric side of the lesion. *Cortex*, 8, 41–55.

GAINOTTI, G., and LEMMO, M. A., (1976) Comprehension of symbolic gesture in aphasia. *Brain and language*, 3, 451–460.

GAINOTTI, G., and TIACCI, C., (1971) The relationships between disorders of visual perception and unilateral spatial neglect. *Neuropsychologia*, 9, 451–458.

GALABURDA, A. M., and EIDELBERG, D., (1982) Symmetry and asymmetry in the human posterior thalamus. II. Thalamic lesions in a case of developmental dyslexia. *Archives of neurology*, 39, 333–336.

GALABURDA, A. M., SANIDES, F., and GESCHWIND, N., (1978) Human brain: cytoarchitectonic left-right asymmetries in the temporal speech region. *Archives of neurology*, 35, 812–817.

GALIN, D., (1974) Implications for psychiatry of left and right cerebral specialisation. *Archives of general psychiatry*, 31, 572–583.

GALIN, D., DIAMOND, R., and BRAFF, D., (1977) Lateralization of conversion symptoms: more frequent on the left. *American journal of psychiatry*, 134, 578–583.

GALIN, D., and ELLIS, R. R., (1975) Asymmetry in evoked potentials as an index of lateralised cognitive processes in relation to EEG alpha asymmetry. *Neuropsychologia*, 13, 45–50.

GALIN, D., JOHNSTONE, J., and HERRON, J., (1978) Effects of task difficulty on EEG measures of cerebral engagement. *Neuropsychologia*, 16, 461–472.

GALIN, D., JOHNSTONE, J., NAKELL, L., and HERRON, J., (1979) Development of the capacity for tactile transfer between hemispheres in normal children. *Science*, 204, 1330–1332.

GALIN, D., and ORNSTEIN, R. E., (1972) Lateral specialization of cognitive mode: an EEG study. *Psychophysiology*, 9, 412–418.

GALIN, D., and ORNSTEIN, R. E., (1974) Individual differences in cognitive style. 1. Reflective eye movements. *Neuropsychologia*, 12, 367–376.

GALPER, R. E., and COSTA, L., (1980) Hemispheric superiority for recognizing faces depends upon how they are learned. *Cortex*, 16, 21–38.

GARDNER, E. B., and BRANSKI, D. M., (1976) Unilateral cerebral activation and perception of gaps: a signal detection analysis. *Neuropsychologia*, 14, 43–53.

GARDNER, E. B., ENGLISH, A G., FLANNERY, B. M., HARTNETT, M. B , McCORMICK, J. K., and WILHELMY, B. B., (1977) Shape recognition accuracy and response latency in a bilateral tactile task. *Neuropsychologia*, 15, 607–616.

GARDNER, H., SILVERMAN, J., WAPNER, W., and ZURIF, E., (1978) The appreciation of antonymic contrasts in aphasia. *Brain and language*, 6, 301–317.

GARRICK, C., (1978) Field dependence and hemispheric specialization. *Perceptual and motor skills*, 47, 631–639.

GASPARRINI, W. G., SATZ, P., HEILMAN, K., and COOLIDGE, F. L., (1978) Hemispheric asymmetries of affective processing as determined by the MMPI. *Journal of neurology, neurosurgery and psychiatry*, 41, 470–473.

GATES, A., and BRADSHAW, J. L., (1977a) Music perception and cerebral asymmetries. *Cortex*, 13, 390–401.

GATES, A., and BRADSHAW, J. L., (1977b) The role of the cerebral hemispheres in music. *Brain and language*, 4, 403–431.

GAZZANIGA, M. S., (1968) Short-term memory and brain-bisected man. *Psychonomic science*, 12, 161–162.

GAZZANIGA, M. S., (1969) Eye position and visual motor co-ordination. *Neuropsychologia*, 7, 379–382.

GAZZANIGA, M. S., (1970) *The bisected brain*. Appleton-Century-Crofts: New York.

GAZZANIGA, M. S. (1974) Cerebral dominance viewed as a decision system. In DIMOND, S. J., and BEAUMONT, J. G., (Eds.) *Hemisphere function in the human brain*. Elek: London, 367–382.

GAZZANIGA, M. S., BOGEN, J. E., and SPERRY, R. W., (1962) Some functional effects of sectioning the cerebral commissures in man. *Proceedings of the national academy of science of the USA*, 48, 1765–1769.

GAZZANIGA, M. S., BOGEN, J. E., and SPERRY, R. W., (1963) Laterality effects in somesthesis following cerebral commissurotomy in man. *Neuropsychologia*, 1, 209–215.

GAZZANIGA, M. S., BOGEN, J. E., and SPERRY, R. W., (1965) Observations on visual perception after disconnexion of the cerebral hemispheres in man. *Brain*, 88, 221–236.

GAZZANIGA, M. S., BOGEN, J. E., and SPERRY, R. W., (1967) Dyspraxia following division of the cerebral commissures. *Archives of neurology*, 16, 606–612.

GAZZANIGA, M. S., and HILLYARD, S. A., (1971) Language and speech capacity of the right hemisphere. *Neuropsychologia*, 9, 273–280.

GAZZANIGA, M. S., and HILLYARD, S. A., (1973) Attention mechanisms following brain bi-section In S. KORNBLUM (Ed.) *Attention and performance* IV. New York: Academic Press, 22:–238.

GAZZANIGA, M. S., and LE DOUX, J. E., (1978) *The integrated mind*. Plenum: New York.

GAZZANIGA, M. S., SIDTIS, J. J., VOLPE, B. T., SMYLIE, C., HOLTZMAN, J., and WILSON, D. H., (1982) Evidence for paracallosal verbal transfer after callosal section: a possible consequence of bilateral language organization, *Brain*, 105, 53–63.

GAZZANIGA, M. S., and SPERRY, R. W., (1966) Simultaneous double discrimination response following brain bisection. *Psychonomic science*, 4, 261–262.

GAZZANIGA, M. S., and SPERRY, R. W., (1967) Language after section of the cerebral commissures. *Brain*, 90, 131–148.

GAZZANIGA, M. S., VOLPE, B. T., SMYLIE, C. S., WILSON, D. H., and LE DOUX, J. E., (1979) Plasticity in speech organisation following commissurotomy. *Brain*, 102, 805–815

GAZZANIGA, M. S., and YOUNG, E. D., (1967) Effects of commissurotomy on the processing of increasing visual information. *Experimental brain research*, 3, 368–371.

GEFFEN, G., (1976) Development of hemispheric specialization for speech. *Cortex*, 12, 337–346.

GEFFEN, G., BRADSHAW, J. L., and NETTLETON, N. C., (1972) Hemispheric asymmetry: verbal and spatial encoding of visual stimuli. *Journal of experimental psychology*, 95, 25–31.

GEFFEN, G., BRADSHAW, J. L., and WALLACE, G., (1971) Interhemispheric effects on reaction time to verbal and non-verbal visual stimuli. *Journal of experimental psychology*, 37, 415–422.

GEFFEN, G., and CAUDREY, D., (1981) Reliability and validity of the dichotic monitoring test for language laterality. *Neuropsychologia*, 19, 413–424.

GEFFEN, G., and TRAUB, E., (1979) Preferred hand and familial sinistrality in dichotic monitoring. *Neuropsychologia*, 17, 527–531.

GEFFEN, G., TRAUB, E., and STIERMAN, I., (1978) Language laterality assessed by unilateral ECT and dichotic monitoring. *Journal of neurology, neurosurgery and psychiatry*, 41, 354–360.

GEFFEN, G., and WALE, J., (1979) Development of selective listening and hemispheric asymmetry, *Developmental psychology*, 15, 138–146.

GEFFEN, G., WALSH, A., SIMPSON, D., and JEEVES, M., (1980) Comparison of the effects of transcortical and transcallosal removal of intraventricular tumours. *Brain*, 103, 773–788.

GEFFNER, D. S., and DORMAN, M. F., (1976) Hemispheric specialization for speech perception in four year old children from low and middle socio-economic classes. *Cortex*, 12, 71–73.

GEFFNER, D. S., and HOCHBERG, I., (1971) Ear laterality performance of children from low and middle socioeconomic levels on a verbal dichotic listening task. *Cortex*, 7, 193–203.

GERSTMANN, J., (1927) Fineragnosie und isolierte. Agraphie: ein neues Syndrom. *Zeitschrift für neurologie und psychiatrie*, 108, 152–177.

GESCHWIND, N., (1964) *The development of the brain and the evolution of language*. In STUART, C. I. J. M., (Ed.) Monograph series on language and linguistics. Georgetown University Press: Washington, 17, 155–169.

GESCHWIND, N., (1965) Disconnexion syndromes in animals and man. *Brain*, 88, 237–294 (Part I), 585–644 (Part II).

GESCHWIND, N., (1967) The apraxias. In *Phenomenology of will and action* (Eds.) STRAUS, E. W., and GRIFFITH, I. R. M., Duquesne University Press: Pittsburgh, 91–102.

GESCHWIND, N., (1969) Problems in the Anatomical Understanding of the Apraxias. In BENTON, A. L., (Ed.) *Contributions to clinical neuropsychology*, Chicago: Aldine Publishing Co., Chap. 4, 107–128.

GESCHWIND, N., (1970) The organization of language and the brain. *Science*, 170, 940–944.

GESCHWIND, N., (1974) The anatomical basis of hemispheric differentiation. In DIMOND, S. J., and BEAUMONT, J. G., (Eds.) *Hemisphere function in the human brain*. Elek: London, 7–24.

GESCHWIND, N., and KAPLAN, E., (1962) A human cerebral deconnection syndrome. *Neurology*, 12, 675–685.

GESCHWIND, N., and LEVITSKY, W., (1968) Human Brain: Left-Right asymmetries in temporal speech region. *Science*, 161, 186–187.

GESELL, A., and AMES, L. B., (1947) The development of handedness. *Journal of genetic psychology*, 70, 155–175.

GHENT, L., (1961) Developmental changes in tactual thresholds on dominant and non-dominant sides. *Journal of comparative and physiological psychology*, 54, 670–673.

GIBSON, A. R., DIMOND, S. J., and GAZZANIGA, M. S., Left field superiority for word matching. *Neuropsychologia*, 10, 463–466.

GIBSON, J. B., (1973) Intelligence and handedness. *Nature*, 243, 482.

GILBERT, C., (1973) Strength of left-handedness and facial recognition ability. *Cortex*, 9, 145–151.

GILBERT, C., (1977) Non-verbal perceptual abilities in relation to left-handedness and cerebral lateralization. *Neuropsychologia*, 15, 779–791.

GILHOOLY, K. J., and HAY, D., (1977) Imagery, concreteness, age of acquisition, familiarity and meaningfulness values for 205 five-letter words having single-solution anagrams. *Behavior research methods and instrumentation*, 9, 12–17.

GILHOOLY, K. J., and GILHOOLY, M. L. M., (1980) The validity of age-of acquisition ratings. *British journal of psychology*, 71, 105–110.

GILLIES, S. M., MACSWEENEY, D. A., and ZANGWILL, O. L., (1960) A note on some unusual handedness patterns. *Quarterly journal of experimental psychology*, 12, 113–116.

GLANVILLE, B. B., BEST, C. T., and LEVENSON, R., (1977) A cardiac measure of cerebral asymmetries in infant auditory perception. *Developmental psychology*, 13, 54–59.

GLASS, A. V., GAZZANIGA, M. S., and PREMACK, D., (1973) Artificial language in global aphasics. *Neuropsychologia*, 11, 95–103.

GLICK, S. D., JERUSSI, T. P., and ZIMMERBERG, B., (1977) Behavioral and neuropharmacological correlates of nigrotriatal asymmetry in rats. In HARNAD, S., DOTY, R. W., GOLDSTEIN, L., JAYNES, J., and KRAUTHAMER, G., (Eds.) *Lateralization in the nervous system*. Academic Press: New York, 213–250.

GLICK, S. D., SCHONFIELD, A. R., and STRUMPF, A. J., (1980) Sex differences in brain asymmetry of the rodent. *The behavioral and brain sciences*, 3, 236.

GLICK, S. D., WEAVER, L. M., and MEIBACH, R. C., (1980) Lateralization of reward in rats: differences in reinforcing thresholds. *Science*, 207, 1093–1095.

GLONING, I., GLONING, K., HAUB, G., and QUATEMBER, R., (1969) Comparison of verbal behavior in right-handed and non-right-handed patients with anatomically verified lesion of one hemisphere. *Cortex*, 5, 43–52.

GOLDEN, C. J., GRABER, B., COFFMAN, J., BERG, R. A., NEWLIN, D. B., BLOCH, S., (1981) Structural brain deficits in schizophrenia. Identification by computed tomographic scan density measures. *Archives of general psychiatry*, 38, 1014–1015.

GOLDSTEIN, G., (1974) The use of clinical neuropsychological methods in the lateralisation of brain lesions. In DIMOND, S. J., and BEAUMONT, J. G., (Eds.) *Hemisphere function in the human brain*. Elek: London, 279–310.

GOLDSTEIN, K., (1939) *The organism*. American Book Publishers: New York.

GOLDSTEIN, L., and LACKNER, J. R., (1974) Sideways look at dichotic listening. *Journal of the acoustical society of America*, 55, Supplement 510 (A).

GOLDSTEIN, M. N., and JOYNT, R. J., (1969) Long term follow-up of a callosal sectioned patient. *Archives of neurology*, 20, 96–102.

GOLDSTEIN, S. G., FILSKOV, S. B., WEAVER, L. A., and IVES, J. O., (1977) Neuropsychological effects of electroconvulsive therapy. *Journal of clinical psychology*, 33, 798–806.

GOODALL, R. J., (1957) Cerebral hemispherectomy: present status and clinical indications. *Neurology*, 7, 151–162.

GOODGLASS, H., and QUADFASEL, F. A., Language laterality in left handed aphasics. *Brain* 1954, 521–548.

GORDON, H. W., (1970) Hemispheric asymmetries in the perception of musical chords. *Cortex*, 6, 387–398.

GORDON, H. W., (1978a) Left hemisphere dominance for rhythmic elements in dichotically-presented melodies. *Cortex*, 14, 58–70.

GORDON, H. W., (1978b) Hemisphere asymmetry for dichotically presented chords in musicians and non-musicians, males and females. *Acta psychologica*, 42, 383–395.

GORDON, H. W., (1980a) Right hemisphere comprehension of verbs in patients with complete forebrain commissurotomy: use of the dichotic method and manual performance. *Brain and language*, 11, 76–86.

GORDON, H. W., (1980b) Cognitive asymmetry in dyslexic families. *Neuropsychologia*, 18, 645–656.

GORDON, H. W., and BOGEN, J. E., (1974) Hemispheric lateralization of singing after intracarotid sodium amylobarbitone. *Journal of neurology, neurosurgery and psychiatry*, 37, 727–738.

GORDON, H. W., FROOMAN, B., and LAVIE, P., (1982) Shift in cognitive asymmetries between wakings from REM and NREM sleep. *Neuropsychologia*, 20, 99–104.

GORDON, H. W., and SPERRY, R. W., (1969) Lateralization of olfactory perception in the surgically separated hemispheres of man. *Neuropsychologia*, 7, 111–120.

GOTT, P. S., (1973) Language after dominant hemispherectomy. *Neurosurgery and psychiatry*, 36, 1082–1088.

GOTT, P. S., and SAUL, R. E., (1978) Agenesis of the corpus callosum: limits of functional compensation. *Neurology*, 28, 1272–1279.

GRAPIN, P., and PERPÈRE, C., (1968) Symetrie et lateralisation du nourrisson. In KOURILSKY, R., and GRAPIN, P., (Eds.) *Main droite et main gauche*. Presses Universitaires de France: Paris.

GOTTLIEB, G., and WILSON, I., (1965) Cerebral dominance: temporary disruption of verbal memory by unilateral electroconvulsive shock treatment. *Journal of comparative and physiological psychology*, 60, 368–372.

GRABOW, J. D., and ELLIOTT, F. K. W., (1974) The electrophysiologic assessment of hemispheric asymmetries during speech. *Journal of speech and hearing research*, 17, 64–72.

GRAVES, R., LANDIS, T., and GOODGLASS, H., (1981) Laterality and sex differences for visual recognition of emotional and non-emotional words. *Neuropsychologia*, 19, 95–102.

GREEN, J. B., and HAMILTON, W. J., (1976) Anosognosia for hemiplegia: Somatosensory evoked potential studies. *Neurology*, 26, 1141–1144.

GREEN, P., (1978) Defective inter-hemispheric transfer in schizophrenia. *Journal of abnormal psychology*, 87, 472–480.

GREEN, P., GLASS, A., and O'CALLAGHAN, M. A. J., (1979) Some implications of abnormal hemisphere interaction in schizophrenia. In GRUZELIER, J., and FLOR-HENRY, P. (Eds.) *Hemisphere asymmetries of function in psychopathology*. Elsevier/North Holland Biomedical Press: Amsterdam, 431–448.

GREENBERG, J. H., REIVICH, M., ALAVI, A., et al., (1981) Metabolic mapping of functional activity in man with 18F-fluorodeoxyglucose technique. *Science*, 212, 678–680.

GREENSTADT, L., SCHUMAN, M., and SHAPIRO, D., (1978) Differential effects of left versus right monaural feedback for heart rate increase. *Psychophysiology*, 15, 233–238.

GREENWOOD, P., WILSON, D. H., and GAZZANIGA, M. S., (1977) Dream report following commissurotomy. *Cortex*, 13, 311–316.

GREGORY, R., and PAUL, J., (1980) The effects of handedness and writing posture on neuropsychological test results. *Neuropsychologia*, 18, 231–235.

GRIFFITH, H., and DAVIDSON, M., (1966) Long-term changes in intellect and behaviour after hemispherectomy. *Journal of neurology, neurosurgery and psychiatry*, 29, 571–577.

GROSS, K., ROTHENBERG, S., SCHOTTENFELD, S., and DRAKE, C., (1978) Duration thresholds for letter identification in left and right visual fields for normal and reading-disabled children. *Neuropsychologia*, 16, 709–716.

GROSS, M. M., (1972) Hemispheric specialization for processing of visually presented verbal and spatial material. *Perception and psychophysics*, 12, 357–363.

GROSS, Y., FRANKO, R., and LEWIN, I., (1978) Effects of voluntary eye movements on hemispheric activity and choice of cognitive mode. *Neuropsychologia*, 16, 653–657.

GROVES, C. P., and HUMPHREY, N. K., Asymmetry in gorilla skulls: evidence for lateralized brain function. *Nature*, 1973, 244, 53–54.

GRUZELIER, J. H., (1973) Bilateral asymmetry of skin conductance orienting activity and levels in schizophrenia. *Journal of biological psychology*, 1, 21–41.

GRUZELIER, J., (1979) Synthesis and critical review of the evidence for hemisphere asymmetries of function in psychopathology. In GRUZELIER, J., and FLOR-HENRY, P., (Eds.) *Hemisphere asymmetries of function in psychopathology*. Elsevier/North Holland Biomedical Press: Amsterdam, 647–672.

GRUZELIER, J. H., (1981) Cerebral laterality and psychopathology: fact and fiction. *Psychological medicine*, 11, 219–227.

GRUZELIER, J., and HAMMOND, N., (1976) Schizophrenia: a dominant hemisphere temporal-limbic disorder? *Research communications in psychology, psychiatry and behavior*, 1, 33–72.

GRUZELIER, J. H., and HAMMOND, N. V., (1979) Gains, losses and lateral differences in the hearing of schizophrenic patients. *British journal of psychology*, 70, 319–330.

GRUZELIER, J. H., and HAMMOND, N. V., (1980) Lateralized deficits on the dichotic listening of schizophrenic patients. *Biological psychiatry*, 15, 759–779.

GRUZELIER, J., and MANCHANDA, R., (1982) The syndrome of schizophrenia: relations between electrodermal response, lateral asymmetries and clinical ratings. *British journal of psychiatry*, 141, 488–495.

GRUZELIER, J., and VENABLES, P., (1974) Bimodality and lateral asymmetry of skin conductance orienting activity in schizophrenics: replication and evidence of lateral asymmetry in patients with depression and disorders of personality. *Biological psychiatry*, 8, 55–73.

GUIARD, Y., (1980) Cerebral hemispheres and selective attention: *Acta psychologica*, 46, 41–61.

GUIARD, Y., and REQUIN, J., (1977) Interhemispheric sharing of signals and responses and the psychological refractory period. *Neuropsychologia*, 15, 427–478.

GUIARD, Y., and REQUIN, J., (1978) Between-hand and within-hand choice RT: a single channel of reduced capacity in the split-brain monkey. In *Attention and performance* VII. (Ed.) REQUIN, J., Erlbaum Associates: New Jersey, Chap. 22, 391–410.

GUR, R. C., (1973) Measurement and imaging of regional brain function: implications for neuropsychiatry. In FLOR-HENRY, P., and GRUZELIER, J., (Eds.) *Laterality and psychopathology*. Elsevier Science Publishers B.V., 589–614.

GUR, R. E., (1975) Conjugate lateral eye movements as an index of hemispheric activation. *Journal of personality and social psychology*, 31, 751–757.

GUR, R. E., (1977) Motoric laterality imbalance in schizophrenia: a possible concomitant of left hemisphere dysfunction. *Archives of general psychiatry*, 34, 33–37.

GUR, R. E., (1978) Left hemisphere dysfunction and left hemisphere over-activation in schizophrenia. *Journal of abnormal psychology*, 87, 225–238.

GUR, R. E., (1979) Hemispheric overactivation in schizophrenia. In GRUZELIER, J., and FLOR-HENRY, P., (Eds.) *Hemisphere asymmetries of function in psychopathology*. Elsevier/North Holland Biomedical Press: Amsterdam, 113–123.

GUR, R. E., and GUR, R. C., (1977) Sex differences in the relations among handedness, sighting dominance and eye-acuity. *Neuropsychologia*, 15, 585–590.

GUR, R. E., GUR, R. C., and HARRIS, L. J., (1975) Cerebral activation as measured by subjects lateral eye movements, as influenced by experimenter location. *Neuropsychologia*, 13, 35–44.

GUR, R. E., GUR, R. C., and MARSHALEK, B., (1975) Classroom seating and functional brain asymmetry. *Journal of educational psychology*, 67, 151–153.

GUR, R. C., GUR, R. E., OBRIST, W. D., HUNGERBUHLER, J. P., YOUNKIN, D., ROSEN, A. D., SKOLNICK, B. E., and REIVICH, M., (1982) Sex and handedness differences in cerberal blood flow during rest and cognitive activity. *Science*, 217, 659–661.

GUR, R. C., GUR, R. E., ROSEN, A. D., et al., (1983) A cognitive-motor network demonstrated by positron emission tomography. *Neuropsychologia*, 21, 601–606.

GUR, R. C., PACKER, I. K., HUNGERBUHLER, J. P., REIVICH, M., OBRIST, W. D., AMARNEK, W. S., and SACKEIM, H. A., (1980) Differences in the distribution of gray and white matter in human cerebral hemispheres. *Science*, 207, 1226–1228.

GUR, R., and REIVICH, M., (1980) Cognitive task effects on hemispheric blood flow in humans: evidence for individual differences in hemispheric activation. *Brain and language*, 9, 78–92.

GUTTMAN, E., (1942) Aphasia in children. *Brain*, 65, 205–219.

HAALAND, K. Y., and DELANEY, H. D., (1981) Motor deficits after left or right hemisphere damage due to stroke or tumour. *Neuropsychologia*, 19, 17–27.

HAGGARD, M. P., and PARKINSON, A. M., (1971) Stimulus and task factors as determinants of ear advantage. *Quarterly journal of experimental psychology*, 23, 168–177.

HALLIDAY, A. M., DAVISON, K., BROWNE, M. W., and KREEGER, A. L. C., (1968) A comparison of the effects on depression and memory of bilateral ECT and unilateral ECT to the dominant and non-dominant hemispheres. *British journal of psychiatry*, 114, 997–1012.

HALPERIN, Y., NACHSON, I., and CARMON, A., (1973) Shift of ear superiority in dichotic listening to temporally patterned non-verbal stimuli. *Journal of the acoustical society of America*, 53, 46–50.

HALSEY, J. H., BLAUENSTEIN, U. W., WILSON, E. M., and WILLS, E. L., (1980) Brain activation in the presence of brain damage. *Brain and language*, 9, 47–60.

HAMILTON, C. R., (1977) An assessment of hemispheric specialisation in monkeys. In DIMOND, S. J., and BLIZARD, D. A., (Eds.) Evolution and Lateralization of the Brain. *Annals of the New York academy of sciences*, 299, 222–232.

HAMMOND, N. V., and GRUZELIER, J. H., (1978) Laterality, attention and rate effects in the auditory temporal discrimination of chronic schizophrenics: the effect of treatment with chlorpromazine. *Quarterly journal of experimental psychology*, 30, 91–103.

HAMSHER, K. de S., (1978) Stereopsis and unilateral brain disease. *Investigative ophthalmology and visual science*, 17, 336–343.

HANNAY, H. J., (1976) Real or imagined incomplete lateralization of function in females? *Perception and psychophysics*, 19, 349–352.

HANNAY, H. J., and BOYER, C. L., (1978) Sex differences in hemispheric asymmetry revisited. *Perceptual and motor skills*, 47, 317–321.

HANNAY, H. J., DEE, H. L., BURNS, J. W., and MASEK, B. S., (1981) Experimental reversal of a left visual field superiority for forms. *Brain and language*, 13, 54–66.

HANNAY, H. J., and MALONE, D. R., (1976a) Visual field effects and short-term memory for verbal material. *Neuropsychologia*, 14, 203–210.

HANNAY, H. J., and MALONE, D. R., (1976b) Visual field recognition memory for right-handed females as a function of familial handedness. *Cortex*, 12, 41–48.

HANNAY, H. J., and ROGERS, J. P., (1979) Individual differences and asymmetry effects in memory for unfamiliar faces. *Cortex*, 15, 257–268.

HANSCH, E. C., and PIROZOLLO, F. J., (1980) Task relevant effects on the assessment of cerebral specialization for facial emotion. *Brain and language*, 10, 51–59.

HARBURG, E., FELDSTEIN, A., and PAPSDORF, J., (1978) Handedness and smoking. *Perceptual and motor skills*, 43, 1171–1174.

HARCUM, E. R., and FILION, R. D. L., (1963) Effects of stimulus reversals on lateral dominance in word recognition. *Perceptual and motor skills*, 17, 779–794.

HARDYCK, C., (1977a) Handedness and part-whole relationships: a replication. Cortex, 13, 177–183.

HARDYCK, C., (1977b) A model of individual differences in hemispheric functioning. In *Studies in neurolinguistics*, 3, (Eds.) WHITAKER, H., and WHITAKER, H. A., Academic Press: New York, Chap. 5, 223–255.

HARDYCK, C., and PETRINOVICH, L. F., (1977) Left-handedness. *Psychological bulletin*, 84, 385–404.

HARDYCK, C., PETRINOVICH, L. F., and GOLDMAN, R. D., (1976) Left handedness and cognitive deficit, *Cortex*, 12, 266–279.

HARDYCK, C., TZENG, O. J. L., and WANG, W. S-Y., (1977) Cerebral lateralization effects in visual half-field experiments. *Nature*, 269, 705–707.

HARDYCK, C., TZENG, O. J. L., and WANG, W. S-Y., (1978) Cerebral lateralization of function and bilingual decision processes: Is thinking lateralized? *Brain and language*, 5, 56–71.

HARMAN, D. W., and RAY, W. J., (1977) Hemispheric activity during affective verbal stimuli: an EEG study. *Neuropsychologia*, 15, 457–460.

HARNAD, S., (1972) Creativity, lateral saccades and the non-dominant hemisphere. *Perceptual and motor skills*, 34, 653–654.

HARNER, R. N., (1977) Agenesis of the corpus callosum and associated effects. In GOLDENSOHN, E. G., and APELL, S. H., (Eds.) *Scientific approaches to clinical neurology*, 1, Lea and Febiger: Philadelphia, 616–627.

HARRIMAN, J., and BUXTON, H., (1979) The influence of prosody on the recall of presented sentences. *Brain and language*, 8, 62–68.

HARRIMAN, J., and CASTELL, L., (1979) Manual asymmetry for tactile discrimination. *Perceptual and motor skills*, 48, 290.

HARRIS, A. J., (1953) Lateral dominance directional confusion and reading disability. *Journal of psychology*, 44, 283–294.

HARRIS, L. J., (1978) Sex differences in spatial ability: possible environmental, genetic and neurological factors. In KINSBOURNE, M., (Ed.) *Asymmetrical function of the brain.* Cambridge University Press: Cambridge, 405–522.

HARRIS, L. J., (1980) Left handedness: early theories, facts and fancies. In HERRON, J., (Ed.) *The neuropsychology of left handedness.* Academic Press: New York, 3–78.

HARRIS, L. J., and GITTERMAN, S. R., (1978) University Professors' self descriptions of left-right confusability: sex and handedness differences. *Perceptual and motor skills*, 47, 819–823.

HARSHMAN, R. A., REMINGTON, R., and KRASHEN, S. D., (1974) Sex, language, and the brain: adult sex differences in lateralization. Paper presented at conference on human brain function. Los Angeles.

HARTER-CRAFT, R., (1981) The relationship between right-handed children's assessed and familial handedness and lateral specialization. *Neuropsychologia*, 19, 697–706.

HARTLINE, H K., and RATLIFF, F., (1957) Inhibitory interaction of receptor units in the eye of limulus. *Journal of general physiology*, 40, 357–376.

HARVEY, L. O., (1978) Single representation of the visual midline in humans. *Neuropsychologia*, 16, 601–610.

HASLAM, R. H. A., DALBY, J. T., JOHNS, R. D., RADEMAKER, A. W., (1981) Cerebral asymmetry in developmental dyslexia. *Archives of neurology*, 38, 679–682.

HATTA, T., (1976) Asynchrony of laterality onset as a factor in difference in visual field. *Perceptual and motor skills*, 42, 163–166.

HATTA, T., (1977) Recognition of Japanese Kanji in the left and right visual fields. *Neuropsychologia*, 15, 685–688.

HATTA, T., and DIMOND, S. J., (1980) Comparison of lateral differences for digit and random form recognition in Japanese and Westerners. *Journal of experimental psychology: human perception and performance*, 6, 368–374.

HAUN, F., (1978) Functional dissociation of the hemispheres using foveal visual input. *Neuropsychologia*, 16, 725–733.

HAY, D. C., and YOUNG, A. W., (1982) The human face. In ELLIS, A. W., (Ed.) *Normality and pathology in cognitive functions*. Academic Press: London, 173–202.

HAYASHI, T., and BRYDEN, M. P., (1967) Ocular dominance and perception asymmetry. *Perceptual and motor skills*, 25, 605–612.

HAYDON, S. P., and SPELLACY, F. J., (1973) Monaural reaction time asymmetries for speech and non-speech sounds. *Cortex*, 9, 288–294.

HEAP, M., and WYKE, M., (1972) Learning of a unimanual motor skill by patients with brain lesions: an experimental study. *Cortex*, 8, 1–18.

HÉCAEN, H., (1962) Clinical symptomatology in right and left hemispheric lesions. In MOUNTCASTLE, V. B., (Ed.) *Interhemispheric relations and cerebral dominance*. Baltimore: Johns Hopkins Press, 215–243.

HÉCAEN, H., (1968) La dominance cérébrale. In KOURILSKY, R., and GRAPIN, P., (Eds.) *Main droite et main gauche*. Presses Universitaires de France: Paris, 25–56.

HÉCAEN, H., (1976) Acquired aphasia in children and the ontogenesis of hemispherical functional specialization. *Brain and language*, 3, 114–134.

HÉCAEN, H., (1978) Right hemisphere contributions to language functions. In BUSER, P. A., and ROUGEL-BUSER, A., (Eds.) *Cerebral correlates of conscious experience*. INSERM Symposium Number 6, 199–213.

HÉCAEN, H., and AJURIAGUERRA, J. DE, (1964) *Left handedness: manual superiority and cerebral dominance*. Grune and Stratton: New York and London.

HÉCAEN, H., AJURIAGUERRA, J. DE, and MASSONNET, J., (1951) Les troubles visuoconstructifs par lésion pariéto-occipitale droite: role des perturbations vestibulaires. *Encéphale*, 6, 533–562.

HECAEN, H., and ALBERT, M. L., (1978) *Human neuropsychology*. Wiley: New York.

HÉCAEN, H., and ANGELERGUES, R., (1962) Agnosia for faces (Prosopagnosia) *Archives of neurology*, 7, 92–100.

HÉCAEN, H., DE AGOSTINI, M., and MONZON-MONTES, A., (1981) Cerebral organization in left handers. *Brain and language*, 12, 261–284.

HÉCAEN, H., GOSNAVE, G., VEDRENNE, C., and SZIKLA, G., (1978) Suppression lateralisé du matériel verbal présenté dichotiquement lors d'une déstruction partielle du corps calleux. *Neuropsychologia*, 16, 233–238.

HÉCAEN, H., and PIERCY, M., (1956) Paroxysmal dysphasia and the problem of cerebral dominance. *Journal of neurology, neurosurgery and psychiatry*, 19, 194–201.

HÉCAEN, H., and SAUGUET, J., (1971) Cerebral dominance in left handed subjects. *Cortex*, 7, 19–48.

HEILMAN, K. M., COYLE, J. M., GONYEA, E. F., and GESCHWIND, N., (1973) Apraxia and Agraphia in a left hander. *Brain*, 96, 21–28.

HEILMAN, K. M., GONYEA, E. F., and GESCHWIND, N., (1974) Apraxia and Agraphia in a right hander. *Cortex*, 10, 284–288.

HEILMAN, K. M., SCHOLES, R., and WATSON, R. T., (1975) Auditory affective agnosia. *Journal of neurology, neurosurgery and psychiatry*, 38, 69–72.

HEILMAN, K. M., and VAN DEN ABELL, T., (1979) Right hemispheric dominance for mediating cerebral activation. *Neuropsychologia*, 17, 315–322.

HEILMAN, K. M., and VAN DEN ABELL, T., (1980) Right hemisphere dominance for attention: the mechanism underlying hemispheric asymmetries of inattention (neglect). *Neurology*, 30, 327–330.

HEILMAN, K. M., and WATSON, R. T., (1977) The neglect syndrome: a unilateral defect of the orienting response. In HARNAD, S., DOTY, R. W., GOLDSTEIN, L., JAYNES, J., and KRAUTHAMER, G., (Eds.) *Lateralisation in the nervous system*. Academic Press: New York, 285–302.

HEIM, A. W., and WATTS, K. P., (1976) Handedness and cognitive bias. *Quarterly journal of experimental psychology*, 28, 35–360.

HELLER, W., and LEVY, J., (1981) Perception and expression of emotion in right handers and left handers. *Neuropsychologia*, 19, 263–272.

HELLIGE, J. B., (1976) Changes in same-different laterality patterns as a function of practice and stimulus quality. *Perception and psychophysics*, 20, 267–273.

HELLIGE, J. B., and COX, P. J., (1976) Effects of concurrent verbal memory on recognition of stimuli from the left and right visual fields. *Journal of experimental psychology: human perception and performance*, 3, 210–221.

HELLIGE, J. B., COX, P. J., and LITVAC, L., (1979) Information processing in the cerebral hemispheres: selective hemispheric activation and capacity limitations. *Journal of experimental psychology: general*, 108, 251–279.

HELLIGE, J. B., and LONGSTRETH, L. E., (1981) Effects of concurrent hemisphere – specific activity on unimanual tapping rate. *Neuropsychologia*, 19, 395–405.

HELLIGE, J. B., and WEBSTER, R., (1979) Right hemisphere superiority for initial stages of letter processing. *Neuropsychologia*, 17, 653–661.

HELLIGE, J. B., ZATKIN, J. L., and WONG, T. M., (1981) Intercorrelation of laterality indices. *Cortex*, 17, 129–134.

HENRY, R. G. J., (1979) Monaural studies eliciting an hemispheric asymmetry: a bibliography. *Perceptual and motor skills*, 48, 335–338.

HERMELIN, B., and O'CONNOR, N., (1971) Functional asymmetry in reading of Braile. *Neuropsychologia*, 9, 431–435.

HERON, W., (1957) Perception as a function of retinal locus and attention. *American journal of psychology*, 70, 38–48.

HERMANN, D. J., and VAN DYKE, K. A., (1978) Handedness and the mental rotation of perceived patterns. *Cortex*, 14, 521–529.

HERRON, J., GALIN, D., JOHNSTONE, J., and ORNSTEIN, R. E., (1979) Cerebral specialization, writing posture, and motor control of writing in left-handers. *Science*, 205, 1285–1289.

HEWES, G. W., (1973) Primate communication and the gestural origin of language. *Current anthropology*, 14, 5–24.

HICKS, R. E., (1975) Intrahemispheric response competition between vocal and unimanual performance in normal adult human males. *Journal of comparative and physiological psychology*, 89, 50–60.

HICKS, R. E., and BARTON, A. K., (1975) A note on left-handedness and severity of mental retardation. *Journal of genetic psychology*, 127, 323–324.

HICKS, R. E., and BEVERIDGE, R., (1978) Handedness and intelligence. *Cortex*, 14, 304–307.

HICKS, R. E., BRADSHAW, G. J., KINSBOURNE, M., and FEIGIN, D. S., (1978) Vocal-manual trade-offs in hemispheric sharing of performance control in normal adult humans. *Journal of motor behavior*, 10, 1–6.

HICKS, R. E., and DUSECK, C., (1980) The handedness distributions of gifted and non-gifted children. *Cortex*, 16, 479–482.

HICKS, R. E., DUSECK, C., LARSEN, F., WILLIAMS, S., and PELLEGRINI, R. J., (1980) Birth complications and the distribution of handedness. *Cortex*, 16, 483–486.

HICKS, R. E., ELLIOTT, D., GARBESI, L., and MARTIN, S., (1979) Multiple birth risk factors and the distribution of handedness. *Cortex*, 15, 135–137.

HICKS, R. E., EVANS, E. A., and PELLEGRINI, R. J., (1978) Correlation between handedness and birth order: compilation of five studies. *Perceptual and motor skills*, 46, 53–54.

HICKS, R. E., and KINSBOURNE, M., (1976) Human handedness: a partial cross-fostering study. *Science*, 192, 908–910.

HICKS, R. E., and KINSBOURNE, M., (1978) Human handedness. In KINSBOURNE, M., (Ed.) *Asymmetrical function of the brain*. Cambridge University Press: Cambridge, 523–549.

HICKS, R. E., and PELLEGRINI, R. J., (1978a) Handedness and locus of control. *Perceptual and motor skills*, 46, 369–370.

HICKS, R. E., and PELLEGRINI, R. J., (1978b) Handedness and anxiety. *Cortex*, 14, 119–121.

HICKS, R. E., PELLEGRINI, R. J., and EVANS, E. A., (1978) Handedness and birth risk, *Neuropsychologia*, 16, 243–245.

HICKS, R. E., PELLEGRINI, R. J., and HAWKINS, J., (1979) Handedness and sleep duration. *Cortex*, 15, 327–330.

HICKS, R. E., PROVENZANO, F. J., and RIBSTEIN, E. D., (1975) Generalized and lateralized effects of concurrent verbal rehearsal upon performance of sequential movements of the fingers by the left and right hands. *Acta psychologica*, 39, 119–130.

HIER, D. B., LE MAY, M., and ROSENBERG, P. B., (1979) Autism and unfavorable left-right asymmetries of the brain. *Journal of autism and developmental disorders*, 9, 153–159.

HIER, D. B., LE MAY, M., ROSENBERGER, P. B., PERLO, V. P., (1978) Developmental dyslexia: evidence for a sub group with a reversal of cerebral asymmetry. *Archives of neurology*, 35, 90–92.

HIGGENBOTTAM, J. A., (1973) Relationships between sets of lateral and perceptual preference measures. *Cortex*, 9, 402–410.

HILDRETH, G., (1949) The development and training of hand dominance: II. Developmental tendencies in handedness. *Journal of genetic psychology*, 75, 221–254.

HILL, J., (1978) Apes and language. *Annual review of anthropology*, 7, 89–112.

HILLIARD, R. E., (1973) Hemispheric laterality effect on a facial recognition task in normal subjects. *Cortex*, 9, 246.

HILLIER, W. F., (1954) Total left cerebral hemispherectomy for malignant glioma. *Neurology*, 4, 718–721.

HILLYARD, J. A., and WOODS, D. L., (1979) Electrophysiological analysis of human brain function. In GAZZANIGA, M. S , (Ed.) *Handbook of behavioral neurobiology*, Vol. 2. *Neuropsychology*, Plenum: New York, Chap. 12, 345–378.

HILLSBERG, B., (1979) A comparison of visual discrimination performance of the dominant and non-dominant hemispheres in schizophrenia. In GRUZELIER, J., and FLOR-HENRY, P., (Eds.) *Hemisphere asymmetries of function in psychopathology*. Elsevier/North Holland Biomedical Press: Amsterdam, 527–538.

HINES, D., (1972) Bilateral tachistoscopic recognition of verbal and non-verbal stimuli. *Cortex*, 8, 315–322.

HINES, D., (1975) Independent functioning of the two cerebral hemispheres for recognizing bilaterally presented tachistoscopic visual half-field stimuli. *Cortex*, 11, 132–143.

HINES, D., (1976) Recognition of verbs, abstract nouns and concrete nouns from the left and right visual half-fields. *Neuropsychologia*, 14, 211–216.

HINES, D., (1977) Differences in tachistoscopic recognition between abstract and concrete words as a function of visual half-field and frequency. *Cortex*, 13, 66–73.

HINES, D., (1978) Visual information processing in the left and right hemispheres. *Neuropsychologia*, 16, 593–600.

HINES, D., and SATZ, P., (1971) Superiority of right visual half field in right handers for recall of digits presented at varying rates. *Neuropsychologia*, 9, 21–25.

HINES, D., and SATZ, P., (1974) Cross-modal asymmetries in perception related to asymmetry in cerebral function. *Neuropsychologia*, 12, 239–247.

HINK, R. F., KAGA, K., and SUZUKI, J., (1980) An evoked potential correlate of reading ideographic and phonetic Japanese scripts. *Neuropsychologia*, 18, 455–464.

HIRSCH, H. V. B., and SPINELLI, D. N., (1970) Visual experience modifies distribution of horizontally and vertically oriented receptive fields in cats. *Science*, 168, 869–871.

HIRSHKOWITZ, M., EARLE, J., and PALEY, B., (1978) EEG Alpha asymmetry in musicians and non-musicians: a study of hemispheric specialization. *Neuropsychologia*, 16, 125–128.

HISCOCK, M., (1977) Effects of examiners location and subjects anxiety on gaze laterality. *Neuropsychologia*, 15, 409–416.

HISCOCK, M., and BERGSTROM, K. J., (1982) The lengthy persistence of priming effects in dichotic listening. *Neuropsychologia*, 20, 43–54.

HISCOCK, M., and KINSBOURNE, M., (1978) Ontogeny of cerebral dominance. Evidence from time-sharing asymmetry in children. *Developmental psychology*, 14, 321–329.

HISCOCK, M., and KINSBOURNE, M., (1980a) Asymmetries of selective listening and attention switching in children. *Developmental psychology*, 16, 70–82.

HISCOCK, M., and KINSBOURNE, M., (1980b) Asymmetry of verbal-manual time sharing in children: a follow-up study. *Neuropsychologia*, 18, 151–162.

HOCHBERG, F. H., and LE MAY, M., (1974) Arteriographic correlates of handedness. *Neurology*, 25, 218–222.

HOFFMAN, C., and KAGAN, S., (1977) Lateral eye movements and field dependence-independence. *Perceptual and motor skills*, 45, 767–778.

HOLMES, D. R., and MCKEEVER, W. F., (1975) Material specific serial memory deficit in adolescent dyslexics. *Cortex*, 15, 51–62.

HOLTZMAN, J. D., SIDTIS, J., VOLPE, B. T., WILSON, D. H., AND GAZZANIGA, M. S., (1981) Dissociation of spatial information for stimulus localization and the control of attention. *Brain*, 104, 861–872.

HOMMES, O. R., and PANHUYSEN, L. H. H. M., (1970) Bilateral intracarotid amytal injection. A study of dysphasia, disturbance of consciousness and paresis. *Psychiatria, neurologia, neurochirurgia*, 73, 447–459.

HOWES, D., and BOLLER, F., (1975) Simple reaction time: evidence for focal impairment from lesions of the right hemisphere. *Brain*, 98, 317–332.

HUPPERT, F., (1981) Memory in split-brain patients: a comparison with organic syndromes. *Cortex*, 17, 303–312.

HURWITZ, L., (1971) Evidence for restitution of function ahd development of new function in cases of brain bisection. *Cortex*, 7, 401–409.

HYND, G. W., and OBRZUT, J. E., (1977) Effects of grade level and sex on the magnitude of the dichotic ear advantage. *Neuropsychologia*, 15, 689–692.

INGLIS, J., (1965) Dichotic listening and cerebral dominance. *Acta otolaryngologica*, 60, 231–238.

INGLIS, J., and LAWSON, J. S., (1981) Sex differences in the effects of unilateral brain damage on intelligence. *Science*, 212, 693–695.

INGRAM, D., (1975a) Motor asymmetries in young children. *Neuropsychologia*, 13, 95–102.

INGRAM, D., (1975b) Cerebral speech lateralisation in young children. *Neuropsychologia*, 13, 103–105.

INGVAR, D. H., and SCHWARTZ, M. S., (1974) Blood flow patterns induced in the dominant hemisphere by speech and reading. *Brain*, 97, 273–288.

INNOCENTI, G. M., and FROST, D. O., (1979) Effects of visual experience on the maturation of the efferent system to the corpus callosum. *Nature*, 280, 231–234.

IWATA, M., SUGISHITA, M., TOYOKURA, Y., YAMADA, R., and YOSHIOKA, M., (1974) Etude sur le syndrome de disconnexion visuo-linguale après la transection du splénium du corps calleux: troubles de la verbalisation des informations visuelles dans l'hémisphère mineur. *Journal of the neurological sciences*, 23, 421–432.

JACKSON, J. H., (1868) Defect of intellectual expression (aphasia) and left hemiplegia. *The lancet*, 1, 457.

JACKSON, J. H., (1874) On the duality of the brain. Medical Press and Circular 1, 19, 14, 63. Reprinted in *Selected writings of John Hughlings Jackson* (Ed.) TAYLOR, J., Vol. II, (1932) London: Hodder and Stoughton, 129–145.

JAMES, W. E., MIFFERD, R. B., and WIELAND, B., (1967) Repetitive psychometric measures: handedness and performance. *Perceptual and motor skills*, 25, 209–212.

JEEVES, M. A., (1979) Some limits to interhemispheric integration in cases of callosal agenesis and partial commissurotomy. In STEELE-RUSSELL, I., VAN HOF, M., and BERLUCCHI, G., (Eds.) *Structure and function of the cerebral commissures* Macmillan: New York, 449–474.

JEEVES, M. A., SIMPSON, D. A., and GEFFEN, G., (1979) Functional consequences of the transcallosal removal of intraventricular tumours. *Journal of neurology, neurosurgery and psychiatry*, 42, 134–142.

JOHNSON, D., and HARLEY, C., (1980) Handedness and sex differences in cognitive tests of brain laterality. *Cortex*, 16, 73–82.

JOHNSON, O., and KOZMA, A., (1977) Effects of concurrent verbal and musical tasks on a unimanual skill. *Cortex*, 13, 11–16.

JOHNSON, P. R., (1977) Dichotically-stimulated ear differences in musicians and non-musicians. *Cortex*, 13, 385–389.

JOHNSON, R. C., COLE, R. E., BOWERS, J. K., FOILES, S. V., NIKAIDO, A. M., PATRICK, J. W., and WOLIVER, R. E., (1979) Hemispheric efficiency in middle and later adulthood. *Cortex*, 15, 109–119.

JONES, B., (1979) Lateral asymmetry in testing long-term memory for faces. *Cortex*, 15, 183–186.

JONES, B., (1980) Sex and handedness as factors in visual-field organization for a categorization task. *Journal of experimental psychology: human perception and performance*, 6, 494–500.

JONES, B., and BELL, J., (1980) Handedness in engineering and psychology students. *Cortex*, 16, 521–625.

JONES, G. H., and MILLER, J. J., (1981) Functional tests of the corpus callosum in schizophrenia. *British journal of psychiatry*, 139, 553–57.

JONES, R. J., (1966) Observations on stammering after localized cerebral injury. *Journal of neurology, neurosurgery and psychiatry*, 29, 192–195.

JONES-GOTMAN, M., and MILNER, B., (1978) Right temporal-lobe contribution to image-mediated verbal learning. *Neuropsychologia*, 16, 61–71.

JONIDES, J., (1979) Left and right visual field superiority for letter classification. *Quarterly journal of experimental psychology*, 31, 428–429.

JORM, A. F., (1979) The cognitive and neurological basis of developmental dyslexia: a theoretical framework and review. *Cognition*, 7, 19–33.

JULESZ, B., BREITMEYER, B., and KROPFL, W., (1976) Binocular-disparity-dependent upper-lower hemifield anisotropy and left-right hemifield isotropy as revealed by dynamic random dot stereograms. *Perception*, 5, 129–141.

JUNG, R., (1962) Summary of the conference. In (Ed.) MOUNTCASTLE, V. B., *Interhemispheric relations and cerebral dominance*. Baltimore: Johns Hopkins Press, 264–277.

KAHNEMAN, D., (1973) *Attention and effort*. Prentice-Hall: Englewood Cliffs, New Jersey.

KAIL, R. V., and SIEGEL, A. W., (1978) Sex and hemispheric differences in the recall of verbal and spatial information. *Cortex*, 14, 557–563.

KALLMAN, H. J., (1978) Can expectancy explain reaction time ear asymmetries? *Neuropsychologia*, 16, 225–228.

KALLMAN, H. J., and CORBALLIS, M. C., (1975) Ear asymmetry in reaction time to musical sounds. *Perception and psychophysics*, 17, 368–370.

KALSBEEK, J. W. H., and SYKES, R. N., (1967) Objective measurement of mental load. *Acta psychologica*, 27, 253–261.

KAPPAUF, W. E., and YEATMAN, F. R., (1970) Visual on-and-off latencies and handedness. *Perception and psychophysics*, 8, 46–50.

KASTE, M., and WALTIMO, O., (1976) Prognosis of patients with middle cerebral artery occlusion. *Stroke*, 7, 482–485.

KATZ, A. N., (1980) Cognitive arithmetic: evidence for right hemisphere mediation in an elementary component stage. *Quarterly journal of experimental psychology*, 32, 69–84.

KAUFER, I., MORAIS, J., and BERTELSON, P., (1975) Lateral differences in tachisto-scopic recognition of bilaterally presented verbal material. *Acta psychologica*, 39, 369–376.

KEEFE, B., and SWINNEY, D., (1979) On the relationship of hemispheric specialization and developmental dyslexia. *Cortex*, 15, 471–481.

KELLAR, L. A., (1978) Words on the right sound louder than words on the left in free field listening. *Neuropsychologia*, 16, 221–223.

KELLAR, L. A., and BEVER, T. G., (1980) Hemispheric asymmetries in the perception of musical intervals as a function of musical experience and family handedness background. *Brain and language*, 10, 24–38.

KELLY, R. R., and ORTON, K. D., (1979) Dichotic perception of word-pairs with mixed image values, *Neuropsychologia*, 17, 363–372.

KELLY, R. R., and TOMLINSON-KEASEY, C., (1981) The effect of auditory input on cerebral laterality. *Brain and language*, 13, 67–77.

KELSO, J. A. S., SOUTHARD, D. L., and GOODMAN, D., (1979) On the coordination of two handed movements. *Journal of experimental psychology: human perception and performance*, 5, 221–238.

KERSHNER, J. R., (1974) Ocular-manual laterality and dual hemisphere specialization. *Cortex*, 10, 293–302.

KERSHNER, J. R., (1977) Cerebral dominance in disabled readers, good readers and gifted children: search for a valid model. *Child development*, 48, 61–67.

KERSHNER, J. R., (1979) Rotation of mental images and asymmetries in word recognition in disabled readers. *Canadian journal of psychology*, 33, 39–50.

KERSHNER, J. R., and JENG, A. G. R., (1972) Dual functional hemispheric asymmetry in visual perception: effects of ocular dominance and postexposural processes. *Neuropsychologia*, 10, 437–445.

KERSHNER, J. R., and KING, A. J., (1974) Laterality of cognitive functions in achieving hemiplegic children. *Perceptual and motor skills*, 39, 1283–1289.

KERSHNER, J. R., THOMAE, R., and CALLAWAY, R., (1977) Non-verbal fixation control in young children induces a left field advantage in digit recall. *Neuropsychologia*, 15, 569–576.

KERTESZ, A., and GESCHWIND, N., (1971) Patterns of pyramidal decussation and their relationship to handedness. *Archives of neurology*, 24, 326–332.

KERTESZ, A., and SHEPPARD, A., (1981) The epidemiology of aphasic and cognitive impairment in stroke: age, sex, aphasia type and laterality differences. *Brain*, 104, 117–128.

KETTERER, M. W., and SMITH, B. D., (1977) Bilateral electrodermal activity, lateralized cerebral processing and sex. *Psychophysiology*, 14, 513–516.

KILSHAW, D., and ANNETT, M., (1983) Right and left hand skill – I. Effects of age, sex and hand preference showing superior skill in left handers. *British journal of psychology*, 74, 253–268.

KIM, Y., ROYER, F., BONSTELLE, C., and BOLLER, F., (1980) Temporal sequencing of verbal and non-verbal materials: the effect of laterality of lesion. *Cortex*, 16, 135–144.

KIMURA, D., (1961a) Some effects of temporal lobe damage on auditory perception. *Canadian journal of psychology*, 15, 156–165.

KIMURA, D., (1961b) Cerebral dominance and the perception of verbal stimuli. *Canadian journal of psychology*, 15, 166–171.

KIMURA, D., (1963a) Right temporal lobe damage. *Archives of neurology*, 8, 264–271.

KIMURA, D., (1963b) Speech lateralization in young children as determined by an auditory test. *Jouranl of comparative and physiological psychology*, 56, 899–902.

KIMURA, D., (1964) Left-right differences in the perception of melodies. *Quarterly journal of experimental psychology*, 16, 355–358.

KIMURA, D., (1966) Dual functional asymmetry of the brain in visual perception. *Neuropsychologia*, 4, 275–285.

KIMURA, D., (1967) Functional asymmetry of the brain in dichotic listening. *Cortex*, 3, 163–178.

KIMURA, D., (1969) Spatial localization in left and right visual fields. *Canadian journal of psychology*, 23, 445–458.

KIMURA, D., (1973a) Manual activity during speaking – I. Right handers. *Neuropsychologia*, 11, 45–50.

KIMURA, D., (1973b) Manual activity during speaking – II. Left handers. *Neuropsychologia*, 11, 51–55.

KIMURA, D., (1977a) Acquisition of a motor skill after left hemisphere damage. *Brain*, 100, 527–542.

KIMURA, D., (1977b) The neural basis of language qua gesture. In WHITAKER, H., and WHITAKER, H. A., (Eds.) *Studies in neurolinguistics* Vol. 3, Academic Press: New York, Chap. 4, 145–156.

KIMURA, D., (1980) Sex differences in intrahemispheric organization of speech. *The behavioral and brain sciences*, 3, 240–241.

KIMURA, D., and ARCHIBALD, Y., (1974) Motor functions of the left hemisphere. *Brain*, 97, 337–350

KIMURA, D., BATTISON, R., and LUBERT, B., (1976) Impairment of non-linguistic hand movements in a deaf aphasic. *Brain and language*, 3, 566–571.

KIMURA, D., and DURNFORD, M., (1974) Normal studies on the function of the right hemisphere in vision. In DIMOND, S. J., and BEAUMONT, J. G., (Eds.) *Hemisphere function in the human brain*. Elek: London, 25–47.

KIMURA, D., and FOLB, S., (1968) Neural processing of backward speech sounds. *Science*, 161, 395–396.

KIMURA, D., and HUMPHREYS, C. A., (1981) A comparison of left and right arm movements during speaking. *Neuropsychologia*, 19, 807–812.

KIMURA, D , and VANDERWOLF, C. H., (1970) The relation between hand preference and the performance of individual finger movements by left and right hands. *Brain*, 93, 769–774.

KINSBOURNE, M., (1970) The cerebral basis of lateral asymmetries in attention. *Acta psychologica*, 33, 193–201.

KINSBOURNE, M., (1971) The minor cerebral hemisphere as a source of aphasic speech. *Archives of neurology*, 25, 302–306.

KINSBOURNE, M., (1972) Eye and head turning indicates cerebral localization. *Science*, 176, 539–541.

KINSBOURNE, M., (1973) The control of attention by interaction between the cerebral hemispheres. In KORNBLUM, S., (Ed.) *Attention and performance IV*, 239–255. Academic Press: London, 239–255.

KINSBOURNE, M., (1974a) Mechanisms of hemispheric interaction in man. In KINSBOURNE, M., and SMITH, W. L., (Eds.) *Hemisphere disconnection and cerebral function*. Thomas: Springfield, Illinois, 260–284.

KINSBOURNE, M., (1974b) Lateral interactions in the brain. In KINSBOURNE, M., and SMITH, W. L., (Eds.) *Hemisphere disconnection and cerebral function*. Thomas: Springfield, Illinois, 239–259.

KINSBOURNE, M., (1975a) The mechanism of hemispheric control of the lateral gradient of attention. In RABBITT, P. M. F., and DORNIC, S., (Eds.) *Attention and performance V*. Chap. 7, 81–97.

KINSBOURNE, M., (1975b) Minor hemisphere language and cerebral maturation. In LENNEBERG, E. H., and LENNEBERG, E., (Eds.) *Foundations of language development: a multidisciplinary approach*, Vol. 2. Academic Press, Chap. 7, 107–116.

KINSBOURNE, M , (1975c) The ontogeny of cerebral dominance. *Annals of the New York academy of sciences*, 263, 244–250.

KINSBOURNE, M., (1978) Evolution of language in relation to lateral action. In KINSBOURNE, M., (ed.) *Asymmetrical function of the brain*. Cambridge University Press: Cambridge, 553–565.

KINSBOURNE, M., (1980) If sex differences in brain lateralization exist, they have yet to be discovered. *The behavioral and brain sciences*, 3, 241–242.

KINSBOURNE, M., and COOK, J., (1971) Generalized and lateralized effects of concurrent verbalization on a unimanual skill. *Quarterly journal of experimental psychology*, 23, 341–345.

KINSBOURNE, M., and FISHER, M., (1971) Latency of uncrossed and crossed reaction in callosal agenesis. *Neuropsychologia*, 9, 471–473.

KINSBOURNE, M., and HICKS, R. E., (1978) Functional cerebral space: a model for overflow, transfer and interference affects in human performance: a tutorial review. In REQUIN, J., (Ed.) *Attention and performance* VII. Erlbaum Associates: New Jersey, Chap. 19, 345–362.

KINSBOURNE, M., and LEMPERT, H., (1979) Does left brain lateralization of speech arise from right-biased orienting to salient perceptions? *Human development*, 22, 270–276.

KIRSNER, K., (1980) Hemisphere specific processes in letter matching. *Journal of experimental psychology: human perception and performance*, 6, 167–179.

KIRSNER, K., and BROWN, H., (1981) Laterality and recency effects in working memory. *Neuropsychologia*, 19, 249–262.

KLATZKY, R. L., (1970) Interhemispheric transfer of test stimulus representations in memory scanning. *Psychonomic science*, 21, 201–203.

KLATZKY, R. L., and ATKINSON, J., (1971) Specialization of the cerebral hemispheres in scanning for information in short term memory. *Perception and psychophysics*, 10, 335–338.

KLEIN, D., MOSCOVITCH, M., and VIGNA, C., (1976) Attentional mechanisms and perceptual asymmetries in tachistoscopic recognition of words and faces. *Neuropsychologia*, 14, 55–66.

KLEIN, R., and ARMITAGE, A., (1979) Rhythms in human performance: 1.5 hour oscillations in cognitive style. *Science*, 204, 1326–1328.

KLEIN, S. P., and ROSENFIELD, W. D., (1980) The hemispheric specialization for linguistic and non-linguistic tactile stimuli in third grade children. *Cortex*, 16, 205–212.

KLEINEMAN, K. M., and CLONINGER, L. H., (1973) Inter-manual transfer of tactual learning as a function of stimulus meaningfulness. *Perceptual and motor skills*, 37, 875–880.

KLICPERA, C., WOLFF, P. H., and DRAKE, C., (1981) Bimanual co-ordination in adolescent boys with reading retardation. *Developmental medicine and child neurology*, 23, 617–625.

KLISZ, D. K., and PARSONS, O. A., (1975) Ear asymmetry in reaction time tasks as a function of handedness. *Neuropsychologia*, 13, 323–330.

KNOX, A. W., and BOONE, D. R., (1970) Auditory laterality and tested handedness. *Cortex*, 6, 164–173.

KNOX, C., and KIMURA, D., (1970) Cerebral processing of non-verbal sounds in boys and girls. *Neuropsychologia*, 8, 227–237.

KOCEL, K. M., (1976) Cognitive abilities: handedness, familial sinistrality and sex. *Annals of the New York academy of sciences*, 280, 233–243.

KOCEL, K., GALIN, D., ORNSTEIN, R., and MERRIN, E. L., (1972) Lateral eye movement and cognitive mode. *Psychonomic science*, 27, 223–224.

KOHN, B., and DENNIS, M., (1974) Selective impairments of visuo-spatial abilities in infantile hemiplegics after right cerebral hemidecortication. *Neuropsychologia*, 12, 505–512.

KOFF, E., BOROD, J C., and WHITE, B., (1981) Asymmetries for hemiface size and mobility. *Neuropsychologia*, 19, 825–830.

KOLB, B., and MILNER, B., (1981a) Performance of complex arm and facial movements after focal brain lesions. *Neuropsychologia*, 19, 491–503.

KOLB, B., and MILNER, B., (1981b) Observations on spontaneous facial expressions after focal cerebral excisions and after intracarotid injection of sodium amytal. *Neuropsychologia*, 19, 505–514.

KRASHEN, S., (1973) Lateralization, language learning and the critical period: some new evidence. *Language learning*, 23, 63–74.

KREUTER, C., KINSBOURNE, M., and TREVARTHEN, C., (1972) Are deconnected cerebral hemispheres independent channels? A preliminary study of the effect of unilateral loading on bilateral finger tapping. *Neuropsychologia*, 10, 453–461.

KROLL, N. E. A, and MADDEN, D. J., (1978) Verbal and pictorial processing by hemisphere as a function of the subject's verbal scholastic aptitude score. In REQUIN, J.,

(Ed.) *Attention and performance* Vol. VII. Lawrence Erlbaum Associates: Hillsdale, New Jersey, 375–390.

KRONFOL, Z., HAMSHER, K. De S., DIGRE, K., and WAZIRI, R., (1978) Depression and hemispheric functions: changes associated with unilateral ECT. *British journal of psychiatry*, 132, 560–567.

KRUPER, D. C., PATTON, R. A., and KOSKOFF, Y. D., (1971) Hand and eye preference in unilaterally brain ablated monkeys. *Physiology and behavior*, 7, 181–185.

KUGLER, B. T., and HENLEY, S. H. A., (1979) Laterality effects in the tactile modality in schizophrenia. In GRUZELIER, J., and FLOR-HENRY, P., (Eds.) *Hemisphere asymmetries of function in psychopathology*. Elsevier/North Holland Biomedical Press: Amsterdam, 475–489.

KUHN, G., (1973) The phi coefficient as an index of ear differences in dichotic listening, *Cortex*, 9, 450–457.

KRYNAUW, R. A., (1950) Infantile hemiplegia treated by removing one cerebral hemisphere. *Journal of neurology, neurosurgery and psychiatry*, 13, 243–267.

KUMAR, S., (1977) Short-term memory for a non-verbal tactual task after cerebral commissurotomy. *Cortex*, 13, 55–61.

KUTAS, M., McCARTHY, G., and DONCHIN, E., (1975) Differences between sinistrals' and dextrals' ability to infer a whole from its parts: a failure to replicate. *Neuropsychologia*, 13, 455–464.

LACROIX, J. M., and COMPER, P., (1979) Lateralization in the electrodermal system as a function of cognitive hemispheric manipulations. *Psychophysiology*, 16, 116–129.

LADAVAS, E., UMILTÀ, C., and RICCI-BITTI, P. E., (1980) Evidence for sex differences in right-hemisphere dominance for emotions. *Neuropsychologia*, 18, 361–366.

LAKE, D. A., and BRYDEN, M. P., (1976) Handedness and sex differences in hemispheric asymmetry. *Brain and language*, 3, 266–282.

LAMBERT, A. J., and BEAUMONT, J. G., (1982) On Kelly and Orton's 'dichotic perception of word pairs with mixed image values'. *Neuropsychologia*, 20, 209–211.

LANDIS, T., ASSAL, G., and PERRET, E., (1979) Opposite cerebral hemispheric superiorities for visual and associative processing of emotional facial expressions and objects. *Nature*, 278, 739–740.

LANSDELL, H., (1961) The effect of neurosurgery on a test of proverbs. *American psychologist*, 16, 448.

LANSDELL, H., (1962) Laterality of verbal intelligence in the brain. *Science*, 135, 922–923.

LANSDELL, H., (1968) Effect of extent of temporal lobe ablations on two lateralized deficits. *Physiology and behavior*, 3, 271–273

LANSDELL, H., (1969) Verbal and non-verbal factors in right-hemisphere speech: relation to early neurological history. *Journal of comparative and physiological psychology*, 69, 734–738.

LANSDELL, H., (1973) Effect of neurosurgery on the ability to identify popular word associations. *Journal of abnormal psychology*, 81, 255–258.

LASHLEY, K. S., (1929) *Brain mechanisms and intelligence*. Chicago: University of Chicago Press.

LASSEN, N. A., and INGVAR, D. H., (1972) Radiotopic assessment of rCBF. *Progress in nuclear medicine*, 1, 376–409.

LASSONDE, M. C., LORTIE, J., PTITO, M., and GEOFFROY, G., (1981) Hemispheric asymmetry in callosal agenesis as revealed by dichotic listening performance. *Neuropsychologia*, 19, 455–458.

LAWSON, N. C., (1978) Inverted writing in right and left handers in relation to lateralization of face recognition. *Cortex*, 14, 207–211.

LEDLOW, A., SWANSON, J. M., and KINSBOURNE, M., (1978) Reaction times and evoked potential as indicators of hemispheric differences for laterally presented name and physical matches. *Journal of experimental psychology: human perception and performance*, 4, 440–454.

LE DOUX, J. E., RISSE, G. L., SPRINGER, S. P., WILSON, D. H., and GAZZANIGA, M. S., (1977) Cognition and commissurotomy. *Brain*, 100, 87–104.

LE DOUX, J. E., WILSON, D. H., and GAZZANIGA, M. S., (1977a) A divided mind: observations on the conscious properties of the separated hemispheres. *Annals of neurology*, 2, 417–421.

LE DOUX, J. E., WILSON, D. H., and GAZZANIGA, M. S., (1977b) Manipulo-spatial aspects of cerebral lateralization: clues to the origin of lateralization. *Neuropsychologia*, 15, 743–750.

LE DOUX, J. E., WILSON, D. H., and GAZZANIGA, M. S., (1979) Beyond commissurotomy: clues to consciousness. In GAZZANIGA, M. S., (Ed.) *Handbook of behavioural neurobiology*, Vol. 2. *Neuropsychology*, Plenum: New York, Chap. 17, 543–554.

LEEHEY, S. C., and CAHN, A., (1979) Lateral asymmetries in the recognition of words, familiar faces and unfamiliar faces. *Neuropsychologia*, 17, 619–635.

LEEHEY, S., CAREY, S., DIAMOND, R., and CAHN, A., (1978) Upright and inverted faces: the right hemisphere knows the difference. *Cortex*, 14, 411–419.

LEFEVRE, E., STARCK, R., LAMBERT, W. E., and GENESE, F., (1977) Lateral eye movements during verbal and non-verbal dichotic listening. *Perceptual and motor skills*, 44, 1115–1122.

LEHMAN, R. A., (1978) The handedness of rhesus monkeys: I. Distribution. *Neuropsychologia*, 16, 33–42.

LEHMAN, R. A. W., and SPENCER, D. D., (1973) Mirror-image shape discrimination. Reversal of responses in the optic chiasm sectioned monkey. *Brain research*, 52, 233–241.

LEIBER, L., (1976) Lexical decisions in the right and left cerebral hemispheres. *Brain and language*, 3, 443–450.

LEIBER, L., and AXELROD, S., (1981a) Intra-familial learning is only a minor factor in manifest handedness. *Neuropsychologia*, 19, 273–288.

LEIBER, L., and AXELROD, S., (1981b) Not all sinistrality is pathological. *Cortex*, 12, 259–272.

LENNEBERG, E. H., (1967) *Biological foundations of language*. John Wiley and Sons: New York.

LEONG, C. K., (1976) Lateralization in severely disabled readers in relation to functional cerebral development and synthesis of information. In KNIGHTS, R. M., and BAKKER, D. J., (Eds ) *The neuropsychology of learning disorders: theoretical approaches*. Baltimore: University Park Press, 221–231.

LE MAY, M., (1976) Morphological cerebral asymmetries of modern man, fossil man, and nonhuman primates. *Annals of the New York academy of sciences*, 349–366.

LE MAY, M., (1977) Asymmetries of the skull and handedness. *Journal of the neurological sciences*, 32, 243–253.

LE MAY, M., and CULEBRAS, A , (1972) Human brain and morphological differences in the hemispheres demonstrated by carotid arteriography. *New England journal of medicine*, 287, 168–170.

LE MAY, M., and GESCHWIND, N., (1975) Hemispheric differences in the brains of great apes. *Brain behavior and evolution*, 11, 48–52.

LE MAY, M., and GESCHWIND, N., (1978) Asymmetries of the human cerebral hemispheres. In CARAMAZZA, A., and ZURIF, E. B., (Eds.) *Language acquisition and language breakdown: parallels and divergencies*. Johns Hopkins University Press: Baltimore, 311–328.

LEMKUHL, G., and POECK, K., (1981) Conceptual organisation and ideational apraxia. *Cortex*, 17, 153–158.

LERNER, J , NACHSON, I., and CARMON, I., (1977) Responses of paranoid and non-paranoid schizophrenics in a dichotic listening task. *Journal of nervous and mental disease*, 164, 247–252.

LEVITON, A., and KILTY, T., (1976) Birth order and left handedness. *Archives of neurology*, 33, 664.

LEVY, C. M., and BOWERS, D., (1974) Hemispheric asymmetry of reaction time in a dichotic discrimination task. *Cortex*, 10, 18-25.

LEVY, J., (1969) Possible basis for the evolution of lateral specialization of the human brain. *Nature*, 224, 614–615.

LEVY, J., (1974a) Psychobiological implications of bilateral asymmetry. In *Hemisphere function in the human brain* (Eds.) BEAUMONT, J. G., and DIMOND, S. J., Elek: London, 121–183.

LEVY, J., (1974b) Cerebral asymmetries as manifested in split-brain man. In *Hemispheric disconnection and cerebral function*. (Eds.) KINSBOURNE, M., and SMITH, W. L., Springfield, Illinois. Charles C. Thomas, 165–183.

LEVY, J., (1976a) A review of evidence for a genetic component in handedness. *Behavior genetics*, 6, 429–453.

LEVY, J., (1976b) Evolution of language lateralization and cognitive function. *Annals of the New York academy of sciences*, 280, 810–820.

LEVY, J., (1976c) Lateral dominance and aesthetic preference. *Neuropsychologia*, 14, 431–446.

LEVY, J., (1977a) A reply to Hudson regarding the Levy-Nagylaki model for the genetics of handedness. *Neuropsychologia*, 15, 187–190.

LEVY, J., (1977b) The origins of lateral asymmetry. In HARNAD, S., DOTY, R. W., GOLDSTEIN, G., JAYNES, J., and KRAUTHAMER, G., (Eds.) *Lateralization in the nervous system*. Academic Press: New York, 195–209.

LEVY, J., (1977c) The correlation of the Φ function of the difference score with performance and its relevance to laterality experiments. *Cortex*, 13, 458–464.

LEVY, J., (1977d) The mammalian brain and the adaptive advantage of cerebral asymmetry. *Annals of the New York academy of science*, 299, 266–272.

LEVY, J., (1978) Lateral differences in the human brain in cognition and behavioural control. In BUSER, P. A., and ROUGEUL-BOUSER, A., (Eds.) *Cerebral correlates of conscious experience*. Elsevier/North Holland Biomedical Press, INSERM Symposium no. 6, 285–298 (Amsterdam), 285–298.

LEVY, J., (1980) Cerebral asymmetry and the psychology of man. In WITTROCK, M. G., (Ed.) *The brain and psychology*. Academic Press: New York, 259–321.

LEVY, J., (1982) Handwriting posture and cerebral organization: how are they related? *Psychological bulletin*, 91, 589–608.

LEVY, J., and LEVY, J. M., (1978) Human lateralization from head to foot. *Science*, 200, 1291–1292.

LEVY, J., and NAGYLAKI, T., (1972) A model for the genetics of the hand. *Genetics*, 72, 117–128.

LEVY, J., NEBES, R. D., and SPERRY, R. W., (1971) Expressive language in the surgically separated minor hemisphere. *Cortex*, 7, 49–58.

LEVY, J., and REID, M. L., (1976) Variations in writing posture and cerebral organization. *Science*, 194, 337–339.

LEVY, J., and REID, M., (1978) Variations in cerebral organization as a function of handedness, hand posture in writing and sex. *Journal of experimental psychology: general*, 107, 119–144.

LEVY, J., and TREVARTHEN, C., (1976) Metacontrol of hemispheric function in human split-brain patients. *Journal of experimental psychology: human perception and performance*, 2, 295–312.

LEVY, J., and TREVARTHEN, C., (1977) Perceptual, semantic and phonetic aspects of elementary language processes in split-brain patients. *Brain*, 100, 105–118.

LEVY, J., TREVARTHEN, C., and SPERRY, R. W., (1972) Perception of bilateral chimeric figures following hemispheric deconnexion. *Brain*, 95, 61–78.

LEVY, R. S., (1978) The question of electrophysiological asymmetries preceding speech. In *Studies in neurolinguistics*, 3, (Eds.) WHITAKER, H., and WHITAKER, H. A., Academic Press: New York, 257–318.

LEVY-AGRESTI, J., and SPERRY, R. W., (1968) Differential perceptual capacities in major and minor hemispheres. *Proceedings of the national academy of sciences*, 61, 1151.

LEY, R. G., and BRYDEN, M. P., (1979) Hemispheric differences in processing emotions and faces. *Brain and language*, 7, 127–138.

LIBERMAN, A. M., (1974) The specialization of the language hemisphere. In *The neurosciences: third study program*. (Eds.) SCHMITT, F. O., and WORDEN, F. G., MIT Press: Cambridge, Massachussets, 43–56.

LIBERMAN, A. M., COOPER, F. S., SHANKWEILER, D. S., and STUDDERT-KENNEDY, M., (1967) Perception of the speech code, *Psychological review*, 74, 431–461.

LIEDERMAN, J., and KINSBOURNE, M., (1980) Rightward motor bias in newborns depends upon parental right handedness. *Neuropsychologia*, 18, 579–584.

LIEPMANN, H., (1908) *Drei aufsatze aus dem apraxiegebeit*, Karger: Berlin.

LIEPMANN, H., and MASS, O., (1907) Fall Von Linksseitiger, Agraphie und Apraxie bei rechtsseitiger Lahmung. *Journal of psychology and neurology*, 10, 214–227.

LINDGREN, S. D., (1978) Finger localisation and the prediction of reading disability. *Cortex*, 87–107.

LISHMAN, W. A., and McMEEKAN, E. R. L., (1976) Hand preference in psychiatric patients. *British journal of psychiatry*, 129, 156–166.

LISHMAN, W. A., and McMEEKAN, E. R. L., (1977) Handedness in relation to direction and degree of cerebral dominance for language. *Cortex*, 13, 30–43.

LISHMAN, W. A., TOONE, B. K., COLBOURN, C. J., MCMEEKAN, E. R. L., and MANCE, R. M., (1978) Dichotic listening in psychotic patients. *British journal of psychiatry*, 132, 333–341.

LOMAS, J., (1980) Competition within the left hemisphere between speaking and unimanual tasks performed without visual guidance. *Neuropsychologia*, 18, 141–149.

LOMAS, J., and KIMURA, D., (1976) Intrahemispheric interaction between speaking and sequential manual activity. *Neuropsychologia*, 14, 23–33.

LONGDEN, K., ELLIS, C., and IVERSEN, S. D., (1976) Hemispheric differences in the discrimination of curvature. *Neuropsychologia*, 14, 195–202.

LOO, R., and SCHNEIDER, R., (1979) An evaluation of the Briggs-Nebes modified version of Annett's handedness inventory. *Cortex*, 15, 683–686.

LORDAHL, D. S., KLEINMAN, K. M., LEVY B., MASSOTH, N. A., PESSIN, M. S., STORANDT, M., TUCKER, R., and VANDERPLAS, J. M., (1965) Deficits in recognition of random shapes with changed visual fields. *Psychonomic science*, 3, 245–246.

LUCHINS, D., POLLIN, W., and WYATT, R. J., (1980) Laterality in monozygotic schizophrenic twins: an alternative hypothesis. *Biological psychiatry*, 15, 87–94.

LUESSENHOP, A. J., BOGGS, J. S., LABOWIT, L. J., and WALLE, E. L., (1973) Cerebral dominance in stutterers determined by Wada testing. *Neurology*, 23, 1190–1192.

LURIA, A. R., (1966) *Higher cortical functions in man*. New York: Basic Books.

LURIA, A. R., (1970) *Traumatic aphasia*. Mouton: The Hague.

LURIA, A. R., SIMERNITSKAYA, E. G., and TYBULEVICH, B., (1970) The structure of psychological processes in relation to cerebral organisation. *Neuropsychologia*, 8, 13–19.

LYNN, R. B., BUCHANAN, D. C., FENICHEL, G. M., and FREEMON, F. R., (1981) Agenesis of the corpus callosum. *Archives of neurology*, 37, 444–445.

McADAM, D. W., and WHITAKER, H A., (1971) Language production: Electroencephalographic localization in the normal human brain. *Science*, 172, 499–502.

McCANN, B., (1981) Hemispheric asymmetries and early infantile autism. *Journal of autism and developmental disorders*, 11, 401–411.

McCARTHY, G., and DONCHIN, E., (1978) Brain potentials associated with structural and functional visual matching. *Neuropsychologia*, 16, 571–585.

McFARLAND, K., and ANDERSON, J., (1980) Factor stability of the Edinburgh Handedness Inventory as a function of test-retest performance, age and sex. *British journal of psychology*, 71, 135–142.

McFARLAND, K. A., and ASHTON, R. A., (1975) The lateralized effect of concurrent cognitive activity on a unimanual skill. *Cortex*, 11, 283–290.

McFARLAND, and ASHTON, R., (1978a) The influence of brain lateralization of function on a unimanual skill. *Cortex*, 14, 102–111.

McFARLAND, K., and ASHTON, R., (1978b) The influence of concurrent task difficulty on manual performance. *Neuropsychologia*, 16, 735–741.

McFARLAND, K., McFARLAND, M. L., BAIN, J. D., and ASHTON, R., (1978) Ear differences of abstract and concrete word recognition. *Neuropsychologia*, 16, 555–561.

McFIE, J., (1961) The effects of hemispherectomy on intellectual functioning in cases of infantile hemiplegia. *Journal of neurology, neurosurgery and psychiatry*, 24, 240–249.

McGEE, M., (1979) Human spatial abilities: psychometric studies and environmental, genetic hormonal and neurological influences. *Psychological bulletin*, 86, 889–918.

McGLONE, J., (1977) Sex differences in the cerebral organisation of verbal functions in patients with unilateral cerebral lesions. *Brain*, 100, 775–794.

McGLONE, J., (1978) Sex differences in functional brain asymmetry. *Cortex*, 14, 122–128.

McGLONE, J., (1980) Sex differences in human brain asymmetry: a critical survey. *The behavioral and brain sciences*, 3, 215–263.

McGLONE, J., and DAVIDSON, W., (1973) The relation between cerebral speech laterality and spatial ability with special reference to sex and hand preference. *Neuropsychologia*, 11, 105–113.

McGLONE, J., and KERTESZ, A., (1973) Sex differences in cerebral processing of visuo-spatial tasks. *Cortex*, 9, 313–320.

MacKAVEY, W., CURCIO, F., and ROSEN, J., (1975) Tachistoscopic word recognition performance under conditions of simultaneous bilateral presentation. *Neuropsychologia*, 13, 27–33

McKEE, G., HUMPHREY, B., and McADAM, D., (1973) Scaled lateralization of alpha activity during linguistic and musical tasks. *Psychophysiology*, 10, 441–443.

McKEEVER, W. F., (1974) Does post-exposural directional scanning offer a sufficient explanation for lateral differences in tachistoscopic recognition? *Perceptual and motor skills*, 38, 43–50.

McKEEVER, W. F., (1979) Handwriting posture in left handers: sex, familial sinistrality and language laterality correlates. *Neuropsychologia*, 17, 429–444.

McKEEVER, W. F., and DIXON, M. S., (1981) Right hemisphere superiority for discriminating memorized from non-memorized faces: affective imagery, sex and perceived emotionality effects. *Brain and language*, 12, 246–260.

McKEEVER, W. F., and GILL, K. M., (1972a) Interhemispheric transfer time for visual stimulus information as a function of the retinal locus of stimulation. *Psychonomic science*, 26, 308–310.

McKEEVER, W. F., and GILL, K. M., (1972b) Visual half-field differences in masking effects for sequential letter stimuli in the right and left handed. *Neuropsychologia*, 10, 111–117.

McKEEVER, W. F., and GILL, K. M., (1972c) Visual half-field differences in the recognition of bilaterally presented single letters and vertically spelled words. *Perceptual and motor skills*, 34, 815–818.

McKEEVER, W. F., and GILL, K. M., (1975) Letter versus dot stimuli as tools for splitting the normal brain with reaction time. *Quarterly journal of experimental psychology*, 27, 363–373.

McKEEVER, W. F., HOEMANN, H. W., FLORIAN, V. A., and VAN DEVENTER, A. D. (1976) Evidence of minimal cerebral asymmetries for the processing of English words and American sign language in the congenitally deaf. *Neuropsychologia*, 14, 413–423.

McKEEVER, W. F., and HULING, M. D., (1970a) Left cerebral hemisphere superiority in tachistoscopic word recognition performance. *Perceptual and motor skills*, 30, 763–766.

McKEEVER, W. F., and HULING, M. D., (1970b) Lateral dominance in tachistoscopic recognition of children at two levels of ability. *Quarterly journal of experimental psychology*, 22, 600–604.

McKEEVER, W. F., and HULING, M. D., (1971a) Lateral dominance in tachistoscopic word recognition performances obtained with bilateral simultaneous input. *Neuropsychologia*, 9, 15–20.

McKEEVER, W. F., and HULING, M. D., (1971b) Bilateral tachistoscopic word recognition as a function of hemisphere stimulated and interhemispheric transfer time. *Neuropsychologia*, 9, 281–288.

McKEEVER, W. F., and JACKSON, T. L., (1979) Cerebral dominance assessed by object and color-naming latencies. *Brain and language*, 7, 175–190.

McKEEVER, W. F., LARRABEE, G. J., SULLIVAN, K. F., and JOHNSON, H. J., (1981) Unimanual tactile anomia consequent to corpus callosotomy: reduction of anomic deficit under hypnosis. *Neuropsychologia*, 19, 179–190.

McKEEVER, W. F., and SUBERI, M., (1974) Parallel but temporarily displaced visual half-field meta-contrast functions. *Quarterly journal of experimental psychology*, 26, 258–265.

McKEEVER, W. F., SUBERI, M., and VAN DEVENTER, A. D., (1972) Fixation control in tachistoscopic studies of laterality effects: comments and data relevant to Hines' experiment. *Cortex*, 8, 473–479.

McKEEVER, W. F., and VAN DEVENTER, A. D., (1977a) Familial sinistrality and degree of left handedness. *British journal of psychology*, 68, 469–471.

McKEEVER, W. F., and VAN DEVENTER, A. D., (1977b) Visual and auditory language processing asymmetries: influence of handedness, familial sinistrality and sex. *Cortex*, 13, 225–241.

McKEEVER, W. F., and VAN DEVENTER, A. D., (1977c) Failure to confirm a spatial ability impairment in persons with evidence of right hemisphere speech capability. *Cortex*, 13, 321–326.

McKEEVER, W. F., SULLIVAN, K. F., FERGUSON, S. M., and RAYPORT, M., (1981) Typical cerebral hemisphere disconnection deficits following corpus callosum section despite sparing of the anterior commissure. *Neuropsychologia*, 19, 745–756.

McKEEVER, W. F., VAN DEVENTER, A. D., (1975) Dyslexic adolescents: evidence of impaired visual and auditory language processing associated with normal lateralization and visual responsivity. *Cortex*, 11, 361–378.

McKEEVER, W. F., VAN DEVENTER, A. D., and SUBERI, M., (1973) Avowed assessed and familial handedness and differential hemispheric processing of brief sequential and non-sequential visual stimuli. *Neuropsychologia*, 11, 235–238.

McKEEVER, W. F., and VAN HOFF, A. L., (1979) Evidence of a possible isolation of left hemisphere visual and motor areas in sinistrals employing an inverted handwriting posture. *Neuropsychologia*, 17, 445–455.

McKINNEY, J. P., (1967) Handedness, eyedness and perceptual stability of the left and right visual fields. *Neuropsychologia*, 5, 339–344.

McMANUS, I. C., (1977) Predominance of left sided breast tumours. *Lancet*, 2, 297–298.

McMANUS, I., (1980a) Left handedness in epilepsy. *Cortex*, 16, 487–491.

McMANUS, I. C., (1980b) Handedness in twins: A critical review. *Neuropsychologia*, 18, 347–355.

McMANUS, I. C., (1981) Handedness and birth stress. *Psychological medicine*, 11, 485–496.

McMEEKAN, E. R. L., and LISHMAN, W. A., (1975) Retest reliabilities and interrelationship of the Annett hand preference questionnaire and the Edinburgh handedness inventory. *British journal of psychology*, 66(1), 53–59.

McNEILAGE, P. F., SUSSMAN, H. M., and STOLZ, W., (1975) Incidence of laterality effects in mandibular and manual performance of dichoptic visual pursuit tracking. *Cortex*, 11, 251–258.

McRAE, D. L., BRANCH, C. L., and MILNER, B., (1968) The occipital horns and cerebral dominance. *Neurology*, 18, 95–98.

MACCOBY, E. E., and JACKLIN, C. N., (1974) *The psychology of sex differences*. Stanford University Press: Stanford, Ca.

MADDESS, R. J., (1975) Reaction time to hemiretinal stimulation. *Neuropsychologia*, 13, 213–218.

MAJKOWSKI, J., BOCHENEK, Z., BOCHENEK, W., KNAPIK-FIJALKOWSKA, D., and KOPEC, J., (1971) Latency of average evoked potentials to contralateral and ipsilateral auditory stimulation in normal subjects. *Brain research*, 25, 416–419.

MALONE, D. R., and HANNAY, H. J., (1978) Hemispheric dominance and normal colour memory. *Neuropsychologia*, 16, 51–59.

MANNING, A. A., GOBLE, W., MARKMAN, R., and LA BRECHE, T., (1977) Lateral cerebral differences in the deaf in response to linguistic and non-linguistic stimuli. *Brain and language*, 4, 309–321.

MARCEL, A. J., and PATTERSON, K. E., (1978) Word recognition and production: reciprocity in clinical and normal studies. In REQUIN, J., (Ed.) *Attention and performance* VII, Lawrence Erlbaum and Associates: New Jersey, Chap. 12, 209–226.

MARCEL, T., KATZ, L., and SMITH, M., (1974) Laterality and reading proficiency. *Neuropsychologia*, 12, 131–139.

MARCEL, T., and RAJAN, P., (1975) Lateral specialization for recognition of words and faces in good and poor readers. *Neuropsychologia*, 13, 489–497.

MARCIE, P., HÉCAEN, H., DUBOIS, J., and ANGELERGUES, R., (1965) Les réalisations du langage chez les malades atteints de lésions de l'hémisphère droit. *Neuropsychologia*, 3, 217–245.

MARIN, O. S., SCHWARTZ, M. F., and SAFFRAN, E. M., (1979) Origins and distribution of language. In *Handbook of behavioral neurobiology*, Vol. 2, *Neuropsychology* (Ed.) GAZZANIGA, M. S., Plenum: New York, Chap. 8, 179–213.

MARKOWITZ, H., and WEITZMAN, D. O., (1969) Monocular recognition of letters and Landolt 'C's in left and right visual hemifields. *Journal of experimental psychology*, 79, 187–189.

MARSH, G., (1978) Asymmetry of electrophysiological phenomena and its relation to behavior in humans. In KINSBOURNE, M., (Ed.) *Asymmetrical function of the brain*. Cambridge University Press: Cambridge, 292–317.

MARSHALL, J. C., (1973) Language, learning and laterality. In HINDE, R. A., and STEVENSON, J , (Eds.) *Constraints on learning*. London: Academic Press, 445–456.

MARSHALL, J. C., (1981) Hemispheric specialization: what, how and why. *The behavioral and brain sciences*, 4, 72–73.

MARSHALL, J., CAPLAN, D., and HOLMES, J. M., (1975) The measure of laterality. *Neuropsychologia*, 13, 315–321.

MARSHALL, J. C., and NEWCOMBE, F., (1966) Syntactic and semantic errors in paralexia. *Neuropsychologia*, 4, 169–176.

MARTIN, C. M., (1978) Verbal and spatial encoding of visual stimuli: the effects of sex, hemisphere and yes-no judgements. *Cortex*, 14, 227–233.

MARTIN, M., (1979) Hemispheric specialization for local and global processing. *Neuropsychologia*, 17, 33–40.

MARZI, C. A., and BERLUCCHI, G., (1977) Right field superiority for accuracy of recognition of famous faces in normals. *Neurospsychologia*, 15, 751–756.

MARZI, C. A., STEFANO, M. Di, TASSINARI, G., and CREA, F., (1979) Iconic storage in the two hemispheres. *Journal of experimental psychology: human perception and performance*, 5, 31–41.

MASCIE-TAYLOR, C. G. N., (1981) Hand preference and personality traits. *Cortex*, 17, 319–322.

MATEER, C., (1978) Impairments of nonverbal oral movements after left hemisphere damage: a follow-up analysis of errors. *Brain and language*, 6, 334–341.

MATEER, C., and KIMURA, D., (1977) Impairment of non-verbal oral movements in aphasia. *Brain and language*, 4, 262–276.

MATSUMIYA, Y., TAGLIASCO, V., LOMBROSO, C. T., and GOODGLASS, H., (1972) Auditory evoked response: meaningfulness of stimuli and interhemispheric asymmetry. *Science*, 175, 780–792.

MAYES, A., and BEAUMONT, J. G., (1977) Does visual evoked potential asymmetry index cognitive activity? *Neuropsychologia*, 15, 249–257.

MEADOWS, J. C., (1974) The anatomical basis of prosopagnosia. *Journal of neurology, neurosurgery and psychiatry*, 37, 489–501.

MEBERT, C. J., and MICHEL, G. F., (1980) Handedness in artists. In HERRON, J., (Ed.) *The neuropsychology of left handedness*. Academic Press: New York, 273–279.

MELEKIAN, B., (1981) Lateralization in the human newborn at birth: asymmetry of the stepping reflex. *Neuropsychologia*, 19, 707–711.

MERRELL, D. J., (1957) Dominance of eye and hand. *Human biology*, 23, 314–328.

MERSKEY, H., and WATSON, G. D., (1979) The lateralization of pain. *Pain*, 7, 271–280.

METZGER, R. L., and ANTES, J. R., (1976) Sex and coding strategy effects on reaction time to hemispheric probes. *Memory and cognition*, 4, 167–171.

MEYER, G., (1976) Right hemispheric sensitivity for the McCullough effect. *Nature*, 264, 751–753.

MEYER, J. S., SAKAI, F., YAMAGUCHI, F., YAMAMOTO, M., and SHAW, T., (1980) Regional changes in cerebral blood flow during standard behavioral activation in patients with disorders of speech and mentation compared to normal volunteers. *Brain and language*, 9, 61–77,

MEYER, V., (1959) Cognitive changes following temporal lobectomy for relief of temporal lobe epilepsy. *Archives of neurology and psychiatry*, 49, 299–309.

MEYER, V., (1960) Psychological effects of brain damage. In *Handbook of abnormal psychology* (Ed.) EYSENCK, H. J., Pitman: London 529–565.

MEYER, V., and JONES, H. G., (1957) Patterns of cognitive test performance as functions of the lateral localization of cerebral abnormalities in the temporal lobe. *Journal of mental science*, 103, 758–772.

MEYER, V., and YATES, A. B., (1955) Intellectual changes following temporal lobectomy for psychomotor epilepsy. *Journal of neurology, neurosurgery and psychiatry*, 18, 44–52.

MICELLI, G., CALTAGIRONE, C., GAINOTTI, G., MASULLO, C., SILVERI, M. C., and VILLA, G., (1981) Influence of age, sex, literacy and pathological lesions on incidence, severity and type of aphasia. *Acta neurologica Scandinavica*, 64, 370–382.

MICHEL, G., (1981) Right-handedness: a consequence of infant supine head-orientation preference? *Science*, 212, 685–697.

MILES, T. R., (1978) *Understanding dyslexia*. Hodder and Stoughton, London.

MILLER, E., (1971) Handedness and the pattern of human ability. *British journal of psychology*, 62, 111–112.

MILLS, L., and ROLLMAN, G. B., (1980) Hemispheric asymmetry for auditory perception of temporal order. *Brain and language*, 18, 41–47.

MILNER, A. D., and DUNNE, J. J., (1977) Lateralised perception of bilateral chimeric faces by normal subjects. *Nature*, 268, 175–176.

MILNER, A. D., and JEEVES, M. A., (1979) A review of behavioural studies of agenesis of the corpus callosum. In STEELE RUSSELL, I., VAN HOF, M. W., and BERLUCCHI, G., (Eds.) *Structure and function of the cerebral commissures*. Macmillan: London, 429–448.

MILNER, B., (1962) Laterality effects in audition. In MOUNTCASTLE, V. B., (Ed.) *Interhemispheric relations and cerebral dominance*. Baltimore: Johns Hopkins Press, 177–195.

MILNER, B., (1965) Visually-guided maze learning in man: effects of bilateral hippocampal, bilateral frontal and unilateral cerebral lesions. *Neuropsychologia*, 3, 317–338.

MILNER, B., (1967) Brain mechanisms suggested by studies of temporal lobes. In *Brain mechanisms underlying speech and language*. (Eds.) MILIKAN, C. H., and DARLEY, F. L., Grune and Stratton: New York.

MILNER, B., (1971) Interhemispheric differences and psychological processes. *British medical bulletin*, 27, 272–277.

MILNER, B., (1974) Some effects of frontal lobectomy in man. In WARREN, J. M., and AKERT, K., (Eds.) *The frontal granular cortex and behavior*. McGraw Hill: New York, 313–334.

MILNER, B., and TAYLOR, L., (1972) Right hemisphere superiority in tactile pattern recognition after cerebral commissurotomy: evidence for non-verbal memory. *Neuropsychologia*, 10, 1–15.

MILNER, B., TAYLOR, L., and SPERRY, R. W., (1968) Lateralized suppression of dichotically presented digits after commissural section in man. *Science*, 164, 184–185.

MISHKIN, M., and FORGAYS, D. G., (1952) Word recognition as a function of retinal locus. *Journal of experimental psychology*, 43, 43–48.

MITCHELL, D. E., and BLAKEMORE, C. B., (1970) Binocular depth perception and the corpus callosum. *Vision research*, 10, 49–54.

MOLFESE, D. L., (1977) Infant cerebral asymmetry. In SEGALOWITZ, S. J., and GRUBER, F. A., (Eds.) *Language development and neurological theory*. Academic Press: London, Chap. 2, 22–33.

MOLFESE, D. L., (1978) Left and right hemisphere involvement in speech perception: electrophysiological correlates. *Perception and psychophysics*, 23, 237–243.

MOLFESE, D. L., FREEMAN, R. B., and PALERMO, D. S., (1975) The ontogeny of brain lateralization for speech and non-speech stimuli. *Brain and language*, 2, 356–368.

MOLFESE, D. L., and HESS, T. M., (1978) Hemispheric specialization for VOT Perception in the pre-school child. *Journal of experimental child psychology*, 26, 71–84.

MOLFESE, D. L., and MOLFESE, V., (1979) Hemisphere and stimulus differences as reflected in the cortical responses of newborn infants to speech stimuli. *Developmental psychology*, 15, 505–511.

MOLFESE, D. L., and NUNEZ, V., (1976) Cerebral asymmetry: changes in factors affecting its development. *Annals of the New York academy of sciences*, 280, 821–833.

MOLLON, J. D., (1976) Review of 'Hemisphere function in the human brain. (Eds.) DIMOND, S. J., and BEAUMONT, J. G.' *British journal of psychology*, 67, 113–115.

MONEY, J., (1972) Studies on the function of sighting dominance. *Journal of experimental psychology*, 24, 454–464.

MONONEN, L. J., and SEITZ, M. R., (1977) An AER analysis of contralateral advantage in the transmission of auditory information. *Neuropsychologia*, 15, 165–173.

MOORE, W. H., (1976) Bilateral tachistoscopic word perception of stutterers and normal subjects. *Brain and language*, 3, 434–442.

MOORE, W. H., and LANG, M. K., (1977) Alpha asymmetry over the right and left hemispheres of stutterers and control subjects preceding massed oral readings: a preliminary investigation. *Perceptual and motor skills*, 44, 223–230.

MORAIS, J., and BERTELSON, P., (1973) Laterality effects in diotic listening. *Perception*, 2, 107–111.

MORAIS, J., and BERTELSON, P., (1975) Spatial position versus ear of entry as determinants of the auditory laterality effect: a stereophonic test. *Journal of experimental psychology: human perception and performance*, 1, 253–262.

MORAIS, J., and LANDERCY, M., (1977) Listening to speech while retaining music: what happens to the right-ear advantage? *Brain and language*, 4, 295–308.

MORGAN, A. H., and MACDONALD, H., and HILGARD, E. R., (1974) EEG alpha: lateral asymmetry related to task and hypnotizability. *Psychophysiology*, 11, 275–282.

MORGAN, A. H., McDONALD, P. J., and MacDONALD, H., (1971) Differences in bilateral alpha activity as a function of experimental task, with a note on lateral eye movements and hypnotizability. *Neuropsychologia*, 9, 459–469.

MORGAN, M., (1977) Embryology and inheritance of asymmetry. In HARNAD, S., DOTY, R. W., GOLDSTEIN, L., JAYNES, J., and KRAUTHAMER, G., (Eds.) *Lateralization in the nervous system*. Academic Press: New York. 173–194.

MORGAN, M. J., (1981) Hemispheric specializations and spatiotemporal interactions. *The behavioral and brain sciences*, 4, 74–75.

MORGAN, M. J., and CORBALLIS, M. C., (1978) On the biological basis of human laterality: II. The mechanism of inheritance. *The behavioral and brain sciences*, 1, 35–62.

MORRELL, L. K., and HUNTINGDON, D. A., (1972) Cortical particals time locked to speech production: evidence for cerebral origin. *Life sciences*, 72, 921–929.

MORRELL, L. K., and SALAMY, J. G., (1971) Hemispheric asymmetry of electrocortical responses to speech stimuli. *Science*, 174, 164–167.

MORROW, L., VRTUNSKI, P. Bart., KIM, Y., and BOLLER, F., (1981) Arousal responses to emotional stimuli and laterality of lesion. *Neuropsychologia*, 19, 65–72.

MOSCOVITCH, M., (1972) Choice reaction-time study assessing the verbal behavior of the minor hemisphere in normal adult humans. *Journal of comparative and physiological psychology*, 80, 66–74.

MOSCOVITCH, M., (1973) Language and the cerebral hemispheres: reaction time studies and their implications for models of cerebral dominance. In PLINER, P., KRAMES, L., and ALLOWAY, T., (Eds.) *Communication and affect: language and thought*. Academic Press: New York, 89–126.

MOSCOVITCH, M., (1976) On the representation of language in the right hemisphere of right-handed people. *Brain and language*, 3, 47–71.

MOSCOVITCH, M., (1979) Information processing and the cerebral hemispheres. In GAZZANIGA, M. S., (Ed.) *Handbook of neurobiology*, 2, *Neuropsychology*. Plenum: New York, Chap. 13, 379–346.

MOSCOVITCH, M., and OLDS, J., (1982) Asymmetries in spontaneous facial expressions and their possible relation to hemispheric specialization. *Neuropsychologia*, 71–82.

MOSCOVITCH, M., SCULLION D., and CHRISTIE, D., (1976) Early versus late stage of processing and their relation to functional hemispheric asymmetries in face recognition. *Journal of experimental psychology: human perception and performance*, 2, 401–416.

MOSCOVITCH, M., and SMITH, L. C., (1979) Differences in neural organisation between individuals with inverted and non-inverted handwriting postures. *Science*, 205, 710–712.

MOUNTCASTLE, V. B., (1962) (Ed.) *Interhemispheric relations and cerebral dominance.* Johns Hopkins University Press: Baltimore.

MURPHY, E. H., and VENABLES, P. H., (1970) The investigation of ear asymmetry by simple and disjunctive reaction-time tasks. *Perception and psychophysics,* 8, 104–106.

MYERS, R. E., (1960) Failure of intermanual transfer in corpus callosum – sectioned chimpanzees. *Anatomical record,* 136, 358.

MYERS, R. E., (1962) Transmission of visual information within and between the hemispheres: a behavioural study. In MOUNTCASTLE, V. B., (Ed.) *Interhemispheric relations and cerebral dominance,* The Johns Hopkins University Press: Baltimore, Chap. IV., 51–74.

MYERS, R. E., (1965) The neocortical commissures and interhemispheric transmission of information. In ETTLINGER, G., (Ed.) *CIBA foundation study group no. 20: Functions of the corpus callosum.* Churchill: London, 1–17.

MYERS, R. E., and SPERRY, R. W., (1953) Interocular transfer of a visual form discrimination habit in cats after section of the corpus callosum and optic chiasm. *Anatomical record,* 115, 351–352.

MYSLOBODSKY, M. S., and RATTOK, J., (1977) Bilateral electrodermal activity in waking man. *Acta Psychologica,* 41, 273–282.

MYSLOBODSLY, M. S., and SHAVIT, Y., (1978) Hemispheric asymmetry of visual evoked potentials with motor imbalance in rats. *Brain research,* 157, 356–359.

NACHSON, I., (1973) Effects of cerebral dominance and attention on dichotic listening. *Journal of life sciences,* 3, 107–14.

NACHSON, I., and CARMON, A., (1975) Hand preference in sequential and spatial tasks. *Cortex,* 11, 123–131.

NAGAFUCHI, M., (1970) Development of dichotic and monaural hearing abilities in young children. *Acta otolaryngologica,* 69, 409–414.

NAGYLAKI, T., and LEVY, J., (1973) The sound of one paw clapping isn't sound. *Behavior genetics,* 3, 279–292.

NAIDOO, S., (1972) *Specific dyslexia.* Pitman: London.

NAIR, K. R., and VIRMANI, V., (1973) Speech and language disturbances in hemiplegics. *Indian journal of medical research,* 61, 1395–1403.

NAKAMURA, R., and GAZZANIGA, M. S., (1977) Processing capacities following commissurotomy in the monkey. *Experimental neurology,* 56, 323–333.

NAKAMURA, R., and TANIGUCHI, R., (1977) Reaction time in patients with cerebral hemiparesis. *Neuropsychologia,* 15, 845–848.

NAKAMURA, R., and SAITO, H., (1974) Preferred hand and reaction time in different movement patterns. *Perceptual and motor skills,* 39, 1275–1281.

NASRALLAH, H., KEELOR, K., VAN SCHROEDER, C., and WHITTERS, M. M., (1981) Motoric lateralization in schizophrenic males. *American journal of psychiatry,* 138, 1114–1115.

NATALE, M., (1977) Perception of non-linguistic auditory rhythms by the speech hemisphere. *Brain and language,* 4, 32–44.

NAVA, P. L., BUTLER, S. R., and GLASS, A., (1975) Asymmetries of the alpha rhythm associated with functions of the right hemisphere. Paper presented to a meeting of the EEG society (June).

NAYLOR, H., (1980) Reading disability and lateral asymmetry: an information-processing analysis. *Psychological bulletin,* 87, 531–545.

NEBES, R. D., (1971a) Superiority of the minor hemisphere in commissurotomized man for the perception of part-whole relations. *Cortex,* 8, 333–349.

NEBES, R. D., (1971b) Handedness and perception of part-whole relations. *Cortex,* 8, 350–356.

NEBES, R. D., (1972) Dominance of the minor hemisphere in commissurotomized man on a test of figural unification. *Brain,* 85, 633–639.

NEBES, R. D., (1973) Perception of spatial relationships by the right and left hemispheres in commissurotomized man. *Neuropsychologia,* 11, 285–289.

NEBES, R. D., (1974) Hemispheric specialization in commissurotomized man. *Psychological bulletin,* 81, 1–14.

NEBES, R. D., and BRIGGS, G. C., (1974) Handedness and the retention of visual material. *Cortex*, 10, 209–214.

NEBES, R. D., MADDEN, D. J., and BERG, W. D., (1981) Lateral asymmetry of auditory attention in hemispherectomized patients. *Neuropsychologia*, 19, 307–310.

NEBES, R. D., and SPERRY, R W., (1971) Hemispheric deconnection syndrome with cerebral birth injury in the dominant arm area. *Neuropsychologia*, 9, 247–259.

NEILL, D. O., SAMPSON, H., and GRIBBEN, J. A., (1971) Hemiretinal effects in tachistoscopic letter recognition. *Journal of experimental psychology*, 91, 129–135.

NEISSER, U., (1967) *Cognitive psychology*. Appleton-Century-Crofts: New York.

NELSON, H. E., and WARRINGTON, E. K., (1980) An investigation of memory functions in dyslexic children. *British journal of psychology*, 71, 487–503.

NETLEY, C., (1977) Dichotic listening of callosal agenesis and Turners Syndrome patients. In SEGALOWITZ, S. J., and GRUBER, F. A., (Eds.) *Language development and neurological theory*. Academic Press: New York and London, Chap. 11, 133–141.

NEVILLE, H. J., (1977) Electroencephalographic testing of cerebral specialization in normal and congenitally deaf children: a preliminary report. In SEGALOWITZ, S. J., and GRUBER, F. A., (Eds.) *Language development and neurological theory*. Academic Press: New York, London, Chap. 10, 121–131.

NEVILLE, H. J., and BELLUGI, U., (1978) Patterns of cerebral specialization in congenitally deaf adults: a preliminary report. In SIPLE, P., (Ed.) *Understanding language through sign language research*. Academic Press: New York, Chap. 9, 239–257.

NEWCOMBE, F., (1969) *Missile wounds of the brain* Oxford University Press: Oxford.

NEWCOMBE, F., (1974) Selective deficits after focal cerebral injury. In DIMOND, S. J., and BEAUMONT, J. G., (Eds.) *Hemisphere function in the human brain*. Elek: London, 311–334.

NEWCOMBE, F., and RATCLIFF, G., (1973) Handedness, speech lateralization and ability. *Neuropsychologia*, 11, 399–408.

NEWCOMBE, F., and RATCLIFF, G., (1979) Long-term psychological consequences of cerebral lesion. In GAZZANIGA, M. S., (Ed.) *Handbook of behavioral neurobiology*, 2, *Neuropsychology*. Plenum: New York, Chap. 16, 495–540.

NEWCOMBE, F. G., RATCLIFF, G. G., CARRIVICK, P. J., HIORNS, R. W., HARRISON, G. A., and GIBSON, J. B., (1975) Hand preference and IQ in a group of Oxfordshire villages. *Annals of human biology*, 2, 235–242.

NIEDERBUHL, J., and SPRINGER, S., (1979) Task requirements and hemispheric asymmetry for the processing of single letters. *Neuropsychologia*, 17, 689–692.

NOBLE, J., (1966) Mirror images and the forebrain commissures of the monkey. *Nature*, 211, 1263–1265.

NOBLE, J., (1968) Paradoxical interocular transfer of mirror-image discriminations in the optic chiasm sectioned monkey. *Brain research*, 10, 127–151.

NOTTEBOHM, F., (1971) Neural lateralization of vocal control in a passerine bird. 1. Song. *Journal of experimental zoology*, 177, 229–261.

NOTTEBOHM, F., (1972) Neural lateralization of vocal control in a passerine bird. 2. Subsong, calls and a theory of vocal learning. *Journal of experimental zoology*, 179, 35–49.

NOTTEBOHM, F., (1977) Asymmetries in neural control of vocalization in the canary. In HARNAD, S. A., DOTY, R. W., GOLDSTEIN, L., JAYNES, J., and KRAUTHAMER, G., (Eds.) *Lateralization in the nervous system*. Academic Press: New York, Chap. 2, 23–44.

NOTTEBOHM, F., (1979) Origins and mechanisms in the establishment of cerebral dominance. In *Handbook of behavioral neurobiology* Vol. 2. *Neuropsychology*, (Ed.) GAZZANIGA, M. S., Plenum: New York, 295–343.

OBLER, L. K., ZATTORE, R. J., GALLOWAY, L., and VAID, J., (1982) Cerebral lateralization in bilinguals: methodological issues. *Brain and language*, 15, 40–54.

OBRIST, W. D., THOMPSON, W. K., WANG, H. S., and WILKINSON, W. D., (1975) Regional cerebral blood flow estimated by 133-Xe inhalation. *Stroke*, 6, 245–256.

O'CONNOR, K. P., and SHAW, J. C., (1978) Field dependence, laterality and the EEG. *Biological psychology*, 6, 93–109.

ODDY, H. C., and LOBSTEIN, T. J., (1972) Hand and eye dominance in schizophrenia. *British journal of psychiatry*, 120, 331–332.

OJEMANN, G. A., (1979) Individual variability in cortical localization of language. *Journal of neurosurgery*, 50, 164–169.

OJEMANN, G. A., and MATEER, C., (1979a) Cortical and subcortical organization of human communication: evidence from stimulation studies. In STEKLIS, H. D., and RALEIGH, M. J., (Eds.) *Neurobiology of social communication in primates.* Academic Press: London, 111–131.

OJEMANN, G. A., and MATEER, C., (1979b) Human language cortex: localisation of memory, syntax and sequential motor-phoneme identification systems. *Science*, 205, 1401–1403.

OKE, A., KELLER, R., MEFFORD, I., and ADAMS, R. N., (1978) Lateralization of norepinephrine in human thalamus. *Science*, 200, 1411–1412.

OLDFIELD, R. C., (1969) Handedness in musicians. *British journal of psychology*, 60, 91–99.

OLDFIELD, R. C., (1971) The assessment and analysis of handedness: the Edinburgh Inventory. *Neuropsychologia*, 9, 97–113.

OLSON, G. G., and LAXAR, K., (1974) Processing the terms right and left: a note on left handers. *Journal of experimental psychology*, 102, 1135–1137.

OLSON, M. E., (1973) Laterality differences in tachistoscopic word recognition in normal and delayed readers in elementary school. *Neuropsychologia*, 11, 343–350.

OLTMAN, P. K., EHRLICHMAN, H., and COX, P. W., (1977) Field independence and laterality in the perception of faces.

O'NEIL, C., STRATTON, H. T. R., INGERSOLL, R. H., and FOUTS, R. S., (1978) Conjugate lateral eye movements in Pan Troglodytes. *Neuropsychologia*, 16, 759–762.

ORBACH, J., (1967) Differential recognition of Hewbrew and English words in right and left visual fields as a function of cerebral dominance and reading habits. *Neuropsychologia*, 5, 127–134.

ORENSTEIN, H. B., and MEIGHAN, W. B., (1976) Recognition of bilaterally presented words varying in concreteness and frequency: lateral dominance or sequential processing? *Bulletin of the psychonomic society*, 7, 179–180.

ORME, J. E., (1970) Left handedness, ability and emotional instability. *British journal of social and clinical psychology*, 9, 87–88.

ORTON, S. T., (1937) *Reading, writing and speech problems in children.* Chapman and Hall: London.

OSAKA, N., (1978) Naso-temporal differences in human reaction-time in the peripheral visual field. Neuropsychologia, 16, 299–304.

OSCAR-BERMAN, M., GOODGLASS, H., and CHERLOW, D. G., (1973) Perceptual laterality and iconic recognition of visual materials by Korsakoff patients and normal adults. *Journal of comparative and physiological psychology*, 82, 316–321.

OSCAR-BERMAN, M., GOODGLASS, H., and DONNENFELD, H., (1974) Dichotic ear order effects with nonverbal stimuli. *Cortex*, 10, 270–277.

OSCAR-BERMAN, M., REHBEIN, L., PORFERT, A., and GOODGLASS, H., (1978) Dichaptic hand-order effects with verbal and non-verbal tactile stimulation. *Brain and language*, 6, 323–333.

OVERMAN, W. H., and DOTY, R. W., (1982) Hemispheric specialization displayed by man but not macaques for analysis of faces. *Neuropsychologia*, 20, 113–125.

OVERTON, W., and WIENER, M., (1966) Visual field position and word-recognition threshold. *Jouranl of experimental psychology*, 17, 249–253.

OXBURY, J. M., and OXBURY, S. M., (1969) Effects of temporal lobectomy on the report of dichotically presented digits. *Cortex*, 5, 3–14.

OLTMAN, P. K., and CAPOBIANCO, F., (1967) Field dependence and eye dominance. *Perceptual and motor skills*, 25, 645–646.

PAIVIO, A., (1971) *Imagery and visual processes.* Holt, Rinehart and Winston: New York.

PAIVIO, A., (1975) Perceptual comparisons through the mind's eye. *Memory and cognition*, 3, 635–647.

PALMER, R. D., (1964) Development of differential handedness. *Psychological bulletin*, 62, 257–272.

PAPCUN, G., KRASHEN, S., TERBEEK, D., REMINGTON, R., and HARSHMAN, R.,

(1974) Is the left hemisphere specialized for speech, language and/or something else? *Journal of the acoustical society of America*, 55, 319–328.

PARADIS, M., (1977) Bilingualism and aphasia. In *Studies in neurolinguistics*, 3, (Eds.) WHITAKER, H., and WHITAKER, H. A., Academic Press: New York, 65–115.

PARLOW, S., (1978) Differential finger movements and hand preference. *Cortex*, 14, 608–611.

PATTERSON, K., BRADSHAW, J. L., (1975) Differential hemispheric mediation of non-verbal visual stimuli. *Journal of experimental psychology: human perception and performance*, 1, 246–252.

PENFIELD, W., and RASMUSSEN, T., (1950) *The cerebral cortex of man*. Macmillan: New York.

PENFIELD, W., and ROBERTS, L., (1959) *Speech and brain mechanisms*. Princeton University Press: Princeton.

PERL, N., and HAGGARD, M., (1975) Practice and strategy in a measure of cerebral dominance. *Neuropsychologia*, 13, 341–351.

PERRIA, L., ROSADINI, G., and ROSSI, G. F., (1961) Determination of side of cerebral dominance with amobarbital. *Archives of neurology*, 4, 173–189.

PERRIS, C., MONAKHOV, K., VON KNORRING, L., BOTSKABEV, V., and NIKI-FOROV, A., (1978) Systemic structural analysis of the electroencephalogram of depressed patients. *Neuropsychobiology*, 4, 207–228.

PERRIS, C., VON KNORRING, L., and MONAKHOV, K., (1979) Functional inter-hemispheric differences in affective disorders. In OBIOLS, J., BALLUS, C., GONZALEZ-MONCLUS, E., and PUJOL, J., (Eds.) *Biological psychiatry today*, 13. Elsevier/North Holland: Amsterdam, 1237–1241.

PETERS, M., (1977) Simultaneous performance of two motor activities: the factor of timing. *Neuropsychologia*, 15, 461–465.

PETERS, M., (1980) Why the preferred hand taps more quickly than the non-preferred hand: three experiments on handedness. *Canadian journal of psychology*, 34, 62–71.

PETERS, M., (1981) Handedness effect of prolonged practice on between hand performance differences. *Neuropsychologia*, 19, 570–590.

PETERS, M., and DURDING, B. M., (1978) Handedness measured by finger tapping: a continuous variable. *Canadian journal of psychology*, 32, 257–261.

PETERS, M., and PEDERSON, K., (1978) Incidence of left handers with inverted writing position in a population of 5910 elementary school children. *Neuropsychologia*, 16, 743–747.

PETERS, M., and PETRIE, B. F., (1979) Functional asymmetries in the stepping reflex of the human neonate. *Canadian journal of psychology*, 33, 198–200.

PETERSEN, M. R., BEECHER, M. D., ZOLOTH, S. R., MOODY, D., and STEBBINS, W. L., (1978) Neural lateralization of species-specific vocalizations by Japanese macaques (Maccaca Fuscata). *Science*, 202, 324–327.

PETERSEN, G. M., (1934) Mechanisms of handedness in the rat. *Comparative psychology monographs*, 9, 1–67.

PETERSON, J. M., and LANSKY, L. M., (1974) Left handedness among architects: some facts and speculations. *Perceptual and motor skills*, 38, 2, 547–550.

PETERSON, J. M., and LANSKY, L. M., (1977) Left handedness among architects: partial replication and some new data. *Perceptual and motor skills*, 45, 1216–1218.

PETTIT, J. M., and NOLL, J. D., (1979) Cerebral dominance in aphasic recovery. *Brain and language*, 7, 191–200.

PHELPS, M. E., KUHL, D. E., and MAZZIOTTA, J. C. (1981) Metabolic mapping of the brain's response to visual stimulation: studies in humans. *Science*, 211, 1445–1448.

PHELPS, M. E., MAZZIOTTA, J. C., and HUANG, S. C., (1982) Study of cerebral function with positron computed tomography. *Journal of cerebral blood flow and metabolism*, 2, 113–162.

PHIPPARD, D., (1977) Hemifield differences in visual perception in deaf and hearing subjects. *Neuropsychologia*, 15, 555–562.

PIAZZA, D., (1977) Cerebral lateralization in young children as measured by dichotic listening and finger tapping tasks. *Neuropsychologia*, 15, 417–425.

PIAZZA, D. M., (1980) The influence of sex and handedness in the hemispheric specialization of verbal and nonverbal tasks. *Neuropsychologia*, 18, 163–176.

PIC'L, A. K., MAGARO, P. A., and WADE, E. A., (1979) Hemispheric functioning in paranoid and non-paranoid schizophrenia. *Biological psychiatry*, 14, 891–903.

PIERCY, M., (1964) The effects of cerebral lesions on intellectual function: a review of current research trends. *British journal of psychiatry*, 110, 310–352.

PINSKY, S. D., and McADAM, D. W., (1980) Electroencephalographic and dichotic indices of cerebral laterality in stutterers. *Brain and language*, 11, 374–379.

PIROT, M., PULTON, T. W., and SUTKER, L. W., (1977) Hemispheric asymmetry in reaction time to colour stimuli. *Perceptual and motor skills*, 45, 1111–1155.

PIROZZOLO, F. J., (1977) Lateral asymmetries in visual perception: a review of tachistoscopic visual half-field studies. *Perceptual and motor skills*, 45, 695–701.

PIROZZOLO, F. J., and RAYNER, K., (1977) Hemispheric specialization in reading and word recognition. *Brain and language*, 4, 248–261.

PIROZZOLO, F. J., and RAYNER, K., (1979) Cerebral organization and reading disability. *Neuropsychologia*, 17, 485–491.

PIROZZOLO, F. J., and RAYNER, K., (1980) Handedness, hemispheric specialization and saccadic eye movement latencies. *Neuropsychologia*, 18, 225–229.

PITBLADO, C., (1979a) Visual field differences in perception of the vertical with and without a visible frame of reference. *Neuropsychologia*, 17, 381–392.

PITBLADO, C. B., (1979b) Cerebral asymmetries in random-dot stereopsis: reversal of direction with changes in dot size. *Perception*, 8, 683–690.

PIZZAMIGLIO, L., (1974) Handedness, ear preference and field dependence-independence. *Perceptual and motor skills*, 38, 700–702.

PIZZAMIGLIO, L., and CHECCHINE, M., (1971) Development of the hemispheric dominance in children from 5 to 10 years of age and their relations with development of cognitive processes. *Brain research*, 31, 363.

PIZZAMIGLIO, L., DE PASCALIS, C., and VIGNATI, A., (1974) Stability of dichotic listening test. *Cortex*, 10, 203–205.

PIZZAMIGLIO, L, and ZOCCOLOTTI, P., (1981) Sex and cognitive influence on visual hemifield superiority for face and letter recognition. *Cortex*, 17, 215–226.

POECK, K., and HUBER, W., (1977) To what extent is language a sequential activity? *Neuropsychologia*, 15, 359–363.

POFFENBERGER, A. T., (1912) Reaction time to retinal stimulation with special reference to the time lost in conduction through nerve centers. *Archives of psychology*, 23, 1–73.

POIZNER, H., BATTISON, R., and LANE, H., (1979) Cerebral asymmetry for American sign language: the effects of moving stimuli. *Brain and language*, 7, 351–362.

POIZNER, H., and LANE, H., (1979) Cerebral asymmetry in the perception of American sign language. *Brain and language*, 7, 210–266.

POLICH, J. M., (1980) Left hemisphere superiority for visual search. *Cortex*, 16, 39–50.

POLYAK, S., (1957) *The vertebrate visual system*. University of Chicago Press: Chicago.

PORAC, C., and COREN, S., (1975) Is eye dominance a part of generalized laterality? *Perceptual and motor skills*, 40, 363–769.

PORAC, C., and COREN, S., (1976) The dominant eye. *Psychological bulletin*, 83, 880–897.

PORAC, C., and COREN, S., (1979a) Individual and familial patterns in four dimensions of lateral preference. *Neuropsychologia*, 17, 543–548.

PORAC, C., and COREN, S., (1979b) Monocular asymmetries in recognition after an eye movement: sighting dominance and dextrality. *Perception and psychophysics*, 25, 55–59.

PORAC, C., COREN, S., STEIGER, J. H., and DUNCAN, P., (1980) Human laterality: a multidimensional approach. *Canadian journal of psychology*, 34, 91–96.

PORTER, R. J., and BERLIN, C. I., (1975) On interpreting changes in the dichotic right-ear advantage. *Brain and language*, 2, 186–200.

POSNER, M. I., and MITCHELL, R. F., (1967) Chronometric analysis of classification. *Psychological review*, 74, 392–409.

PRATT, R. T. C., and WARRINGTON, E. K., (1972) The assessment of cerebral dominance with unilateral ECT. *British journal of psychiatry*, 121, 327–328.

PRATT, R. T. C., WARRINGTON, E. K., and HALLIDAY, A. M., (1971) Unilateral ECT as a test for cerebral dominance with a strategy for treating left handers. *British journal of psychiatry*, 119, 79–83.

PREILOWSKI, B., (1972) Possible contribution of the anterior forebrain commissures to bilateral motor co-ordination. *Neuropsychologia*, 10, 267-277

PREILOWSKI, B., (1975) Bilateral motor interaction: perceptual-motor performance of partial and complete split-brain patients. In *Cerebral localization*. (Eds.) ZULCH, K. J., CREUTZFELDT, D., and GALBRAITH, G. C., Springer Verlag: New York, 115-132.

PRING, T. R., (1981) The effect of stimulus size and exposure duration on visual field asymmetries. *Cortex*, 17, 227-240.

PRIOR, M. R., and BRADSHAW, J. L., (1979) Hemisphere functioning in autistic children. *Cortex*, 15, 73-81.

PROVINS, K. A., (1967) Motor skills, handedness and behaviour. *Australian journal of psychology*, 19, 137-150.

PROVINS, K. A., and CUNLIFFE, P., (1972) The relationship between EEG activity and handedness. *Cortex*, 8, 136-146.

PUCETTI, R., (1981) The case for mental duality: evidence from split-brain data and other considerations. *The behavioral and brain sciences*, 4, 93-124.

PYLYSHYN, Z. W., (1973) What the mind's eye tells the mind's brain: a critique of mental imagery. *Psychological bulletin*, 80, 1-24.

RABBITT, P. M. A, (1965) Response facilitation on repetition of a limb movement. *British journal of psychology*, 56, 303-304.

RABBITT, P. M. A., (1978) Hand dominance, attention and the choice between responses. *Quarterly journal of experimental psychology*, 30, 407-416.

RACZOWSKI, D., KALAT, J. W., and NEBES, R., (1974) Reliability and validity of some handedness questionnaire items. *Neuropsychologia*, 12, 43-47.

RAMSAY, D. S., (1979) Manual preference for tapping in infants. *Developmental psychology*, 15, 437-442.

RAMSAY, D. S., CAMPOS, J. J., and FENSON, L., (1979) Onset of bimanual handedness in infants. *Infant behavior development*, 2, 69-76.

RANKIN, J. M., ARAM, D. M., and HORWITZ, S. J. (1981) Language ability in right and left hemiplegic children. *Brain and language*, 14, 292-306.

RAO, S. M., ROURKE, D., and WHITMAN, R. D., (1981) Spatio-temporal discrimination of frequency in the right and left visual fields: a preliminary report. *Perceptual and motor skills*, 53, 311-376.

RASMUSSEN, T., and MILNER, B., (1975) Clinical and surgical studies of the cerebral speech areas in man. In *Cerebral localization*. (Eds.) ZULCH, K. J., CREUTZFELDT, O., and GALBRAITH, G. C., Springer Verlag: New York, 238-255.

RATCLIFF, G., DILA, C., TAYLOR, L., and MILNER, B., (1980) The morphological asymmetry of the hemispheres and cerebral dominance for speech: a possible relationship. *Brain and language*, 11, 87-98.

RAY, W J., GEORGIOU, S., and RAVIZZA, R., (1979) Spatial abilities, sex differences and lateral eye movements. *Developmental psychology*, 15, 455-457.

RAY, W. M., MORELL, M., FREDIANI, A. W., and TUCKER, D., (1976) Sex differences and lateral specialization of hemispheric functioning. *Neuropsychologia*, 14, 391-393.

REBERT, C., and MAHONEY, R., (1978) Functional cerebral asymmetry and performance. III. Reaction time as a function of task, hand, sex and EEG asymmetry. *Psychophysiology*, 15, 9-16.

REITAN, M., (1955) Certain differential effects of left and right cerebral lesions in human adults. *Journal of comparative and physiological psychology*, 48, 474-477.

REIVICH, M., GUR, R., and ALAVI, A., (1983) Positron emission tomographic studies of sensory stimuli, cognitive processes and anxiety. *Human neurobiology*, 2, 25-33.

REMINGTON, R., KRASHEN, S., and HARSHMAN, R., (1974) A possible sex difference in degree of lateralization of dichotic stimuli? *Journal of the acoustical society of America*, 55, 434.

REPP, B., (1977) Dichotic competition for speech sounds: the role of acoustic stimulus structure. *Journal of experimental psychology: human performance and perception*, 3, 37-50.

REUTER-LORENZ, P., and DAVIDSON, R. J., (1981) Differential contributions of the two cerebral hemispheres to the perception of happy and sad faces. *Neuropsychologia*, 19, 609-614.

REYNOLDS, D. McQ, and JEEVES, M. A., (1977) Further studies of tactile perception and motor co-ordination in agenesis of the corpus callosum. *Cortex*, 13, 257–272.

RICHARDSON, J. T. E, (1975) The effect of word imageability in acquired dyslexia. *Neuropsychologia*, 13, 281–288.

RICHARDSON, J. T. E., (1976) How to measure laterality. *Neuropsychologia*, 14, 135–136.

RICHARDSON, J. T. E., (1978) A factor-analysis of self-reported handedness. *Neuropsychologia*, 16, 747–748.

RICHARDSON, J. T. E., and FIRLEJ, M. D. E., (1979) Laterality and reading attainment. *Cortex*, 15, 581–596.

RIFE, D. C., (1940) Handedness, with special reference to twins. *Genetics*, 25, 178–186.

RISBERG, J., HALSEY, J. H., BLAUENSTEIN, V. W., et al., (1975) Bilateral measurements of the rCBF during mental activation in normals and in dysphasic patients. In HARPER, A. M., et al., (Eds.) *Blood flow and metabolism in the brain*. London: Churchill Livingston.

RISSE, G. L., and GAZZANIGA, M. S., (1976) Verbal retrieval of right hemisphere memories established in the absence of language. *Neurology*, 26, 354.

RISSE, G. L., LE DOUX, J., SPRINGER, S. P., WILSON, D. H., and GAZZANIGA, M. S., (1978) The anterior commissure in man: functional variation in a multisensory system. *Neuropsychologia*, 16, 23–32.

RIZZOLATTI, G., BERTOLINI, G., and BUCHTEL, H. A., (1979) Interference of concomitant motor and verbal tasks on simple reaction time: a hemispheric difference. *Neuropsychologia*, 17, 323–320.

RIZZOLATTI, G., and BUCHTEL, H. A., (1977) Hemispheric superiority in reaction time to faces: a sex difference. *Cortex*, 13, 300–305.

RIZZOLATTI, G., UMILTÀ, C., and BERLUCCHI, G., (1971) Opposite superiorities of the right and left cerebral hemispheres in discriminative reaction time to physiognomic and alphabetical material. *Brain*, 94, 431–442.

ROBERTS, L., (1969) Aphasia, Apraxia and Agnosia in abnormal states of cerebral dominance. In *Handbook of clinical neurology*, 4, (Eds.) VINKEN, P. J., and BRUYN, G. W., North Holland Publishing Co.: Amsterdam, 312–326.

ROBERTSHAW, S., and SHELDON, M., (1976) Laterality effects in judgement of the identity and position of letters: a signal detection analysis. *Quarterly journal of experimental psychology*, 28, 115–122.

ROBINSON, G. M., and SOLOMON, D. J., (1974) Rhythm is processed by the speech hemisphere. *Journal of experimental psychology*, 102(3), 508–511.

ROBINSON, J. S., and VONEIDA, T. J., (1970) Quantitative differences in performance on abstract discrimination using one or both hemispheres. *Experimental neurology*, 26, 72–83.

ROBINSON, J. S., and VONEIDA, T. J., (1973) Hemisphere differences in cognitive capacity in the split-brain cat. *Experimental neurology*, 38, 123–134.

ROBINSON, R. G., (1979) Differential behavioral and biochemical effects of right and left hemispheric cerebral infarction in the rat. *Science*, 205, 707–710.

ROBINSON, R. G., and BLOOM, F. E., (1977) Pharmacological treatment following experimental infarction: implications for understanding psychological symptoms of human stroke, *Biological psychiatry*, 12, 669–680.

ROBINSON, R. G., and BENSON, D. F., (1981) Depression in aphasic patients: frequency, severity, and clinico-pathological correlations. *Brain and language*, 14, 282–291.

ROBINSON, T. E., BECKER, J. B., and RAMIREZ, V. D., (1980) Sex differences in amphetamine-elicited rotational behavior and the lateralization of striatal dopamine in rats. *Brain research bulletin*, 5, 539–545.

ROGERS, L. J., (1981) Environmental influences on brain lateralization. *The behavioral and brain sciences*, 4, 35–36.

ROGERS, L. J., and ANSON, J. M., (1979) Lateralisation of function in the chicken forebrain. *Pharmacology, biochemistry and behavior*, 10, 679–686.

ROSADINI, G., and ROSSI, G. F., (1967) On the suggested cerebral dominance for consciousness. *Brain*, 80, 101–112.

ROSENFIELD, D. B., and GOODGLASS, H., (1980) Dichotic testing of cerebral dominance in stutterers. *Brain and language*, 11, 170–180.

ROSENTHAL, R., and BIGELOW, L. B., (1972) Quantitative brain measurements in chronic schizophrenia. *British journal of psychiatry*, 121, 259–264

ROSENZWEIG, M. R., (1951) Representations of the two ears at the auditory cortex. *American journal of physiology*, 167, 147–158.

ROSS, E. D., and MESULAM, M. M., (1979) Dominant language functions of the right hemisphere? Prosody and emotional gesturing. *Archives of neurology*, 36, 144–148.

ROSS, P., PERGAMENT, L., and ANISFIELD, M., (1979) Cerebral lateralization of deaf and hearing individuals for linguistic comparison judgements. *Brain and language*, 8, 69–80.

ROSS, P., and TURKEWITZ, G., (1981) Individual differences in cerebral asymmetries for facial recognition. *Cortex*, 17, 199–214.

ROSSI, G. F., and ROSADINI, G., (1967) Experimental analysis of cerebral dominance in man. In *Brain mechanisms underlying speech and language*. (Eds.) MILLIKAN, C. H., and DARLEY, F. L., Grune and Stratton: New York, 167–184.

ROTTER, J. B., (1966) Generalized expectancies for internal versus external control of reinforcement. *Psychological monographs*, 80, Whole No. 609.

RUBENS, A. B., (1977) Asymmetries of human cerebral cortex. In HARNAD, S., DOTY, R. W., GOLDSTEIN, L., JAYNES, J., and KRAUTHAMER, (Eds.) *Lateralization in the nervous system*, Academic Press, Chap. 26, 503–516.

RUBENS, A. B., MAHOWALD, M. W., and HUTTON, J. T., (1976) Asymmetry of the lateral (Sylvian) fissures in man. *Neurology*, 26, 620–624.

RUDEL, G., DENCKLA, M. B., and HIRSCH, S., (1977) The development of left-handed superiority for discriminating Braille configurations. *Neurology*, 27, 160–164.

RUDEL, R. G., DENCKLA, M. B., and SPALTEN, E., (1974) The functional asymmetry of Braille letter learning in normal sighted children. *Neurology*, 24, 733–738.

RUDEL, R. G., TEUBER, H. L., and TWITCHELL, T. E., (1974) Levels of impairment of sensori-motor functions in children with early brain damage. *Neuropsychologia*, 12, 95–108.

RUGG, M. D., (1982) Electrophysiological studies. In BEAUMONT, J. G. (Ed.) *Divided visual field studies of cerebral organization*. Academic Press: London, Chap. 12, 249–252.

RUGG, M. D., and BEAUMONT, J. G., (1978) Interhemispheric asymmetries in the visual evoked response: effect of stimulus lateralisation and task: *Biological psychology*, 6, 283–292.

RUSSELL, J. R., and REITAN, R. M., (1955) Psychological abnormalities in agenesis of the corpus callosum. *Journal of nervous and mental disease*, 121, 675–685.

RUSSELL, W. R., and ESPIR, M. L. E., (1961) *Traumatic aphasia*. Oxford University Press: London.

SACKHEIM, H. A., GREENBERG, M. S., WEIMAN, A. L., GUR, R. C., HUNGERBUHLER, J. P., and GESCHWIND, N., (1982) Hemispheric asymmetry in the expression of positive and negative emotions. *Archives of neurology*, 39, 210–218.

SACKEIM, H. A., and GUR, R. C., (1978) Lateral intensity of emotional expression. *Neuropsychologia*, 16, 473–482.

SACKEIM, H. A., and GUR, R. C., (1980) Asymmetry in facial expression. *Science*, 209, 834–836.

SACKEIM, H. A., GUR, R. C., SAUCY, M. C., (1978) Emotions are expressed more intensely on the left side of the face. *Science*, 202, 434–435.

SADICK, T. L., and GINSBURG, B. E., (1978) The development of the lateral functions and reading ability. *Cortex*, 14, 3–11.

SAFER, M. A., (1981) Sex and hemisphere differences in access to codes processing emotional expressions and faces. *Journal of experimental psychology: general*, 110, 86–100.

SAFER, M. A., and LEVENTHAL, H., (1977) Ear difference in evaluating emotional tones of voice and verbal content. *Journal of experimental psychology: human perception and performance*, 3, 75–82.

SAFFRAN, M., BOGYO, L. C., SCHWARTZ, M. F., and MARIN, O. S. M., (1980) Does deep dyslexia reflect right-hemisphere reading? In *Deep dyslexia*. (Eds.) COLTHEART, M., PATTERSON, K., and MARSHALL, J. C., Routledge and Kegan Paul: London, 381–403.

SAMPSON, H., (1969) Recall of digits projected to temporal and nasal hemi-retinas. *Quarterly journal of experimental psychology*, 21, 39–42.

SARING, W., and VON CARMON, D., (1980) Is there an interaction between cognitive activity and lateral eye movements? *Neuropsychologia*, 18, 591–596.

SASANUMA, S., (1975) KANA and KANJI processing in Japanese aphasics. *Brain and language*, 2, 369–383.

SASANUMA, S., ITOH, M., MORI, K., and KOBAYASHI, Y., (1977) Tachistoscopic recognition of Kana and Kanji words. *Neuropsychologia*, 15, 547–553.

SASANUMA, S., and KOBAYASHI, Y., (1978) Tachistoscopic recognition of line orientation. *Neuropsychologia*, 16, 239–242.

SATZ, P., (1972) Pathological left handedness: an explanatory model. *Cortex*, VIII, 121–135.

SATZ, P., (1973) Left handedness and early brain insult: an explanation. *Neuropsychologia*, 11, 115–117.

SATZ, P., (1976) Cerebral dominance and reading disability: an old problem revisited. In KNIGHT, R. M., and BAKKER, D. J., (Eds.) *The neuropsychology of learning disorders.* University Park Press: Baltimore, 273–294.

SATZ, P., (1977) Laterality tests: an inferential problem. *Cortex*, 13, 208–212.

SATZ, P., (1979) A test of some models of hemispheric speech organization in the left and right handed. *Science*, 203, 1131–1133.

SATZ, P., (1980) Incidence of aphasia in left handers: a test of some hypothetical models of cerebral speech organization. In HERRON, J., (Ed.) *The neuropsychology of left handedness.* Academic Press: New York, 189–198.

SATZ, P., ACHENBACH, K., and FENNELL, E., (1967) Correlations between assessed manual laterality and predicted speech laterality in a normal population. *Neuropsychologia*, 5, 295–310.

SATZ, P., ACHENBACH, K., PATTISHALL, E., and FENNELL, E., (1965) Order of report, ear asymmetry and handedness in dichotic listening. *Cortex*, 1, 377–396.

SATZ, P., BAKKER, D. J., TEUNISSEN, J., GOEBEL, R., and VAN DER VLUGT, H., (1975) Developmental parameters of the ear asymmetry: a multivariate approach. *Brain and language*, 2, 171–185.

SATZ, P., BAYMUR, L., and VAN DER VLUGT, H., (1979) Pathological left handedness: cross cultural tests of a model. *Neuropsychologia*, 17, 77–82.

SATZ, P., FENNELL, E., and JONES, M. W., (1968) Comments on: a model of the inheritance of handedness and cerebral dominance. *Neuropsychologia*, 7, 101–103.

SATZ, P., LEVY, C. M., and TYSON, M., (1970) Effects of temporal delays on the ear asymmetry in dichotic listening. *Journal of experimental psychology*, 84, 372–374.

SATZ, P., RARDIN, D., and ROSS, J., (1971) An evaluation of a theory of specific developmental dyslexia. *Child development*, 42, 2009–2021.

SATZ, P., and SPARROW, S. S., (1970) Specific developmental dyslexia: a theoretical formulation. In BAKKER, D. J., and SATZ, P., (Eds.) *Specific reading disability: advances in theory and method.* Rotterdam University Press: Rotterdam, 17–40.

SAUERWEIN, H. C., LASSONDE, M. C., CARDU, B., and GEOFFREY, G., (1981) Interhemispheric integration of sensory and motor functions in agenesis of the corpus callosum. *Neuropsychologia*, 19, 445–454.

SAUL, R., and SPERRY, R. W., (1965) Absence of commissurotomy symptoms with agenesis of the corpus callosum. *Neurology*, 18, 307.

SCHACTER, S., (1959) *The psychology of affiliation.* Stanford University Press: Stanford, Ca.

SCHACHTER, S., (1975) Cognition and peripheralist-centralist controversies in motivation and emotion. In GAZZANIGA, M. S., and BLAKEMORE, C., (Eds.) *Handbook of psychobiology*, Academic Press: New York, 529–564.

SCHAFER, E. W. P, (1967) Cortical activity preceding speech: semantic specificity. *Nature*, 216, 1338–1339.

SCHALLER, G. B., (1963) *The mountain gorilla.* Chicago University Press: Chicago.

SCHLANGER, B. B., SCHLANGER, P., and GERSTMANN, L. J., (1976) The perception of emotionally toned sentences by right-hemisphere-damaged and aphasic subjects. *Brain and language*, 3, 396–403.

SCHMIT, V., and DAVIS, R., (1974) The role of hemispheric specialisation in the analysis of Stroop stimuli. *Acta psychologica*, 38, 149–158.

SCHMULLER, J., (1980) Stimulus repetition in studies of laterality. *Brain and language*, 10, 205–207.

SCHMULLER, J., and GOODMAN, R., (1979) Bilateral tachistoscopic perception, handedness and laterality. *Brain and language*, 8, 81–91.

SCHMULLER, J., and GOODMAN, R., (1980) Bilateral tachistoscopic perception, handedness and laterality. II. Non-verbal stimuli. *Brain and language*, 11, 12–18.

SCHOLES, R. J., and FISCHLER, I , (1979) Hemispheric function and linguistic skill in the deaf. *Brain and language*, 7, 336–350.

SCHULHOFF, C , and GOODGLASS, H , (1969) Dichotic listening, side of brain injury and cerebral dominance. *Neuropsychologia*, 7, 149–160.

SCHULMAN-GALAMBOS, C., (1977) Dichotic listening performance in elementary and college students. *Neuropsychologia*, 15, 577–584.

SCHWARTZ, B., (1967) Hemispheric dominance and consciousness. *Acta Neurologica Scandinavica*, 43, 513–525.

SCHWARTZ, G. E., AHERN, G. L., and BROWN, S. L., (1979) Lateralized facial muscle response to positive and negative emotional stimuli. *Psychophysiology*, 16, 561–571.

SCHWARTZ, G. E., BROWN, S. L., and AHERN, G. L., (1980) Facial muscle patterning and subjective experience during affective imagery: sex differences. *Psychophysiology*, 17, 75–82.

SCHWARTZ, G. E., DAVIDSON, R. J., and MAER, F., (1975) Right hemisphere lateralization for emotion in the human brain: interactions with cognition. *Science*, 190, 286–288.

SCHWARTZ, M., (1977) Left-handedness and high-risk pregnancy. *Neuropsychologia*, 15, 341–344.

SCHWEITZER, L., (1979) Differences of cerebral lateralization among schizophrenic and depressed patients. *Biological psychiatry*, 14, 721–733.

SCHWEITZER, L., BECKER, E., and WELSH, M., (1978) Abnormalities of cerebral lateralization in schizophrenia patients. *Archives of general psychiatry*, 35, 982–985.

SCHWEITZER, L., and CHACKO, R., (1980) Cerebral lateralization: relation to subject's sex. *Cortex*, 16, 559–566.

SEAMON, J. G., and GAZZANIGA, M. S., (1973) Coding strategies and cerebral laterality effects. *Cognitive psychology*, 1973, 5, 249–256.

SEARLEMAN, A., (1977) A review of right hemisphere linguistic capabilities. *Psychological bulletin*, 84, 503–528.

SEARLEMAN, A., (1980) Subject variables and cerebral organisation for language. *Cortex*, 16, 239–254

SEARLEMAN, A., TWEEDY, J., and SPRINGER, S. P., (1979) Inter-relationships among subject variables believed to predict cerebral organisation. *Brain and language*, 7, 267–276.

SEGALOWITZ, S. J., and GRUBER, F. A., (1977) The development of cerebral dominance: why is language lateralized to the left? In *Language development and neurological theory*. (Eds ) SEGALOWITZ, S. I., and GRUBER, F. A., Academic Press: New York and London, 165–169.

SEGALOWITZ, S. J., and STEWART, C., (1979) Left and right lateralization for letter matching: strategy and sex differences. *Neuropsychologia*, 17, 521–525.

SELNES, O. A , (1974) The corpus callosum: some anatomical and functional considerations with special reference to language. *Brain and language*, 1, 111–139.

SEMMES, J., (1968) Hemispheric specialisation: a possible clue to mechanism. *Neuropsychologia*, 6, 11–26.

SEMMES, J., WEINSTEIN, S., GHENT, L., and TEUBER, H. L., (1955) Spatial orientation in man after cerebral injury I. Analysis by locus of lesion. *The journal of psychology*, 39, 227–244.

SEMMES, J., WEINSTEIN, S., GHENT, L., and TEUBER, H. L., (1960) *Somatosensory changes after penetrating brain wounds in man*. Cambridge: Harvard University Press.

SERAFETINIDES, E. A., (1973) Voltage laterality in the EEG of psychiatric patients. *Diseases of the nervous system*, 34, 190–191.

SERAFETINIDES, E A., HOARE, R. D., and DRIVER, M., (1965) Intracarotid sodium amylobarbione and cerebral dominance for speech and consciousness. *Brain*, 88, 107–130.

SERGENT, J., and BINDRA, D., (1981) Differential hemispheric processing of faces: methodological considerations and reinterpretation. *Psychological bulletin*, 89, 541–554.

SERON, X., VAN DER KAA, M. A., REMITZ, A., and VAN DER LINDEN, M., (1979) Pantomime interpretation and aphasia, *Neuropsychologia*, 17, 661–668.

SETH, G., (1973) Eye-hand co-ordination and handedness: a developmental study of visuo-motor behaviour in infancy. *British journal of educational psychology*, 43, 35–49.

SHAGASS, C., JOSIASSEN, R. C., ROEMER, R. A., STRAUMANIS, J. J., and SLEPNER, S. M., (1983) Failure to replicate evoked potential observation suggesting corpus callosum dysfunction in schizophrenia. *British journal of psychiatry*, 142, 471–476.

SHANKWEILER, D., (1966) Effects of temporal lobe damage on perception of dichotically presented melodies. *Journal of comparative and physiological psychology*, 62, 115–119.

SHANKWEILER, D., (1971) An analysis of laterality effects in speech perception. In *The perception of language*. (Eds.) HORTON, D. L., and JENKINS, J. J., Columbus, Ohio: Chas. E. Merrill, 185–200.

SHANKWEILER, D., and STUDDERT-KENNEDY, M., (1967) Identification of conso-nants and vowels presented to left and right ears. *Quarterly journal of experimental psychology*, 19, 59–63.

SHANKWEILER, D., and STUDDERT-KENNEDY, M., (1975) A continuum of laterali-zation for speech perception? *Brain and language*, 2, 212–225.

SHANON, B., (1978) Writing positions in Americans and Israelis. *Neuropsychologia*, 16, 587–591.

SHANON, B., (1979a) Graphological patterns as a function of handedness and culture. *Neuropsychologia*, 17, 457–465.

SHANON, B., (1979b) Lateralization effects in response to words and non-words. *Cortex*, 15, 541–550.

SHAPIRO, B. E., GROSSMAN, M., and GARDNER, H., (1981) Selective musical processing deficits in brain damaged populations. *Neuropsychologia*, 19, 161–169.

SHAW, J. C., BROOKS, S., COLTER, N., and O'CONNOR, K. P., (1979) A comparison of schizophrenic and neurotic patients using EEG power and coherence spectra. In GRUZELIER, J., and FLOR-HENRY, P., (Eds.) *Hemisphere asymmetries of function in psychopathology*. Elsevier/North Holland Biomedical Press: Amsterdam, 257–284.

SHEREMATA, W. A., DEONNA, T. W., and ROMANUL, F. C. A., (1973) Agenesis of the corpus callosum and inter-hemispheric transfer of information. *Neurology*, 23, 390.

SHEPPARD, G., GRUZELIER, J., MANCHANDA, R., et al., (1983) 15-0 positron emission tomographic scanning in predominantly never-treated acute schizophrenic patients. *The Lancet*, 2.2, 1448–1452.

SHERMAN, J. L., KULHAVY, R. W., and BURNS, K., (1976) Cerebral laterality and verbal processes. *Journal of experimental psychology: human language and memory*, 2, 720–727.

SHERWIN, I., and EFRON, R., (1980) Temporal ordering deficits following anterior temporal lobectomy. *Brain and language*, 11, 195–203.

SHIMKUNAS, A., (1978) Hemispheric asymmetry and schizophrenic thought disorder. In SCHWARTZ, S., (Ed.) *Language and cognition in schizophrenia*. Lawrence Erlbaum Associates: New Jersey, Chap. 7, 193–235.

SHLAER, R., (1971) Shift in binocular disparity causes compensatory change in the cortical structure of kittens. *Science*, 173, 638–641.

SHUCARD, D. W., CUMMINS, K. R., THOMAS, D. G., and SHUCARD, J. L., (1981) Evoked potentials to auditory probes as indices of cerebral specialisation of func-tion: replication and extensiuon.*Electroencephalography and clinical neurophysiology*, 52, 389–393.

SHUKLA, G. D., and KATIYAR, B. C., (1980) Psychiatric disorders in temporal lobe epilepsy: the laterality effect. *British journal of psychiatry*, 137, 181–182.

SHUKLA, G. D., SAHU, S. C., TRIPATHI, R. P., and GUPTA, D. K., (1982) A psychiatric study of amputees. *British journal of psychiatry*, 141, 50–53.

SIDTIS, J. J., (1980) On the nature of the cortical function underlying right hemisphere auditory perception. *Neuropsychologia*, 18, 321–330.

SIDTIS, J. J., and BRYDEN, M. P., (1978) Asymmetrical perception of language and music: evidence for independent processing strategies. *Neuropsychologia*, 16, 627–632.

SILVA, D. A., and SATZ, P., (1979) Pathological left handedness: evaluation of a model. *Brain and language*, 7, 8–16.

SILVERBERG, R., BENTIN, S., GAZIEL, T., OBLER, L. K., and ALBERT, M. L., (1979) Shift of visual field preference for English words in native Hebrew speakers. *Brain and language*, 8, 184–190.

SILVERBERG, R., GORDON, H. W., POLLACK, S., and BENTIN, S., (1980) Shift of visual field preference for Hebrew words in native speakers learning to read. *Brain and language*, 11, 99–105.

SILVERBERG, R., OBLER, L. K., and GORDON, H. W., (1979) Handedness in Israel. *Neuropsychologia*, 17, 83–88.

SILVERMAN, A. J., ADEVAI, G., and McGOUGH, W. E., (1966) Some relationships between handedness and perception. *Journal of psychosomatic research*, 10, 151–158.

SIMERNITSKAYA, E. G., (1974) On two forms of writing defect following brain lesions. In DIMOND, S. J., and BEAUMONT, J. G., (Eds.) *Hemisphere function in the human brain*. Elek: London, 335–344.

SIMION, F., BAGNARA, S., BISIACCHI, P., RONCATO, S., and UMILTÀ, C., (1980) Laterality effects, levels of processing and stimulus properties. *Journal of experimental psychology: human perception and performance*, 6, 184–195.

SIMON, J. R., (1967) Ear preference in a simple reaction time task. *Journal of experimental psychology*, 75, 49–55.

SKLAR, B., HANLEY, J., and SIMONS, W. W., (1972) An EEG experiment aimed toward identifying dyslexic children. *Nature*, 240, 414–416.

SLORACH, N., and NOEHR, B., (1973) Dichotic listening in stuttering and dyslabic children. *Cortex*, 9, 295–300.

SMITH, A., (1966) Speech and other functions after left (dominant) hemispherectomy. *Journal of neurology, neurosurgery and psychiatry*, 29, 467–471.

SMITH, A., (1972) Dominant and non-dominant hemispherectomy. In W. LYNN SMITH, (Ed.) *Drugs, development and cerebral function.*, Chap. 3, Thomas, C. Springfield, Illinois, 37–68.

SMITH, A., (1974) Dominant and non-dominant hemispherectomy. In KINSBOURNE, M., and SMITH, W. L., (Eds.) *Hemispheric disconnection and cerebral function*. Charles C. Thomas: Springfield, Illinois, 5–33.

SMITH, A., and BURKLAND, C. W., (1966) Dominant hemispherectomy: preliminary report on neuropsychological sequelae. *Science*, 153, 1280–1282.

SMITH, A., and SUGAR, O., (1975) Development of above normal language and intelligence 21 years after left hemispherectomy. *Neurology*, 25, 813–818.

SMITH, K. N., and AKELAITIS, A. J., (1942) Studies on the corpus callosum 1. Laterality in behavior and bilateral motor organization in man before and after section of the corpus callosum. *Archives of neurology and psychiatry*, 47, 519–543.

SMITH, L. C., and MOSCOVITCH, M., (1979) Writing posture, hemispheric control of movement and cerebral dominance in individuals with inverted and non-inverted hand postures during writing. *Neuropsychologia*, 17, 637–644.

SMITH, M. O., CHU, J., and EDMONTON, W. E., (1977) Cerebral lateralization of haptic perception: interaction of responses to Braille and music reveals a functional basis. *Science*, 197, 689–690.

SNYDER, S. H., (1977) Opiate receptors and internal opiates. *Scientific American*, 236, 44–56.

SPARKS, R., and GESCHWIND, N., (1968) Dichotic listening in man after section of neocortical commissures. *Cortex*, 4, 3–16.

SPARKS, R., GOODGLASS, H., and NICKEL, B., (1970) Ipsilateral versus contralateral extinction in dichotic listening resulting from hemisphere lesions. *Cortex*, 6, 249–260.

SPARKS, R., HELM, N., and ALBERT, M., (1974) Aphasia rehabilitation resulting from melodic intonation therapy. *Cortex*, 10, 303–316.

SPARROW, S. S., and SATZ, P., (1970) Dyslexia, laterality and neuropsychological development. In *Specific Reading Disability: advances in theory and method*, (Eds.) BAKKER, D. J., and SATZ, P., Chap. 2, Rotterdam University Press, 41–60.

SPELLACY, F., and BLUMSTEIN, S., (1970) The influence of language set on ear preference in phoneme recognition. *Cortex*, 6, 430–439.

SPERLING, G., (1960) The information available in brief visual presentations. *Psychological monographs*, 74 whole no. 498.

SPERLING, G., (1963) A model for visual memory tasks. *Human factors*, 5, 19–31.

SPERLING, G., (1967) Successive approximations to a model for short-term memory. *Acta psychologica*, 27, 285–292.

SPERRY, R. W., (1958) Corpus callosum and interhemispheric transfer in the monkey (Macaca Mulatta). *Anatomical record*, 131, 297.

SPERRY, R. W., (1968) Hemisphere deconnection and unity in conscious awareness. *American psychologist*, 23, 723–733.

SPERRY, R. W., (1974) Lateral specialization in the surgically separated hemispheres. In *The neurosciences: third study program*. (Eds.) SCHMITT, F. D., and WORDEN, F. G., Cambridge, Massachusetts: MIT Press, 5–19.

SPERRY, R. W., and GAZZANIGA, M. S., (1967) Language following disconnection of the hemispheres. In MILLIKAN, C. H., and DARLEY, F. L., (Eds.) *Brain mechanisms underlying speech and language*. Grune and Stratton: New York and London, 108–116.

SPERRY, R. W., GAZZANIGA, M. S., and BOGEN, J. E., (1969) Interhemispheric relationships: the neocortical commissures. Syndromes of hemisphere disconnection, In VINKEN, P. J., and BRUYN, G. W., (Eds.) *Handbook of clinical neurology*, 4, *Disorders of speech, perception and symbolic behaviour*. North Holland Publishing Co.: Amsterdam, 273–290.

SPERRY, R. W., STAMM, J. S., and MINER, N., (1956) Relearning tests for interocular transfer following division of optic chiasma and corpus callosum in cats. *Journal of comparative and physiological psychology*, 49, 529–533.

SPERRY, R. W., ZAIDEL, E., and ZAIDEL, D., (1979) Self recognition and awareness in the deconnected minor hemisphere. *Neuropsychologia*, 17, 153–166.

SPRINGER, S., (1971) Ear asymmetry in a dichotic detection task. *Perception and psychophysics*, 10, 239–241.

SPRINGER, S. P., (1973) Hemispheric specialization for speech opposed by contralateral noise: *Perception and psychophysics*, 13, 391–393.

SPRINGER, S. P., (1979) Speech perception and the biology of language. In *Handbook of behavioral neurobiology*, Vol. 2, *Neuropsychology*, (Ed.) GAZZANIGA, M. S., Plenum: New York, Chap. 7, 153–176.

SPRINGER, S. P., and GAZZANIGA, M. S., (1975) Dichotic testing of partial and complete split-brain subjects. *Neuropsychologia*, 13, 341–346.

SPRINGER, S. P., and SEARLEMAN, A., (1978) The ontogeny of hemispheric specialization: evidence from dichotic listening in twins. *Neuropsychologia*, 16, 269–282.

SPRINGER, S. P., and SEARLEMAN, A., (1980) Left handedness in twins: implications for the mechanisms underlying asymmetry of function. In HERRON, J., (Ed.) *The neuropsychology of left handedness*. Academic Press: New York, 139–158.

SPRINGER, S. P., SIDTIS, J., WILSON, D., and GAZZANIGA, M. S., (1978) Left ear performance in dichotic listening following commissurotomy. *Neuropsychologia*, 16, 305–312.

STAMM, J. S., ROSEN, S. C., and GADOTTI, A., (1977) Lateralization of functions in the monkey's frontal cortex. In HARNAD, S., DOTY, R. W., GOLDSTEIN, L., JAYNES, J., and KRAUTHAMER, G., (Eds.) *Lateralization in the nervous system*. Academic Press: New York, 385–402.

STAMM, J. S., and SPERRY, R. W., (1957) Functions of corpus callosum in contralateral transfer of somesthetic discrimination in cats. *Journal of comparative and physiological psychology*, 50, 138–143.

STARK, R., GENESE, F., LAMBERT, W. E., and SEITZ, M., (1977) Multiple language experience and the development of cerebral dominance. In SEGALOWITZ, S. J., and GRUBER, F. A., (Eds.) *Language development and neurological theory*. Academic Press: New York, 47–55.

STEINGRUBER, H. J., (1975) Handedness as a function of test complexity. *Perceptual and motor skills*, 40, 263–266.

STERN, D., (1971) Handedness and the lateral distribution of conversion reactions. *Journal of nervous and mental disease*, 164, 122–128.

STERNBERG, S., (1966) High-speed scanning in human memory. *Science*, 153, 652–654.

ST. JAMES-ROBERTS, I., (1981) A re-interpretation of hemispherectomy data without functional plasticity of the brain. *Brain and language*, 13, 31–53.

ST. JOHN, R. C., (1981) Lateral asymmetry in face perception. *Canadian journal of psychology*, 35, 213–223.

STONE, J., LEICESTER, J., and SHERMAN, S. M., (1973) The naso-temporal division of the monkey's retina. *Journal of comparative neurology*, 150, 333–348.

STONE, M. A., (1980) Measures of laterality and spurious correlation. *Neuropsychologia*, 18, 339–345.

STRONGMAN, K. T., (1973) *The psychology of emotion*. Wiley: London.

STUDDERT-KENNEDY, M., (1975) Dichotic studies: two questions. *Brain and language*, 2, 123–130.

SUBERI, M., and McKEEVER, W. F., (1977) Differential right hemisphere storage of emotional and non-emotional faces. *Neuropsychologia*, 15, 757–768.

SUBIRANA, A., (1958) The prognosis in aphasia in relation to cerebral dominance and handedness. *Brain*, 81, 415–425.

SUBIRANA, A., (1969) Handedness and cerebral dominance. In *Handbook of clinical neurology*, 4, *Disorders of speech, perception and symbolic behaviour*. (Eds.) VINKEN, P. J., and BRUYN, G. W., North Holland Publishing Co.: Amsterdam, 248–272.

SUGISHITA, M., IWATA, M., TOYOKURA, Y., YOSHIOKA, M., and YAMADA, R., (1978) Reading of ideograms and phonograms in Japanese patients after partial commissurotomy. *Neuropsychologia*, 16, 417–426.

SUMMERS, J. J., and SHARP, C. A., (1979) Bilateral effects of concurrent verbal and spatial rehearsal on complex motor sequencing. *Neuropsychologia*, 17, 331–344.

SUSSMAN, H. M., (1971) The laterality effect in lingual-auditory tracking. *Journal of the acoustical society of America*, 49, 1874–1880.

SUSSMAN, H. M., FRANKLIN, P., and SIMON, T., (1982) Bilingual speech: bilateral control? *Brain and language*, 15, 125–142.

SUSSMAN, H. M., and MACNEILAGE, P. F., (1975a) Hemispheric specialization for speech production and perception in stutterers. *Neuropsychologia*, 13, 19–26.

SUSSMAN, H. M., and MACNEILAGE, P. F., (1975b) Studies of hemispheric specialization for speech production. *Brain and language*, 2, 131–151.

SUTTON, P. R., (1963) Handedness and facial asymmetry: lateral position of the nose in two racial groups. *Nature*, 198, 909.

SWANSON, J., LEDLOW, A., and KINSBOURNE, M., (1978) Lateral asymmetries revealed by simple reaction time. In KINSBOURNE, M., (Ed.) *Asymmetrical function of the brain*. Cambridge University Press: Cambridge, Chap. 8, 274–291.

SWISHER, L., and HIRSH, I. J., (1972) Brain damage and the ordering of two temporally successive stimuli. *Neuropsychologia*, 10, 137–152.

SYMANN-LOUETT, N., GASCON, G. G., MATSUMIYA, Y., and LOMBROSO, C. T., (1977) Wave form differences in visual evoked responses between normal and reading disabled children. *Neurology*, 27, 156–159.

TALLAL, P., and NEWCOMBE, F., (1978) Improvement of auditory perception and language comprehension in dysphasia. *Brain and language*, 5, 13–24.

TAN, L. E., and NETTLETON, N. C., (1980) Left handedness birth order and birth stress. *Cortex*, 16, 363–374.

TAYLOR, D. C., (1969) Differential rates of cerebral maturation between sexes and between hemispheres. Evidence from epilepsy. *The lancet*, 2, 140–142.

TAYLOR, D. C., (1977) Epilepsy and the sinister side of schizophrenia. *Developmental medicine and child neurology*, 19, 402–412.

TAYLOR, D. C., and OUNSTED, C., (1972) The nature of gender differences explored through ontogenetic analyses of sex ratio in disease. In OUNSTED, C., and TAYLOR, D. C., (Eds.) *Gender differences: their ontogeny and significance*. Churchill Livingstone: Edinburgh, 215–240.

TAYLOR, H. G., and HEILMAN, K. M., (1980) Left hemisphere motor dominance in right handers. *Cortex*, 16, 587–603.

TAYLOR, P. J., DALTON, R., and FLEMINGER, J. J., (1980) Handedness in schizophrenia. *British journal of psychiatry*, 136, 375–383.

TAYLOR, P., DALTON, R., FLEMINGER, J. J., and LISHMAN, W. A., (1982) Differences between two studies of hand preference in psychiatric patients. *British journal of psychiatry*, 140, 166–173.

TEI, B. E., and OWEN, D. H., (1980) Laterality differences in sensitivity to line orientation as a function of adaptation duration. *Perception and psychophysics*, 28, 479–483.

TENG, E. L., (1981) Dichotic ear difference is a poor index for the functional asymmetry between the cerebral hemispheres. *Neuropsychologia*, 18, 235–240.

TENG, E. L., LEE, P. H., YANG, K. S., and CHANG, P. C., (1976) Handedness in a chinese population: biological social and pathological factors. *Science*, 193, 1148–1150.

TENG, E. L., LEE, P., YANG, K., and CHANG, P. C., (1979) Lateral preferences for hand, foot and eye, and their lack of association with scholastic achievement, in 4143 Chinese. *Neuropsychologia*, 17, 41–48.

TENG, E. L., and SPERRY, R., (1973) Interhemispheric interaction during simultaneous bilateral presentation of letters in commissurotomized patients. *Neuropsychologia*, 11, 131–140.

TENG, E. L., and SPERRY, R. W., (1974) Interhemispheric rivalry during simultaneous bilateral task presentation in commissurotomized patients. *Cortex*, 10, 111–120.

TERRACE, H. S., (1959) The effects of retinal locus and attention on the perception of words. *Journal of experimental psychology*, 58, 382–385.

TERZIAN, H., and CECOTTO, C., (1959) Un nuovo metodo per la determinazione e lo studio della dominanza emisferica. *Giornale di psychiatria e di neuropatologia*, 87, 889–924.

TEUBER, H. L., (1955) Physiological psychology. *Annual review of psychology*, 6, 267–294.

TEYLER, T. J., ROEMER, R. A., HARRISON, T. F., and THOMPSON, R. F., (1973) Human scalp-recorded evoked potential correlates of linguistic stimuli. *Bulletin of the psychonomic society*, 1, 333–334.

TEZNER, D., TZAVARAS, A., GRUNER, I., and HÉCAEN, H., (1972) L'asymetrie droite-gauche du planum temporale a propos de l'étude anatomique de 100 cerveaux. *Revue neurologique*, 126, 444.

THATCHER, R. W., (1977) Evoked potential correlates of hemispheric lateralization during semantic information processing. In HARNAD, S. A., DOTY, R. W., GOLDSTEIN, L., JAYNES, J., and KRAUTHAMER, G., (Eds.) *Lateralization in the nervous system*. Academic Press: New York, Chap. 22, 429–448.

THOMAS, D. G., and CAMPOS, J. J., (1978) The relationship of handedness to a lateralized task. *Neuropsychologia*, 16, 511–517.

THOMPSON, A. L., and MARSH, J. F., (1976) Probability sampling of manual asymmetry. *Neuropsychologia*, 14, 217–223.

THOMPSON, M., (1975) Laterality and reading attainment. *British journal of educational psychology*, 45, 317–321.

THOMPSON, M. E., (1976) A comparison of laterality effects in dyslexics and controls using verbal dichotic listening tasks. *Neuropsychologia*, 14, 243–246.

TODOR, J. I., (1980) Sequential motor ability of left handed inverted and non-inverted writers. *Acta psychologica*, 44, 165–173.

TODOR, J. I., and DOANE, T., (1977) Handedness classification: preference versus proficiency. *Perceptual and motor skills*, 45, 1041–1042.

TOMLINSON-KEASEY, C., and KELLY, R. R., (1979) Is hemispheric specialization important to scholastic achievement? *Cortex*, 15, 97–107.

TOONE, B. K., COOKE, E., and LADER, M. H., (1979) The effect of temporal lobe surgery on electrodermal activity: implications for an organic hypothesis in the aetiology of schizophrenia. *Psychological medicine*, 9, 281–285.

TRANKELL, A., (1955) Aspects of genetics in psychology. *American journal of human genetics*, 7, 264–276.

TRAVIS, L. E., (1931) *Speech pathology*. Appleton-Century-Crofts: New York.

TRESCHER, J. H., and FORD, F. R., (1937) Colloid cyst of the third ventricle. *Archives of neurology and psychiatry*, 37, 959–973.

TRESS, K. H., and KUGLER, B. T., (1979) Inter-ocular transfer of movement after-effects in schizophrenia. *British journal of psychology*, 70, 389–392.

TRESS, K. H., and KUGLER, B. T., and CAUDREY, D. J., (1979) Interhemispheric integration in schizophrenia. In GRUZELIER, J., and FLOR-HENRY, P., (Eds.) *Hemisphere asymmetries of function in psychopathology*. Elsevier/North Holland Biomedical Press: Amsterdam, 449–462.

TREVARTHEN, C., (1974a) Analysis of cerebral activities. In *Hemisphere function in the human brain*. (Eds.) DIMOND, S. J., and BEAUMONT, J. G., Elek: London, 235–263.

TREVARTHEN, C., (1974b) Functional relations of disconnected hemispheres with the brain stem and with each other: monkey and man. In *Hemisphere disconnection and cerebral function*. (Eds.) KINSBOURNE, M., and SMITH, W. L., Chap. X, 187–207.

TREVARTHEN, C., and SPERRY, R. W., (1973) Perceptual unity of the ambient visual field in human commissurotomy patients. *Brain*, 96, 547–570.

TROTHMAN, G. C. A., and HAMMOND, G. R., (1979) Sex differences in task-dependent EEG asymmetries. *Psychophysiology*, 16, 429–431.

TSAI, L., JACOBY, C. G., STEWART, M. A., and BEISLER, J. M., (1982) Unfavourable left-right asymmetries of the brain and autism: a question of methodology. *British journal of psychiatry*, 140, 312–319.

TSUNODA, T., (1965) Functional differences between right and left cerebral hemispheres detected by the key-tapping method. *Brain and language*, 2, 152–170.

TSUNODA, T., (1971) The difference of the cerebral dominance of vowel sounds among different languages. *The journal of auditory research*, 11, 305–314.

TUCKER, D., (1976) Sex differences in hemispheric specialization for synthetic visuospatial functions. *Neuropsychologia*, 14, 447–454.

TUCKER, D. M., (1981) Lateral brain function, emotion and conceptualization. *Psychological bulletin*, 19, 19–46.

TUCKER, D. M., ROTH, R. S., ARNESON, B. A., and BUCKINGHAM, V., (1977) Right hemisphere activation during stress. *Neuropsychologia*, 15, 697–700.

TUCKER, D. M., WATSON, R. T., and HEILMAN, K. M., (1977) Discrimination and evocation of affectively intoned speech in patients with right parietal disease. *Neurology*, 27, 947–950.

TURKEWITZ, G., (1977) The development of lateral differences in the human infant. In HARNAD, S., DOTY, R. W., GOLDSTEIN, L., JAYNES, J., and KRAUTHAMER, G., (Eds.) *Lateralization in the nervous system*. Academic Press: New York, 251–259.

TURNER, S., and MILLER, L. K., (1975) Some boundary conditions for laterality effects in children. *Developmental psychology*, 11, 342–352.

TURVEY, M., (1973) On peripheral and central processes in vision: inferences from an information processing analysis of masking with patterned stimuli. *Psychological review*, 80, 1–52.

TZAVARAS, A., HÉCAEN, H., and LE BRAS, H., (1971) Troubles de la reconnaissance du visage humain et latéralisation hémisphèrique lesionelle chez les sujets gauchers. *Neuropsychologia*, 9, 475–477.

TZAVARAS, A., KAPRINIS, G., and GATZOYAS, A., (1981) Literacy and hemispheric specialization for language: digit dichotic listening in illiterates. *Neuropsychologia*, 19, 565–570.

TZENG, O. J. L., HUNG, D. L., COTTON, B., and WANG, W. S. Y., (1979) Visual lateralisation effect in reading Chinese characters. *Nature*, 282, 499–501.

ULRICH, G., (1978) Interhemispheric functional relationships in auditory agnosia. *Brain and language*, 5, 286–300.

UMILTÀ, C., BRIZZOLARA, D., TABOSSI, P., and FAIRWEATHER, H., (1978) Factors affecting face recognition in the cerebral hemispheres: familiarity and naming. In REQUIN, J., (Ed.) *Attention and performance*, VII, Erlbaum: Hillsdale, New Jersey, 363–374.

UMILTÀ, C., RIZZOLATTI, G., MARZI, C. A., ZAMBONI, G., FRANZINI, C., CAMARDA, R., and BERLUCCHI, G., (1974) Hemispheric differences in the discrimination of line orientation. *Neuropsychologia*, 12, 165–174.

VAID, J., and LAMBERT, W. E., (1979) Differential cerebral involvement in the cognitive functioning of bilinguals. *Brain and language*, 8, 111–129.

VAN DER STAAK, C., (1975) Intra and inter-hemispheric visual-motor control of human arm movements. *Neuropsychologia*, 13, 439–448.

VAN DUYNE, H. J., and SASS, E., (1979) Verbal logic and ear-asymmetry in third and fifth grade males and females. *Cortex*, 15, 173–182.

VAN DUYNE, H. J., and SCANLAN, D., (1974) Left-right ear differences in auditory perception of verbal instruction for non-verbal behaviour: a preliminary report. *Neuropsychologia*, 4, 545–548.

VAN LANCKER, D., and FROMKIN, V. A., (1973) Hemispheric specialisation for pitch and 'tone': evidence from Thai. *Journal of phonetics*, 1, 101–109.

VAN WAGENEN, W. P., and HERREN, R. Y., (1940) Surgical division of commissural pathways in the corpus callosum: relation to spread of an epileptic attack. *Archives of neurology and psychiatry*, 44, 740.

VARGHA-KHADEM, F., and CORBALLIS, M. C., (1979) Cerebral asymmetry in infants. *Brain and language*, 8, 1–9.

VARNEY, N. R., and BENTON, A. L., (1975) Tactile perception of direction in relation to handedness and familial handedness. *Neuropsychologia*, 13, 449–454.

VELLUTINO, F. R., STEGER, J. A., HARDING, C. J., and PHILLIPS, F., (1975) Verbal vs. non-verbal paired-associates learning in poor and normal readers. *Neuropsychologia*, 13, 75–82.

VERNON, M. D., (1957) *Backwardness in reading: a study of its nature and origin.* Cambridge University Press: Cambridge.

VOLPE, B. T., SIDTIS, J. J., and GAZZANIGA, M. S., (1981) Can left-handed writing posture predict cerebral language laterality? *Archives of neurology*, 38, 637–639.

VON BONIN, G., (1962) Anatomical asymmetries of the cerebral hemispheres. In MOUNT-CASTLE, V. B., (Ed.) *Interhemispheric relations and cerebral dominance.* Johns Hopkins Press: Baltimore, 1–6.

WABER, D. P. (1976) Sex differences in cognition: a function of maturation rate? *Science*, 192, 572–574.

WABER, D., (1977) Sex differences in mental abilities, hemisphere lateralization and rate of physical growth at adolescence. *Developmental psychology*, 13, 29–38.

WADA, J. A., (1949) A new method for the determination of the side of cerebral speech dominance: a preliminary report on the intracarotid injection of sodium amytal in man. *Medical biology* (Tokyo), 14, 221–222.

WADA, J. A., CLARKE, R., and HAMM, A., (1975) Cerebral hemispheric asymmetry in humans. Cortical speech zones in 100 adults and 100 infant brains. *Archives of neurology*, 32, 239–246.

WADA, J. A., and RASMUSSEN, T., (1960) Intracarotid injection of sodium amytal for the lateralization of cerebral speech dominance: *Journal of neurosurgery*, 17, 262–282.

WAHL, O. F., (1976) Handedness in schizophrenia. *Perceptual and motor skills*, 42, 944–946.

WALE, J., and GEFFEN, G., (1981) Dichotic monitoring as a test of hemispheric dominance in cases of epilepsy. *Cortex*, 17, 135–140.

WALKER, E., and McGUIRE, M., (1982) Intra and inter-hemispheric information processing in schizophrenia. *Psychological bulletin*, 92, 701–725.

WALKER, H. A., and BIRCH, H. G., (1970) Lateral preference and right-left awareness in schizophrenic children. *Journal of nervous and mental disease*, 151, 341–351.

WALKER, P., (1978) Binocular rivalry: central or peripheral selective processes. *Psychological bulletin*, 85, 376–389.

WALKER, S. F., (1980) Lateralization of functions in the vertebrate brain: a review. *British journal of psychology*, 71, 329–367.

WAPNER, W., HAMBY, S., and GARDNER, H., (1981) The role of the right hemisphere in the apprehension of complex linguistic materials. *Brain and language*, 14, 15–33.

WARD, T. B., and ROSS, L. E., (1977) Laterality differences and practice effects under central backward masking conditions. *Memory and cognition*, 5, 221–226.

WARREN, J. M., (1977) Handedness and cerebral dominance in monkeys. In HARNAD, S. A., DOTY, R. W., GOLDSTEIN, L., JAYNES, J., and KRAUTHAMER, G., (eds.) *Lateralization in the nervous system*, Academic Press: New York, Chap. 10, 151–172.
WARREN, J. M., (1981) Laterality and natural selection. *The behavioral and brain sciences*, 4, 36–37.
WARRINGTON, E. K., (1969) Constructional apraxia. In *Handbook of clinical neurology*, Vol. 4, *Disorders of speech, perception and symbolic behaviour.* (Eds.) VINKEN, P. J., and BRUYN, G. W., North Holland Publishing Co.: Amsterdam, 67–83.
WARRINGTON, E. K., and JAMES, M., (1967) An experimental investigation of facial recognition in patients with unilateral cerebral lesions. *Cortex*, 3, 317–326.
WARRINGTON, E. K., and PRATT, R. T. C., (1973) Language laterality in left handers assessed by unilateral ECT. *Neuropsychologia*, 11, 423–428.
WARRINGTON, E. K., and RABIN, P., (1970) Perceptual matching in patients with cerebral lesions. *Neuropsychologia*, 8, 475–487.
WARRINGTON, E. K., and SHALLICE, T., (1979) Semantic access dyslexia. *Brain*, 102, 43–63.
WARSHAL, D., and SPIRDUSO, W. W., (1981) Concurrent verbal-manual performance in inverted and non-inverted writers. *Perceptual and motor skills*, 53, 123–126.
WEBER, A. M., and BRADSHAW, J. L., (1981) Levy and Reid's neurological model in relation to writing hand posture: an evaluation. *Psychological bulletin*, 90, 74–88.
WEBSTER, W. G., (1977a) Hemispheric asymmetry in cats. In HARNAD, S. A., DOTY, R. W., GOLDSTEIN, L., JAYNES, J., and KRAUTHAMER, G., (Eds.) *Lateralization in the nervous system*. Academic Press: New York, 471–480.
WEBSTER, W. G., (1977b) Territoriality and the evolution of brain asymmetry. In Evolution and lateralization of the Brain. (Eds.) DIMOND, S. J., and BLIZARD, D., *Annals of the New York academy of sciences*, 229, 213–221.
WECHSLER, A. F., (1972) The effect of organic brain disease on recall of emotionally charged versus neutral narrative texts. *Neurology*, 23, 130–135.
WECHSLER, A. F., (1976) Crossed aphasia in an illiterate dextral. *Brain and language*, 3, 164–172.
WEINSTEIN, S., (1969) Neuropsychological studies of the phantom. In BENTON, A. L., (Ed.) *Contributions to clinical neuropsychology*. Aldine Publishing Co.: Chicago, 73–106.
WELFORD, A. T., (1967) Single channel operation in the brain. *Acta psychologica*, 27, 5–22.
WELLER, M., and KUGLER, B. T., (1979) Tactile discrimination in schizophrenic and affective psychoses. In GRUZELIER, J., and FLOR-HENRY, P., (Eds.) *Hemisphere asymmetries of function in psychopathology*. Elsevier/North Holland Biomedical Press: Amsterdam, 463–474.
WELLER, M., and MONTAGU, J. D., (1979) Electroencephalographic coherence in schizophrenia: a preliminary study. In GRUZELIER, J., and FLOR-HENRY, P., (Eds.) *Hemisphere asymmetries of function in psychopathology*. Elsevier/North Holland Biomedical Press: Amsterdam, 285–292.
WERNICKE, C., (1874) *Der aphasische symtomencomplex*. Max Cohn and Weigert: Breslau.
WEXLER, B. E., (1980) Cerebral laterality and psychiatry: a review of the literature. *American journal of psychiatry*, 137, 279–291.
WEXLER, B. E., and HALWES, T., (1981) Right ear bias in the perception of loudness of pure tones. *Neuropsychologia*, 19, 147–150.
WEXLER, B. E., HALWES, T., and HENINGER, G. R., (1981) Use of a statistical significance criterion in drawing inferences about hemispheric dominance for language function from dichotic listening data. *Brain and language*, 13, 13–18.
WEXLER, B. E., and HENINGER, G. R., (1979) Alterations in cerebral laterality during acute psychotic illness. *Archives of general psychiatry*, 36, 278–284.
WHITE, K., and ASHTON, R., (1976) Handedness assessment inventory. *Neuropsychologia*, 14, 261–264.
WHITE, M. J., (1969a) Laterality differences in perception: a review. *Psychological bulletin*, 72, 387–405.

WHITE, M. J., (1969b) Order of report and letter structure in tachistoscopic recognition. *Psychonomic science*, 17, 364–365.

WHITE, M. J., (1971) Visual hemifield differences in the perception of letters and contour orientation. *Canadian journal of psychology*, 25, 207–212.

WHITE, M. J., (1972) Hemispheric asymmetries in tachistoscopic information processing. *British journal of psychology*, 63, 497–508.

WHITE, M. J., (1977) Does cerebral dominance offer a sufficient explanation for laterality differences in tachistoscopic recognition? *Perceptual and motor skills*, 36, 479–485.

WHITE, M. J., and WHITE, K. G., (1975) Parallel-serial processing and hemispheric function. *Neuropsychologia*, 13, 377–381.

WHITEHOUSE, P. J., (1981) Imagery and verbal encoding in left and right hemisphere damaged patients. *Brain and language*, 14, 315–322.

WHITELEY, A. M., and WARRINGTON, E. K., (1977) Prosopagnosia: a clinical psychological and anatomical study of three patients. *Journal of neurology, neurosurgery and psychiatry*, 40, 395–403.

WHITFIELD, I. C., DIAMOND, I. T., CHINERALLS, K., and WILLIAMSON, T. G., (1978) Some further observations on the effects of unilateral cortical ablation on sound lateralization in the cat. *Experimental brain research*, 31, 221–234.

WIENRICH, A. M., WELLS, P. A., and McMANUS, C., (1982) Handedness, anxiety and sex differences. *British journal of psychology*, 73, 69–72.

WILKINS, A., and STEWART, A., (1974) The time course of lateral asymmetries in visual perception of letters. *Journal of experimental psychology*, 102, 905–908.

WILLIS, S. G., WHEATLEY, G. H., and MITCHELL, O. R., (1979) Cerebral processing of spatial and verbal analytic tasks: an EEG study. *Neuropsychologia*, 17, 473–484.

WILSON, D. H., REEVES, H., GAZZANIGA, M., and CULVER, C., (1977) Cerebral commissurotomy for control of intractable seizures. *Neurology*, 27, 708–715.

WILSON, P. J. E., (1970) Cerebral hemispherectomy for infantile hemiplegia: a report of 50 cases. *Brain*, 93, 147–180.

WINNER, E., and GARDNER, H., (1977) The comprehension of metaphor in brain-damaged patients. *Brain*, 100, 717–729.

WITELSON, S. F., (1974) Hemispheric specialization for linguistic and non-linguistic tactual perception using a dichotomous stimulation technique. *Cortex*, 10, 3–17.

WITELSON, S. F., (1976a) Sex and the single hemisphere: specialization of the right hemisphere for spatial processing. *Science*, 193, 425–427.

WITELSON, S. F., (1976b) Abnormal right hemisphere specialization in developmental dyslexia. In *The neuropsychology of learning disorders*. (Eds.) KNIGHT, R. M., and BAKKER, D. J., University Park Press: Baltimore, 233–255.

WITELSON, S. F., (1977a) Early hemisphere specialization and interhemisphere plasticity: an empirical and theoretical review. In SEGALOWITZ, S., and GRUBER, F., (Eds.) *Language development and neurological theory*. Academic Press: New York, Chap. 16, 213–287.

WITELSON, S. F., (1977b) Developmental dyslexia: two right hemispheres and none left. *Science*, 195, 309–311.

WITELSON, S. F., (1977c) Neural and Cognitive correlates of developmental dyslexia: age and sex differences. In SHAGASS, C., GERSHON, S., and FRIEDHOFF, A. J., *Psychopathology and brain dysfunction*. Raven Press: New York, 15–49.

WITELSON, S. F., (1980) Neuroanatomical asymmetry in left handers: a review and implications for functional asymmetry. In HERRON, J., (Ed.) *The neuropsychology of left handedness*. Academic Press: New York, 79–113.

WITELSON, S. F., and PALLIE, W., (1973) Left hemisphere specialization for language in the newborn. *Brain*, 96, 641–646.

WITELSON, S. F., and RABINOVITCH, M. S., (1972) Hemispheric speech lateralization in children with auditory linguistic deficits. *Cortex*, 8, 412–426.

WITKIN, H. A., DYK, R. B., FATERSON, H. F., GOODENOUGH, D. R., and KARP, S. A., (1962) *Psychological differentiation*. Wiley.

WOLFF, P. H., and COHEN, C., (1980) Dual task performance during bimanual co-ordination. *Cortex*, 16, 119–134.

WOLFF, P. H., and HURWITZ, I., (1976) Sex differences in finger tapping: a developmental study. *Neuropsychologia*, 14, 25–41.

WOLFF, P. H., HURWITZ, I., and MOSS, H., (1977) Serial organization of motor skills in left- and right-handed adults. *Neuropsychologia*, 15, 539–546.

WOOD, C. C., GOFF, W. R., and DAY, R. S., (1971) Auditory evoked potentials during speech perception. *Science*, 173, 1248–1251.

WOOD, F., STUMP, D., McKEEHAN, A., SHELDON, S., and PROCTOR, J., (1980) Patterns of regional cerebral blood flow during attempted reading aloud by stutterers both on and off haloperidol medication: evidence for inadequate left frontal activation during stuttering. *Brain and language*, 9, 141–144.

WOODS, B. T., (1980) The restricted effects of right hemisphere lesions after age one: Wechsler test data. *Neuropsychologia*, 18, 65–70.

WOODS, B. T., and TEUBER, H. L., (1978) Changing patterns of childhood aphasia. *Annals of neurology*, 3, 273–279.

WYKE, M., (1967) Effect of brains lesions on the rapidity of arm movement. *Neurology*, 17, 1113–1120.

WYKE, M., (1968) The effect of brain lesions in the performance of an arm-hand precision task. *Neuropsychologia*, 6, 125–134.

WYKE, M., (1971a) The effects of lesions on the performance of bilateral arm movements. *Neuropsychologia*, 9, 33–42.

WYKE, M., (1971b) The effects of brain lesions on the learning performance of a bi-manual coordination task. *Cortex*, 7, 59–72.

WYKE, M. A., (1977) Musical ability: a neuropsychological interpretation. In CRITCHLEY, M., and HENSON, R. A., (Eds.) *Music and the brain*. The Camelot Press: Southampton, Chap. 10, 156–173.

WYKE, M., and ETTLINGER, G., (1961) Efficiency of recognition in left or right visual fields. *Archives of neurology*, 5, 659–665.

YAKOVLEV, P. I., and LECOURS, A. R., (1967) The myelogenetic cycles of regional maturation of the brain. In MONKOWSKI, A., (Ed.) *Regional development of the brain in early life*. Blackwell: Oxford, 3–70.

YAKOLEV, P. I., and RAKIC, P., (1966) Patterns of decussation of bulbar pyramids and distribution of pyramidal tracts on two sides of the spinal cord. *Transactions of the American Neurological Association*, 91, 366–367.

YAMADORI, A., (1975) Ideogram reading in alexia. *Brain*, 98, 231–238.

YAMADORI, A., OSUMI, Y., MASUHARA, S., and OKUBO, M., (1977) Preservation of singing in Broca's aphasia. *Journal of Neurology, Neurosurgery and Psychiatry*, 40, 221–224.

YENI-KOMSHIAN, G. H., and BENSON, D. A., (1976) Anatomical study of cerebral asymmetry in the temporal lobe of humans, chimpanzees and rhesus monkeys. *Science*, 192, 387–389.

YIN, R. K., (1970) Face recognition by brain-injured patients: a dissociable ability? *Neuropsychologia*, 8, 395–402.

YOUNG, A. W., (1982) Asymmetry of cerebral hemispheric function during development. In DICKERSON, J. W. T., and McGURK, H., (Eds.) *Brain and behavioural development*, Blackie: Glasgow, Chap. 6, 168–202.

YOUNG, A. W., and BION, P. J., (1979) Hemispheric laterality effects in the enumeration of visually presented collections of dots by children. *Neuropsychologia*, 17, 99–102.

YOUNG, A. W., and BION, P. J., (1980a) Absence of any developmental trend in right hemisphere superiority for face recognition. *Cortex*, 16, 213–221.

YOUNG, A. W., and BION, P. J., (1980b) Hemifield differences for naming bilaterally presented nouns varying in age of acquisition. *Perceptual and Motor Skills*, 50, 366.

YOUNG, A. W., and BION, P. J., (1981) Accuracy of naming laterally presented known faces by children and adults. *Cortex*, 13, 97–106.

YOUNG, A. W., BION, P. J., and ELLIS, A. W., (1982) Age of reading acquisition does not affect visual hemifield asymmetries for naming imageable nouns. *Cortex*, 18, 477–482.

YOUNG, A. W., and ELLIS, H. D., (1976) An experimental investigation of developmental differences in ability to recognise faces presented to left and right cerebral hemispheres. *Neuropsychologia*, 14, 495–498.

YOUNG, A. W., and ELLIS, H. D., (1980) Ear asymmetry for the perception of monaurally presented words accompanied by binaural white noise. *Neuropsychologia*, 18, 107–110.

YOUNG, A. W., and ELLIS, H. D., (1981) Asymmetry of cerebral hemispheric function in normal and poor readers. *Psychological Bulletin*, 89, 183–190.

YOUNG, G., (1977) Manual specialization in infancy. Implications for lateralization of brain function. In SEGALOWITZ, S. J., and GRUBER, F. A., (Eds.) *Language development and neurological theory*. Academic Press: New York and London, Chap. 17, 289–311.

YOUNG, J. Z., (1962) Why do we have two brains? in MOUNTCASTLE, V. B., (Ed.) *Cerebral dominance and interhemispheric relations*. Johns Hopkins Press: Baltimore, 7–24.

YOZAWITZ, A., BRUDER, G., SUTTON, S., SHARPE, L., GURLAND, B., FLEISS, J., and COSTA, L., (1979) Dichotic perception: evidence for right hemisphere dysfunction in affective psychosis. *British journal of psychiatry*, 135, 224–237.

YULE, W., and RUTTER, M., (1976) Epidemiology and social implications of specific reading retardation. In KNIGHTS, R. M., and BAKKER, D. J., (Eds.) *The neuropsychology of learning disorders*. University Park Press: Baltimore, 25–39.

ZAIDEL, E., (1976) Auditory vocabulary of the right hemisphere following brain bisection or hemidecortication. *Cortex*, 12, 191–211.

ZAIDEL, E., (1977) Auditory language in the right hemisphere following cerebral commissurotomy and hemispherectomy: a comparison with child acquisition. In CARAMAZZA, A., and ZURIF, E., (Eds.) *Language acquisition and language breakdown: parallels and divergences*. Johns Hopkins University Press: Baltimore, 229–275.

ZAIDEL, E., (1978) Lexical organization in the right hemisphere. In BUSER, P., and ROUGEL-BOUSER, A., (Eds.) *Cerebral correlates of conscious experience*. Elsevier/North Holland Biomedical Press: Amsterdam, 177–197.

ZAIDEL, E., and PETERS, A. M., (1981) Phonological encoding and ideographic reading by the disconnected right hemisphere: two case studies. *Brain and language*, 14, 205–234.

ZAIDEL, D., and SPERRY, R. W., (1974) Memory impairment after commissurotomy in man. *Brain*, 97, 263–272.

ZAIDEL, D., and SPERRY, R. W., (1977) Some long term motor effects of cerebral commissurotomy in man. *Neuropsychologia*, 15, 193–204.

ZAMORA, E. N., and KAELBLING, R., (1965) Memory and electroconvulsive therapy. *American journal of psychiatry*, 122, 546–554.

ZANGWILL, O. L., (1960) *Cerebral dominance and its relation to psychological function*. Oliver and Boyd: Edinburgh.

ZANGWILL, O. L., (1962) Dyslexia in relation to cerebral dominance. In MONEY, J., (Ed.) *Reading disability: progress and research needs in dyslexia*. Johns Hopkins Press: Baltimore, 103–113.

ZANGWILL, O. L., (1967) Speech and the minor hemisphere. *Acta neurologica et psychiatrica Belgica*, 67, 1013–1020.

ZANGWILL, O. L., (1974) Consciousness and the cerebral hemispheres. In *Hemisphere function in the human brain*. DIMOND, S. J., and BEAUMONT, J. G., (Eds.) Elek: London, 264–278.

ZANGWILL, O. L., (1979) Two cases of crossed aphasia in dextrals. *Neuropsychologia*, 17, 167–173.

ZATTORE, R. J., (1979) Recognition of dichotic melodies by musicians and non-musicians. *Neuropsychologia*, 17, 607–617.

ZIMMERBURG, B., GLICK, S. D., and JERUSSI, T., (1974) Neurochemical correlate of a spatial preference in rats. *Science*, 185, 623–625.

ZOCCOLOTTI, P., and OLTMAN, P. K., (1978) Field dependence and lateralization of verbal and configurational processing. *Cortex*, 14, 155–163.

ZOLLINGER, R., (1935) Removal of the left cerebral hemisphere. *Archives of neurology and psychiatry*, 34, 1055–1064.

ZURIF, E. B., (1974) Auditory lateralization: prosodic and syntactic factors. *Brain and language*, 1, 391–404.

ZURIF, E. B., and BRYDEN, M. P., (1969) Familial handedness and left-right differences in auditory and visual perception. *Neuropsychologia*, 7, 179–187.

ZURIF, E. B., and CARSON, G., (1970) Dyslexia in relation to cerebral dominance and temporal analysis. *Neuropsychologia*, 8, 351–361.

ZURIF, E. B., and MENDELSOHN, S., (1972) Hemispheric specialization for the perception of speech sounds: the influence of intonation and structure. *Perception and psychophysics*, 11, 329–332.

ZURIF, E. B., and SAIT, P. E., (1970) The role of syntax in dichotic listening. *Neuropsychologia*, 8, 239–244.

# Index

Aesthetic preference 290
Agraphia 2, 133
Amytal
   (*See* Wada test)
Animal
   Cerebral anatomic asymmetry 1, 151
   Functional asymmetry 152 f
Anterior commissure 37, 46, 85, 246
Anxiety
   Handedness 232
   Lateral eye movements 98
Aphasia (dysphasia) 25, 109–136 *passim*
   Bilingualism 170
   Childhood aphasia 158
   Crossed aphasia 110, 123, 135
   Emotion 133
   Handedness 22, 25 ff, 110–127
      *passim*
   Illiteracy 196
   Recovery 112, 136
   Signing aphasia 132, 174
Arousal
   Brain damage 238
   EEG 94–96 *passim*
   Schizophrenia 252 f, 260
   Split-brain 41, 281
   Warning stimuli 282 f
Attention
   (*See* Arousal)
Auditory pathways 83–85 *passim*
   (*See also* Dichotic listening)
   Schizophrenia 257
   Split-brain 41 f

Bilingualism
   Aphasia 170

Split-brain 170
   Visual field effect 170 f
Braille 93 f, 164, 192
Brainedness 2, 25, 122–125 *passim*
Breast tumours 271

Callosal agenesis 45–47
Categorical perception 159
Cerebral anatomic asymmetry
   Animal 151
   Infant 148, 159
   Man 147–150 *passim*
Cerebral vascular supply 143
Channel capacity 273–281 *passim*
Childhood autism 269
Consciousness
   Sodium amytal 112
   Split-brain 56, 245
Corpus callosum 2, 35–54 *passim*
   Attention 41, 281
   Inter-ocular transfer 254 f
   Schizophrenia 253–256 *passim*
   Vertical meridian 218 f

Deafness
   Visual field asymmetry 172–174
      *passim*
   Signing aphasia 132, 174
Depression 237, 249–253 *passim*, 257 f,
   264
   Hand usage 267
Dichotic listening 83–89 *passim*
   Autism 269
   Callosal agenesis 45 f
   Developmental aspects 165 f, 172,
      213

Dichotic listening – *cont.*
  Emotion 239 f
  Handedness/hand    posture  117–121
    *passim*, 125
  Reading 207 f
  Reliability/validity 91 f, 122
  Schizophrenia 257–261 *passim*, 264
  Sex differences 179–182
  Split-brain 41 f, 47
  Stuttering 128
Directional (left-right) confusion 216,
  226
Dominance 2, 198
  (*See also* Aphasia; Brainedness; Han-
  dedness; Eye dominance)
Dopamine 153
Dot enumeration and localisation 69,
  169, 185, 261
Dreaming 283 f
Dual task performance 101–104 *passim*
  Bilingualism 171
  Dual-channel studies 276, 280
  Dyslexia 216
  Sex differences 188
Dyslexia 196–220 *passim*
  Deep dyslexia 140 f
Dyspraxia 128 f

Electro-convulsive therapy
  Depression 249
  Language laterality 115 f
Electrodermal response 238, 252 f, 260
  (Galvanic skin response)
Electrophysiology/electroencephalo-
  graphy (EEG)
  Handedness 96
  Hypnosis 96
  Infants 160 f
  Language 116
  Methodology 94–97 *passim*
  Orgasm 97
  Schizophrenia 262–264 *passim*
  Warning stimuli 283
Emotion 234–248 *passim*
  Aphasia 133
  Split-brain patients 56, 245
Epilepsy
  Commissurotomy 20
  Handedness 20
  EEG 20
  Psychotic symptoms 236, 250

Evolution
  Language 128
  Asymmetry 288–290 *passim*
Eye dominance 11–13 *passim*, 63–65
  *passim*, 198 f, 231

Face recognition
  Adults 71
  Animals 152
  Children 168, 202
  Emotion 241 ff
  Sex differences 184
Field dependence-independence 230 f
Fixation control 67 f, 168, 203
Foot size 195
Form recognition 68 f, 74, 80, 186
Fornix 57

Galvanic skin response
  (*See* Electrodermal response)
Genie 172

Hallucinations 259 f
Handedness 1, 6–33 *passim*, 65, 92,
  134 f, 144, 162, 195, 226–228
  *passim*, 290
  Adoption 31
  Aesthetic response 290
  Age 267
  Anxiety 232 f
  Apraxia 129 f
  Cross-cultural studies 32
  Depression 267
  Developmental aspects 163 f
  Dichotic listening 180–182
  EEG
  Evolution 3 f, 288 f
  Field dependence 230 f
  Genetic theories
  Illusions 13, 226
  IQ 221–224 *passim*
  Language laterality 110–125 *passim*
  Musicians 224
  Neuroanatomical 149 f
  Occupation 228–230 *passim*
  Questionnaires 8–11
  Reading 198–200 *passim*
  Regional cerebral blood flow 146
  Schizophrenia 265
  Tactile perception 92
  Twins 28, 267 f

Visual field differences 182–185 *passim*

Hand posture 13–15 *passim*, 125–127 *passim*, 200

Hemispherectomy 134–138 *passim*
Attention 86
Eye dominance 12
Information processing 278
Plasticity of function 156 ff
Hemisphericity 228–230 *passim*
Hypnosis 97, 99, 270
Hysteria 270

Illiteracy 172, 196
Illusions
Auditory 226
Visual 13
Imageability 40, 57, 169
Deep dyslexia 140 f
Paired associate learning 56
Temporal lobectomy 57, 149
Infant
Brain damage 156 f
Cerebral anatomic asymmetry 159
Convulsions 189
Electrophysiological responses 160 f
Stepping reflex 163
Turning reflex 161–163 *passim*
Information processing theory 75–82 *passim*
Analytic vs. holistic processing 82, 89, 285 ff
Serial vs. parallel processing 81, 285 ff
Inter-hemispheric transmission time 73
Inter-ocular transfer
Callosal agenesis 254
Schizophrenia 254
Split-brain animals 37, 279
IQ 221–224 *passim*
Early brain damage 156 f
Split-brain patients 38, 57

Japanese 49, 133

Kana/kanji
Aphasia 133
Split-brain 49
Visual field effects 133

Lateral eye movements 97–99 *passim*, 188, 229, 253

Laterality coefficient 106 f, 262, 291–297 *passim*
Lexical decision 139 f, 183
Line orientation 69 f, 93

Mastectomy 271
Melodic intonation therapy 5
Memory
Dichotic listening 86
Split-brain patients 56 f
Visual field differences 73 f
Wada amytal test 245, 270
Morse 89
Music
Dichotic listening 89 f

Neuroticism 232

Occupation 228–230 *passim*
Ocular dominance
(*See* Eye dominance)

Phantom limb 270 f
Phoneme
Discrimination by children 160
Identification 132, 257
Planum temporale
(*See* Cerebral anatomic asymmetry)
Prosody 88, 136
Prosopagnosia 71
Psychological refractory period 275

Reading
Disability 196–220 *passim*
Handedness 225
Hand posture 200
Regional cerebral blood flow 144–146 *passim*
Arousal 283
Cerebral anatomic asymmetry 149
Recovery from aphasia 136
Stuttering 128

Schizophrenia 250–268 *passim*
Corpus callosum 254
Dichotic listening 257–260 *passim*
Electrophysiology 262–265
Handedness 265 ff
Interhemispheric transfer 253 ff
Twins 267
Visual field effects 261 f

Semantic priming 139
Sex differences 146, 176–195 *passim*,
  222, 231, 243, 266 f
  Brain damage 176–178 *passim*
  Developmental aspects 190–192
    *passim*
  Dichotic listening 179–182 *passim*
  Dual task 188
  Electrophysiology 187
  Handedness 13, 19, 28, 189
  Lateral eye movements 101
  Reading 199, 210, 213, 215
  Visual field differences 182–187
    *passim*
Singing 133, 135
Skull asymmetry
  Fossils 150
  Gorillas 151
Sleeping 232
Smoking 232
Socio-economic status
  Dichotic listening 172
  Handedness 18
  Visual field effects 172
Sodium amytal/amylobarbitone
  (*See* Wada amytal test)
Split-brain 3, 34–61 *passim*, 138, 224,
  257
  Attention 281
  Bilingualism 170
  Channel capacity 277–280
  Consciousness 245
  Emotion 245 f
  Hypnosis 270
S-R compatibility 15, 73 f
Stereopsis 70, 254
Stroop interference 227
Stuttering 127 f

Tachistoscopic hemifield presentation
  63–83 *passim*, 105, 133, 273 f, 282 f
  Autism 269
  Bilingualism 170 f
  Callosal agenesis 45 f
  Deafness 172–174
  Developmental aspects 167–170
  Emotion 241 f
  Handedness/hand posture 117,
    119 f, 125 f, 226 f
  Reading 172, 200–206 *passim*, 210 f
  Right hemisphere language 136–141
  Schizophrenia 254, 261 f, 264
  Sex differences 182–186 *passim*, 191 f
  Split-brain 39–41 *passim*, 48, 50,
    279–281
  Stuttering 128

Tactile perception 92–94 *passim*, 164 f,
  192
Testicle 1
Thalamus 109, 149, 176
Twins
  Dichotic listening 120 f
  Handedness 27 ff, 267 f
  Schizophrenia 267

Unilateral neglect 235, 238, 282

Vigilance 58, 283
Visual pathways 39 f, 64, 263, 274 ff

Wada sodium amytal test
  Emotion 23 f
  Language laterality 92, 112 f, 136,
    156
  Memory 270
  Stuttering 127 f

www.ingramcontent.com/pod-product-compliance
Ingram Content Group UK Ltd.
Pitfield, Milton Keynes, MK11 3LW, UK
UKHW041118271225
466389UK00001B/4